Advances in Intelligent Systems and Computing

Volume 223

Series editor

Janusz Kacprzyk, Polish Academy of Sciences, Warsaw, Poland
e-mail: kacprzyk@ibspan.waw.pl

For further volumes:
http://www.springer.com/series/11156

About this Series

The series “Advances in Intelligent Systems and Computing” contains publications on theory, applications, and design methods of Intelligent Systems and Intelligent Computing. Virtually all disciplines such as engineering, natural sciences, computer and information science, ICT, economics, business, e-commerce, environment, healthcare, life science are covered. The list of topics spans all the areas of modern intelligent systems and computing.

The publications within “Advances in Intelligent Systems and Computing” are primarily textbooks and proceedings of important conferences, symposia and congresses. They cover significant recent developments in the field, both of a foundational and applicable character. An important characteristic feature of the series is the short publication time and world-wide distribution. This permits a rapid and broad dissemination of research results.

Advisory Board

Václav Snášel · Pavel Krömer
Mario Köppen · Gerald Schaefer
Editors

Soft Computing in Industrial Applications

Proceedings of the 17th Online World Conference on Soft Computing in Industrial Applications

Editors
Václav Snášel
Pavel Krömer
Faculty of Electrical Engineering
and Computer Science
Department of Computer Science
VŠB-TUO
Ostrava-Poruba
Czech Republic

Gerald Schaefer
Department of Computer Science
Loughborough University
Lougborough
UK

Mario Köppen
Network Design and Research Center
Kyushu Institute of Technology
Fukuoka
Japan

ISSN 2194-5357 ISSN 2194-5365 (electronic)
ISBN 978-3-319-00929-2 ISBN 978-3-319-00930-8 (eBook)
DOI 10.1007/978-3-319-00930-8
Springer Cham Heidelberg New York Dordrecht London

Library of Congress Control Number: 2013953229

Printed on acid-free paper

Springer is part of Springer Science+Business Media (www.springer.com)

Preface

This volume of Advances in Intelligent Systems and Computing contains accepted papers presented at WSC17, the 17th Online World Conference on Soft Computing in Industrial Applications, held from December 2012 to January 2013 on the Internet. A tradition started over a decade ago by the World Federation of Soft Computing (http://www.softcomputing.org/) again brought together researchers from over the world interested in the ever advancing state of the art in the field. Continuous technological improvements make this online forum a viable gathering format for a world class conference.

The 2012 edition of the Online World Conference on Soft Computing in Industrial Applications consisted of general track and two special sessions, namely special session on Continuous Features Discretization for Anomaly Intrusion Detectors Generation and special session on Emerging Theories and Applications in Transportation Science. The program committee received a total of 70 submissions from 25 countries, which reflects the worldwide nature of this event. Each paper was peer reviewed by typically 3 referees, culminating in the acceptance of 33 papers for publication. The organization of the WSC17 conference is entirely voluntary. The review process required an enormous effort from the members of the International Technical Program Committee, and we would therefore like to thank all its members for their contribution to the success of this conference. We would like to express our sincere thanks to the special session organizers, to the host of WSC17, VŠB-Technical University of Ostrava, and to the publisher, Springer, for their hard work and support in organizing the conference. Finally, we would like to thank all the authors for their high quality contributions. The friendly and welcoming attitude of conference supporters and contributors made this event a success!

February 2013

Václav Snášel
Pavel Krömer
Mario Köppen
Gerald Schaefer

Organization

General Chair

Václav Snášel	VŠB-Technical University of Ostrava

Program Chair

Pavel Krömer	VŠB-Technical University of Ostrava

Special Event Chair

Gerald Schaefer	Loughborough University, UK

International Advisory Board

Kalyanmoy Deb	Indian Institute of Technology Kanpur, India
Carlos M. Fonseca	University of Algarve, Portugal
Thomas Stützle	University Libre de Bruxelles, Belgium
Jörn Mehnen	Cranfield University, UK
Mario Köppen	Kyushu Institute of Technology, Japan
Xiao-Zhi Gao	Aalto University, Finland
Gerald Schaefer	Loughborough University, UK

International Technical Program Committee

Janos Abonyi	University of Pannonia, Hungary
Mohanad Alfiras	Gulf University, Bahrain
Sudhirkumar V Barai	IIT Kharagpur, India

Suhail Owais	Applied Science University, Jordan
Ana Pereira	Polytechnic Institute of Bragança, Portugal
Jan Platoš	VŠB-Technical University of Ostrava, Czech Republic
Sebastian Polak	Jagiellonian University Medical College, Poland
Sg Ponnambalam	Monash University Sunway Campus, Malaysia
Petrica Claudiu Pop	North University of Baia Mare, Romania
Radu-Emil Precup	Politehnica University of Timisoara, Romania
Ana Maria A. C. Rocha	University of Minho, Portugal
Riccardo Rossi	University of Padova, Italy
Nicola Sacco	DIME-University of Genoa, Italy
Gerald Schaefer	Loughborough University, UK
Giovanni Semeraro	University of Bari, Italy
Sara Silva	INESC-ID, Portugal
Piotr Sobolewski	Wrocław University of Technology, Poland
Marcone Souza	Universidade Federal de Ouro Preto, Brazil
Thomas Stützle	IRIDIA, ULB, Belgium
Eiji Uchino	Yamaguchi University, Japan
Juan Velasquez	University of Chile, Chile
Milorad Vidovic	University of Belgrade, Serbia
Michał Woźniak	Wrocław University of Technology, Poland

Sponsoring Institutions

World Federation on Soft Computing
VŠB Technical University of Ostrava

Contents

Part I
Soft Computing in Industrial Applications

Advanced Methods for 3D Magnetic Localization in Industrial Process Distributed Data-Logging with a Sparse Distance Matrix

Abhaya Chandra Kammara and Andreas König

Abstract Wireless sensor networks/data-logging devices are increasingly applied for distributed measurement and acquiring additional contextual data. These have been applied in large scale indoor and outdoor systems with solutions based on RF, light based and ultra sound based systems. Data-loggers in liquid filled containers pose new challenges for localization because of the high reflectivity of containers and high attenuation due to the liquids obstructing communication between wireless nodes. Magnetic localization techniques have been used in many places including military research [14]. This approach was adapted for use in liquid filled containers. In this project, two prototypes, a laboratory and an industrial installation have been conceived and served for acquisition of experimental data for localization. In our paper, we exploit the sparsity met in the particular magnetic MEMS sensor swarm localization concept by introducing NLMR which is a simplified form of Sammon's mapping (NLM) and we combine it with different meta-heuristics and soft-computing techniques, e.g., gradient descent, Simulated Annealing and PSO. We compare this with Multilateration and conventional NLM localization technique. Our approach has improved the localization from a mean error of 20 cm in the first cut analysis for the industrial setup using conventional NLM down to 11 cm without and to 9 cm with apriori knowledge. Future improvements are to be expected from a thorough calibration of all system components. in [5]. The modified algorithm is capable of distributed localization producing mean localization error of 10 cm for the Warstein experiment data.

A. C. Kammara (✉) · A. König
Institute of Integrated Sensor Systems, TU Kaiserslautern, 67663 Kaiserslautern, Germany
e-mail: abhay@eit.uni-kl.de

A. König
e-mail: koenig@eit.uni-kl.de

V. Snášel et al. (eds.), *Soft Computing in Industrial Applications*,
Advances in Intelligent Systems and Computing 223, DOI: 10.1007/978-3-319-00930-8_1,

1 Introduction

Sensors and Sensing systems are getting ubiquitous in industry and homes alike. Providing contextual information is essential and advantageous in industry and in ambient intelligence systems. Data-loggers have become prominent in such setups. Localization of such mobile sensors are traditionally done using RF, ultrasonic, IR and other methods. This spectrum of approaches is not suitable in our case, because of the high attenuation and reflections caused in liquid containers of industrial processes. Magnetic localization on the contrary can be a feasible method for localization in this situation.

Magnetic localization techniques have been in use for a long period of time from their introduction in 1962 [1]. They have been used in head tracking applications [3] in military [14], for silent localization of underwater sensors [9], in location and orientation tracking [8], in medical systems [12, 13]. The approaches vary from using Internal (mostly in medical approaches) and External (Using artificially generated magnetic fields or earth magnetic field). In this project quasi-DC fields with artificially generated magnetic fields in coils is used. The idea behind such a localization approach for data-logging had its patent filed [21] on 18.05.2010. Similar idea has been used recently for data-logging in [20]. Interestingly an approach similar to this project is also being used in Indoor positioning systems [10, 11].

In the following section the particular project approach, providing the data for the localization algorithm experiments reported in this work, will be described in detail. Based on the data from measurement, localization algorithms are employed to estimate the coordinates of WSN nodes. Commonly, algorithms from, e.g., multi-dimensional-scaling (MDS), are employed. These basically are fine, but in their majority base on the assumption of a densely populated distance matrix and require substantial post-processing for the final coordinate determination. In the regarded research project, inter node communication is practically unavailable, so

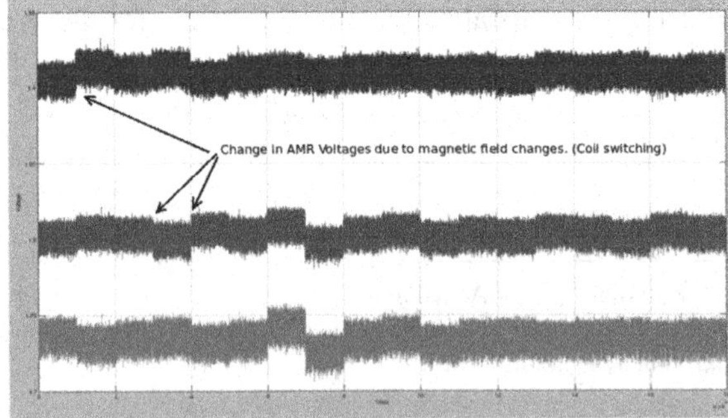

Fig. 1 3D AMR sensor node (*left*), noisy raw data obtained from 3D AMR node in a scaled down ISE lab setup (*right*)

that the resulting distance matrix is sparse, i.e., having only anchor to sensor node non-zero entries. Thus, in this paper we investigate the recall extension of Sammon's non-linear mapping (NLMR) for localization purposes and compare it with multilateration based on data acquired in the industrial target environment. Also, we enhance NLMR with different soft computing techniques and show that we can get more than competitive results, than original Sammon's mapping (NLM) with much less computational effort.

2 The Localization System Setup

In our localization concept and implementation triaxial Anisotropic Magneto-resistive (AMR) sensors complementing the data-loggers which will be deployed in the liquid containers. A magnetic coils system has been mounted on the container hull producing quasi-DC magnetic fields controlled by a central control unit synchronized to the sensors clock. This is used to produce voltage values in the sensors, which can be converted to distances. These distances can be used to compute the location of the sensor node.

2.1 Sensor Node

In the project, artificially generated magnetic fields are sensed by a triaxial Anisotropic Magneto-resistive (AMR) sensor. AMR sensor makes use of a magneto-resistive effect to detect magnetic fields. In the project, AMR sensor type AFF755B from Sensitec Gmbh was used to design a proprietary 3D sensor. The Fig. 1 shows a PCB AMR sensor. There is currently a MEMS prototype which will be used in future experiments.

2.2 Magnetic Field Generation

There are different types of coils that could be used for magnetic field generation. In this project, circular coils of container specific diameters are used to generate the fields. The coils should be positioned in such a way that there are at least 4 magnetic fields observed by the sensor at any point in the cylinder. Generally localization is done in such a way the distance between the coil and the sensor is much greater that the radius of the coil.

Magnetic fields are not simultaneously produced from all the coils. The magnetic field are generated one coil at a time. Each coil is switched-on in both directions. Ternary switching allows a differential and energy-aware measurement, eliminating static magnetic offsets and reducing the flipping of AMR sensors to a

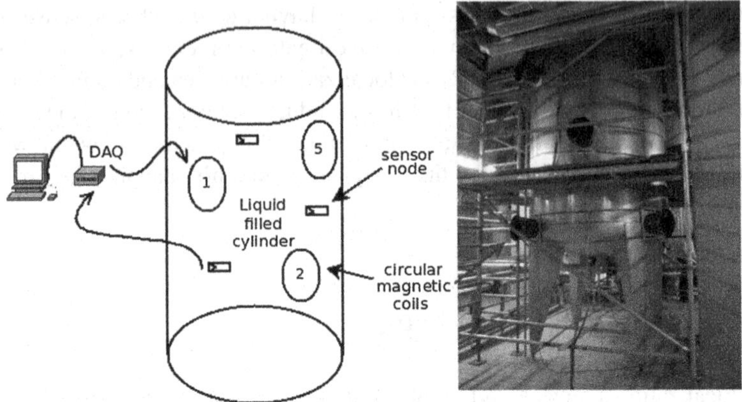

Fig. 2 Sketch of the measurement system with sensor and coils (*left*), photo of industrial container with coils (*right*)

minimum. In the experiments a DAQ board (DT9816) was used to control the coils and in data acquisition as shown in the Fig. 2. In the final planned setup the control circuitry is different for the coil switching and reading from AMR. The clock synchronization errors can be resolved by using synchronization methods described in [19].

2.3 Distance Calculation

Three distances are obtained for each activation of the coil. These voltages are converted into distances using the formulae given below. The angle between the node and the sensor is not considered since it is an unknown. However there are techniques which could be used as mentioned in [18].

$$d = \left(\left(\frac{\frac{1}{2} \times \mu_0 \times n \times R^2 \times I}{B_M} \right) - R^2 \right)^{\frac{1}{2}} \tag{1}$$

where,

$$Bm = \frac{V_M}{S \times V_S \times G}, V_M = \frac{V_i^p - V_i^n}{2}, i = \{x, y, z\}$$

where S is the sensitivity, V_S is sensor voltage, G is the gain of the amplifier, n is the number of windings, R denotes the radius of the coil and $\mu_0 = 4 \times \pi \times 1e^{-7}$.

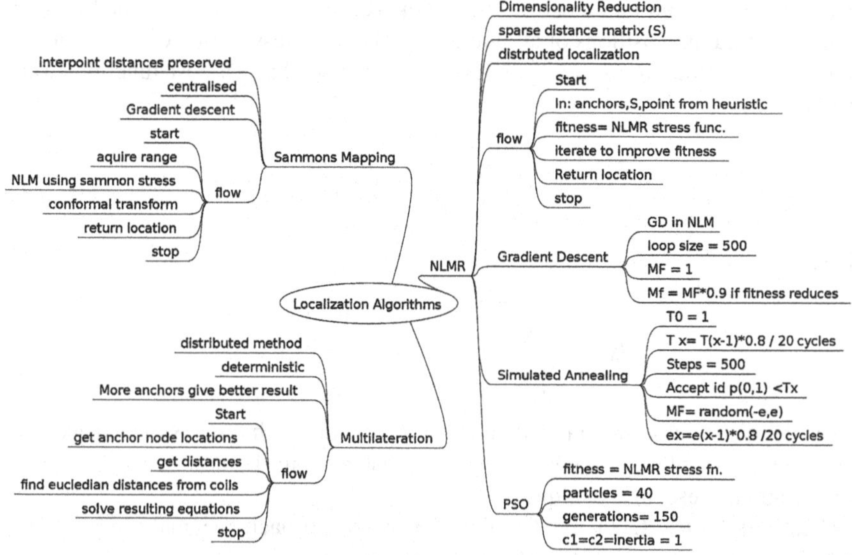

Fig. 3 Survey of localization algorithms and parameter settings

3 Localization

3.1 *MDS Mapping*

MDS mapping is a popular technique used in localization. There are many types of MDS mapping, we are interested in the non linear Sammon's mapping (NLM). In NLM localization is done based on reducing the cost function E(m)

$$E = \frac{1}{\sum_{i=1}^{N-1} \sum_{j=i+1} N d_{ij}^{*}} \sum_{i=1}^{N-1} \sum_{j=i+1}^{N} \frac{(d_{ij}^{*} - d_{ij})^2}{d_{ij}^{*}} \tag{2}$$

where N is number of nodes, d_{ij}^{*} is Euclidean distance between X_i and X_j in Higher dimension and d_{ij} is Euclidean distance between Y_i and Y_j in lower dimension. In the localization problem there is no dimensionality reduction. The main disadvantage of this method in our scenario is that we do not have inter sensor distances. A modification to Sammon's mapping that limits the involved number of distances seems promising [5] Fig. 3.

3.2 *NLMR*

Sammons recall or NLMR was described in [5] for dimensionality reduction. This technique is a simplification of Sammons mapping (NLM) where inter point distances are ignored and a set of code-book vectors (previously mapped points) are used to find the positions of all points. This is much less resource consuming as compared to Sammons mapping (NLM) [5]. Here we have a cost function E(m) described by,

$$E_i(m) = \frac{1}{c}\sum_{j=1}^{K}\frac{(d_{Xij} - d_{Yij}(m))^2}{d_{Xij}} \tag{3}$$

where,

$$d_{Xij} = \sqrt{\sum_{q=1}^{m}(v_{iq}^{r} - v_{jq}^{t})^2}, c = \sum_{j=1}^{K} d_{Xij}$$

d_{Xij} is the distance between recall and training data in high dimensional space and K is the number of code book vectors. The distances in the new space are found using gradient descent technique

Gradient Descent: In the Gradient Descent approach we make use of the NLMR cost function and the following equations.

$$y_{iq}(m+1) = y_{iq}(m) - MF \times \Delta y_{iq}(m) \tag{4}$$

with,

$$\Delta y_{iq}(m) = \frac{\left(\frac{\partial E_i(m)}{\partial y_{iq}(m)}\right)}{\left(\frac{\partial E_i^2(m)}{\partial y_{iq}(m)^2}\right)}, 0 < MF \leq 1 \tag{5}$$

Where $y_{iq}(m+1)$ is the new position, MF is the magic factor which reduces with time, $y_{iq}(m)$ is the current position and E(m) is the cost function at the current position.

Modification: The specialty of NLMR approach is that it matches perfectly with our requirements. We do not obtain an inter-point distance between the nodes. We have a set of mapped positions (coils). We have to make some changes to the original algorithm to make it work for localization.

In our case we do not require a dimensionality reduction. So the known locations (magnetic coils) will act as the trained data set. We do not have any location for the unknown value, however we have the distance information from the known locations which can be directly given to the algorithm.

Unlike Sammon's mapping we do not have to do a reverse mapping (conformal transform) after we get our output (Table 1).

After making these changes, we can make use of any soft-computing technique to generate the values to be given to the Sammon's recall function. In our approach we tried Gradient descent, simulated annealing and PSO.

NLMR - Gradient descent: The gradient descent approach makes use of Quasi- Newton method similar to the method described in Eq. 4, [5]. We make use of Magic factor initialized with 1 and reduced when the new stress is not as good as the old stress. Only better solutions are accepted in this approach.

NLMR - Simulated Annealing: We use the basic simulated annealing where we start with a relatively high temperature ($T_0 = 1$) which is reduced ($Tx = T_{(x-1)} * 0.8$) over the number of cycles and reduce the chances of choosing a bad solution as the temperature decreases (Accept any solution if p(0,1)$leq\ t_x$). The new solutions are found by a markov chain shown in Eq. 4 with a random MF between $-e_0$ and e_0 where e (energy factor) reduces over time. The algorithm runs for 500 iterations to get the best solutions. The number of iterations required was found heuristically.

NLMR - Particle swarm optimization: Standard particle swarm optimization described in [4] is used with $C_1 = C_2 = 1$ and the algorithm has no inertia. Having no inertia helped in faster convergence of the algorithm. 40 particles were used with 150 generations found the best results in the experiments.The standard PSO was unable to converge to a solution using Sammon's mapping and required special approaches to reach convergence [15]. However, in the modified NLMR method the standard PSO is able to converge easily.

Multilateration: Multilateration is used in wireless sensor networks and a standard technique for efficient and effective localization. It serves as a reference in our work. Traditionally the time difference of arrival of the signal is used in multilateration technique, we use the distances obtained with our magnetic localization system. In our approach we make use of the linear least squares method (Moore-Penrose pseudo-inverse) to get the best fit solution.

Coil Selection: In the previous works of this project coil selection was used by taking a heuristic of the closest distances (effectively closest coils (anchors)) available. However in our data analysis we found that erroneous distances detected by the sensor was also present in closer coils. In our analysis we compared the distances from the sensor and the distances calculated from the ground truth value. In Fig. 4, three coils with their distance errors are plotted for all our data. The peaks correspond to distance errors due to angle (which is not honored in our current formula for distance calculation). In Fig. 4, we can see the original positions where the sensors were placed. Even though coil 12 was furtherest from

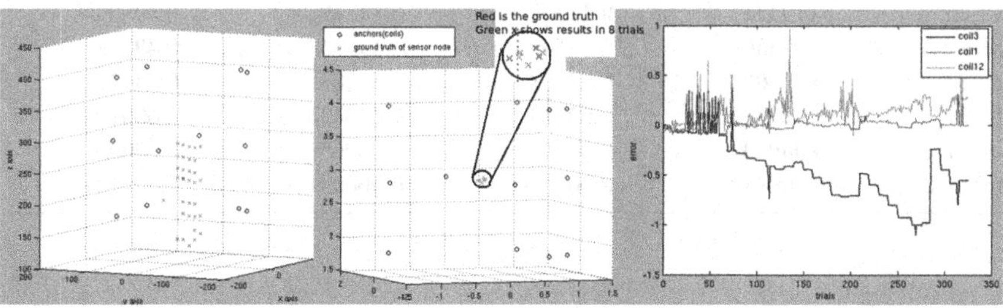

Fig. 4 Plot showing the ground truth and anchor (coils) locations (*left*), results for 8 trials on one ground truth location (*middle*), variations between expected distances and acquired distances for 325 experiments (*right*)

Table 1 Parameters used in the industrial setup

trials	pos	trials/pos	coils	coil dia.	windings	V-Sensor	I-coils
325	40	2 to 20	12	0.25 m	230	3.7V	3A

Table 2 Results making use of all available anchors (coils) (top), results using anchors (coils) selected by apriori knowledge (1,2,4,5,6,7,8,9) (bottom), all values are in meters

All Coils	**NLMR-GD**	**NLMR-SA**	**NLMR-PSO**	**Multilateration**
LEμ	0.17951	0.17914	**0.11429**	0.16702
LEσ	0.13097	0.12557	**0.06580**	0.15194
Max LE	0.76031	**0.62029**	0.69258	1.31961
Min LE	0.02230	0.01720	**0.00911**	0.01302
μ X	−0.09350	−0.09780	**−0.00121**	−0.05915
μ Y	0.08262	0.08639	**0.01846**	−0.02365
μ Z	0.02610	0.02201	0.02336	**−0.00190**
σ X	0.14338	0.13490	**0.06336**	0.12351
σ Y	0.08881	0.08591	**0.07171**	0.11815
σ Z	0.06942	**0.06904**	0.08571	0.13306
Max err X	0.67247	0.55810	0.50856	**0.48092**
Max err Y	0.58066	0.55335	**0.50618**	0.60638
Max err Z	**0.20882**	0.21477	0.30020	1.15277
Min err X	0.00017	**0.00001**	0.00018	0.00043
Min err Y	0.00140	0.00060	0.00065	**0.00020**
Min err Z	0.00026	0.00026	0.00021	**0.00011**
Sel. Coils	**NLMR-GD**	**NLMR-SA**	**NLMR-PSO**	**Multilateration**
LE μ	0.10106	**0.09345**	0.09534	0.13912
LE σ	**0.05822**	0.05861	0.06129	0.11206
Max LE	0.69430	**0.68370**	0.70647	0.80797
Min LE	0.01774	0.01126	0.01239	**0.00573**
μ X	**−0.00238**	0.01458	0.00813	−0.03728
μ Y	**0.00126**	0.00150	0.00686	−0.03760
μ Z	**0.00323**	0.02655	−0.00776	0.03375
σ X	0.04970	**0.04145**	0.04564	0.07312
σ Y	0.06513	0.06770	**0.06290**	0.09209
σ Z	0.08290	**0.07033**	0.08144	0.11892
Max err X	0.19471	**0.13390**	0.17319	0.19803
Max err Y	0.52350	0.53103	0.49736	**0.49735**
Max err Z	**0.50800**	0.57788	0.58183	0.78424
Min err X	0.00043	**0.00010**	0.00043	0.00012
Min err Y	**0.00003**	0.00007	0.00065	**0.00003**
Min err Z	0.00033	0.00021	**0.00009**	0.00018

Table 3 Parameters used in the ISE lab experimental setup

trials	pos	trials/pos	coils	coil dia.	windings	V-Sensor	I-coils
56	56	1	6	0.13 m	100	5V	5A

Table 4 First-cut results from ISE lab orthogonal demonstrator

	NLMR-GD	NLMR-SA	NLMR-PSO	Multilateration
LE μ	0.07169	**0.07137**	0.08715	0.12526
LE σ	**0.03312**	0.03429	0.03428	0.07147
μ X	−0.05512	0.05434	**0.05112**	0.10509
μ Y	0.03315	**0.03307**	0.03383	−0.05274
μ Z	0.00532	−0.00639	−0.00744	**−0.00212**
σ X	**0.02857**	0.02953	0.03727	0.07489
σ Y	0.02062	**0.02057**	0.02612	0.03253
σ Z	0.02881	0.02978	0.05372	**0.01735**

any of the data points while coil 3 was close to most of them coil 3 has a higher error as compared to coil 12. The reason behind this maybe due to an unfortunate angle with respect to the data points. Here, we remove 4 coils depending on such a study (coils 3,10,11,12) and observe the improvement in results. More research on coil selection will be done in our future work (Tables 2, 3).

4 Results

The data acquisition based on the container of Fig. 2 (right), was confined to a cubic volume in the container of 350 cm × 350 cm × 250 cm. The acquired data was first-cut analyzed in prior project work, employing the conventional NLM localization method. These previous results serve as the baseline to comparatatively evaluate the suggested new methods. In the tables below we use Localization Error (LE) which is the distance between the ground truth and obtained values. The parameter settings of all conducted experiments are summarized in Fig. 3. In addition to the *stale* database from the industrial environment, we extracted fresh data from the ISE lab setup, which has just six smaller scaled coils in an orthogonal arrangement of 150 cm × 150 cm × 150 cm. Raw data in Fig. 2 was obtained here. In a first step, 56 different locations with just one trial each were sampled in a single Z-plane and the data was subject to multilateration and the suggested methods. The results given in Table 4 in comparison to Multilateration confirm the viability of our approach. The ISE lab demonstrator is reshaped to cylindrical shape and campaigns with different sensor heads including MEMS and wireless sensor will follow-up.

5 Conclusion

In our work, we have presented an adaptation of NLMR, a simplified Sammon's mapping for dimensionality reduction and used it for localization with a sparse distance matrix, e.g., for liquid filled containers or general indoor localization.

This was applied to industrial data available as benchmark and obtained comparable results to Multilateration. The results were also better than those achieved by conventional NLM in first-cut analysis with the same data. We also presented simulated annealing and particle swarm optimization to reduce the NLMR cost function and these methods provide better results than the commonly used gradient descent method [2,5]. The standard NLMR PSO even provides results for all coils selected comparable to results of the other methods when coils giving unreliable information are removed.

Acknowledgments This work was partly supported by the Federal Ministry of Education and Research (BMBF) in the program mst-AVS, in the project ROSIG grant no. 16SV3604 of the PAC4PT consortium (Partners were UST GmbH, IMST GmbH, Krohne, Warsteiner, and microTEC Ges. für Mikrotechn. GmbH (Coord.). The industrial environment data, employed here for the algorithmic studies, has been acquired and first-cut analysed by conventional NLM in the project work by S. Carrella and D. Groben. All algorithms were implemented using python (numpy, scipy, and matplotlib packages) and Matlab.

References

1. Kalmus, H.P.: A new guiding and tracking system. IRE Trans. Aerosp. Navig. Electron. **9**, 7–10 (1962)
2. Sammon, J.W.: A nonlinear mapping for data structure analysis, IEEE Trans. Comput. C-18, 401–409 (1969)
3. Raab, F.E., et al.: Magnetic position and orientation tracking system, IEEE Trans. Aerosp. Electron. Syst. **5**, 709–717 (1979). ISSN 18237843
4. Kennedy, J., Eberhart, R.C.: Particle swarm optimization. Proc, IEEE International Conference on Neural Networks, IJCNN (1995)
5. König, A.: Interactive visualisation and analysis of hierarchical neural projections for data mining. In IEEE TNN, Special Issue on Neural Networks for, Data Mining and Knowledge Discovery, pp. 615–624, 11(3), May, 2000
6. König, A.K.: "Dimensionality reduction techniques for interactive visualization, exploratory data analysis, and classification", Pattern Recognition in Soft Computing Paradigm, World Scientific, N.R. Pal (eds.), 2: 1–37, 2001
7. Paperno, E., Sasada, I., Leonovich, E. A new method for magnetic position and orientation tracking. IEEE Trans. Magn. **37**(4), JULY 2001
8. Prigge, E.A.: A positioning system with no line-of-sight restrictions for cluttered environments. Stanford University, Dissertation (2004)
9. Callmer, J., Skoglund, M.: F. Silent localization of underwater sensors using magnetometers. EURASIP J. Adv. Sig. Process, Gustafsson (2010)
10. Blankenbach, J., Norrdine, A.: Position estimation using artificial generated magnetic fields, : International Conference on Indoor Positioning and Indoor Navigation (IPIN), 15–17 September 2010. Zürich, Switzerland (2010)
11. Blankenbach, J., Norrdine, A., Hellmers, H.: Adaptive signal processing for a magnetic indoor positioning system geodetic institute, Technische Universität Darmstadt - Short paper IPIN
12. Placidi, G., Franchi, D., Maurizi, A., Sotgiu, A.: Review on patents about magnetic localisation systems for in-vivo catheterizations INFM c/o Dept. of Health Sci., Uni. of LAquila, Via Vetoio Coppito 2, 67100 LAquila, Italy

13. Ascension Technology Corporation Products Application: [Online]. http://www.ascension-tech.com/medical/pdf/TrakStarWRTSpecSheet.pdf, checked Nov. 5 2012
14. Polhemus: [Online]. Available: http://www.polhemus.com/?page=Military_Why_Magnetic_Tracking, checked Nov. 5 2012
15. Iswandy, K., önig, A.K.: Soft-computing techniques to advance non-linear mappings for multi-variate data visualization and wireless sensor localization. e-Newsletter IEEE SMC Soc., Issue #29, Dec. 2009
16. Carrella, S., Iswandy, K., Lutz, K., önig, A.K.: 3D-Localization of low-power wireless sensor nodes based on AMRSensors in industrial and AmI applications, VDE Verlag GmbH Berlin Offenbach, pp. 522–529, 2010
17. Iswandy, K., Carrella, S., önig, A.K.: Localization system for low power sensor nodes deployed in liquid-filled industrial containers based on magnetic sensing. Tagungsband XXIV Messtechnisches Symposium des AHMT, 23.-25. September, pp. 108–121, 2010
18. Carrella, S., Iswandy, K., önig, A.K.: A system for localization of wireless sensor nodes in industrial applications based on sequentially emitted magnetic fields sensed by tri-axial AMR sensors
19. Carrella, S., Iswandy, K., önig, A.K.: System for 3D localization and synchronization of embedded wireless sensor nodes based on AMR sensors in industrial environments, Proceedings Sensor+test 2011
20. Reinecke, S., öpping, U.P., Hampel, U.: Autonome sensorpartikel zur räumlichen Parametererfassung in gro*ß*skaligen Behältern, Sensor & Test 2012
21. Pending Patent Application: Method and apparatus for determining the spatial coordinates of at least one sensor node in a container, filing date: 18.05.2010, Int. Pub. 24.11.2011, Int.No.: WO 2011/144325 A2.

Neural Network Ensemble Based on Feature Selection for Non-Invasive Recognition of Liver Fibrosis Stage

Bartosz Krawczyk, Michał Woźniak, Tomasz Orczyk, Piotr Porwik, Joanna Musialik and Barbara Błońska-Fajfrowska

Abstract Contemporary medicine concentrates on providing high quality diagnostic services, yet it should not be forgotten that the comfort of the patient during the examination is also of high importance. Therefore non-invasive methods that allows to precisely predict the state of the disease are currently one of the key issues in the medical business. The paper presents a novel ensemble of neural networks applied to recognition of liver fibrosis stage from indirect examination method. Several neural network models are build on the basis of outputs of different feature selection algorithms. Then an ensemble pruning procedure with the usage of diversity measures is conducted in order to eliminate redundant predictors from the pool. Finally the weights of classifiers in the fusion process are assessed to establish their influence on the output of the whole ensemble. Proposed method is compared with several state-of-the-art ensemble methods. Extensive experimental investigations, carried out on a dataset collected by authors, show that the proposed method achieve a satisfactory level of the

B. Krawczyk (✉) · M. Woźniak
Department of Systems and Computer Networks, Wroclaw University of Technology, Wyb Wyspianskiego 27, 50-370 Wroclaw, Poland
e-mail: bartosz.krawczyk@pwr.wroc.pl

M. Woźniak
e-mail: michal.wozniak@pwr.wroc.pl

T. Orczyk · P. Porwik
Institute of Computer Science, University of Silesia, Bedzinska 39, 41-200 Sosnowiec, Poland
e-mail: tomasz.orczyk@us.edu.p

P. Porwik
e-mail: piotr.porwik@us.edu.pl

J. Musialik
Department of Gastroenterology and Hepatology, Medical University of Silesia, Katowice, Poland

B. Błońska-Fajfrowska
Department of Basic Biomedical Science, Medical University of Silesia, Sosnowiec, Poland

V. Snášel et al. (eds.), *Soft Computing in Industrial Applications*, Advances in Intelligent Systems and Computing 223, DOI: 10.1007/978-3-319-00930-8_2,

fibrosis level recognition, outperforming other machine learning algorithms and thus may be used as a real-time medical decision support system for this task.

1 Introduction

Liver fibrosis is a condition where fibrous scare tissue accumulates in the liver. It is a common complication of many diseases but in this research we use medical data of patients infected with liver hepatitis type B and C virus (HBV/HCV). Early detection of liver fibrosis is very important as the condition may stay completely hidden for months or even years, but untreated may lead to liver cirrhosis and in consequence to patient's decease.

In most cases the condition stays in so called compensated state, so no visible changes nor dysfunctions might be observed, but although most medical examination results are within their normal results, some slight discrepancies may be observed and used to evaluate the liver fibrosis stage [13]. In this research we will be using blood test results.

Mentioned results coming from the non-invasive biomedical examination are then treated as an input for the machine learning algorithms. Our aim is to create an accurate medical decision support tool that will allow for an automatic classification of patients under the observation.

Among many machine learning techniques it is impossible to select the best one for the task at hand without any a priori knowledge about the data [18]. Therefore standard procedure concentrates on building several predictors, testing their performance and selecting the single best model according to some criterion. Yet it should be beard in mind that in many cases different models may contribute uniquely to the analysed task. That is why the Multiple Classifier Systems (MCS) are one of the major research directions in machine learning [14]. They propose to utilize more than one model in hope to exploit their strengths while reducing their drawbacks. It has been shown many times that a ensemble may achieve a better accuracy than any of the individual members taking part in it.

In this work we propose an ensemble based on neural networks, as those classifiers were one of the first to highlight the effectiveness of the MCS approach [6]. We address three main issues in our compound model—assuring base diversity among individual models, discarding the redundant predictors and creating a fusion methodology that exploits the classifier local competencies. Individual models are build on the basis of outputs of different feature selection algorithms. This way we achieve both—the initial diversity (as different methods return different reduced feature spaces) and the decreased complexity of the model (due to the reduction of the feature space). Then ensemble pruning, based on the diversitymeasure, is conducted to discard predictors that are similar to each other and therefore add no value to the committee. Finally a trained fusion procedure, based on individual discriminant functions, is performed in order to boost the quality of the proposed MCS.

The proposed method is compared to other state-of-the-art ensembles in order to asses its quality for the task of liver fibrosis recognition.

2 Liver Fibrosis Recognition

The only method of liver fibrosis stage recognition giving a 100 % accuracy is an autopsy and this is due to the fact that the condensation of scare tissue within the liver may vary in different regions of the organ. For the same reason the most common examination method—liver biopsy doses not guarantee the correct diagnosis. This method is unfortunately not only inaccurate, but also may lead to serious health complications including risk of patient's death.

There are two (or three) common description methods for liver biopsy samples. One used in the article is METAVIR [2] (4 stages of fibrosis) and the other are: Histological Activity Index (HAI Score) also known as Knodell Score [9] (3 stages of fibrosis) and it's modified version called Ishak Score [7] (6 stages of fibrosis). The METAVIR has been specifically designed and validated for patients with hepatitis C. All these systems rely on a histological image of the liver and in consequence their quality depends on a sample size and doctor's experience.

Due to facts presented in the first paragraph also some non invasive examination methods have been developed. Most common one is APRI-test [17], but also ELF-Test [12] and FIBRO-Test [4] have been developed by medical companies. All these methods are blood test based, but the first one is very general and can be used only to detect advanced fibrosis or cirrhosis. Two other are more specific, but also more expensive for patients. All blood test based methods try to detect some dependencies between liver functionality and blood test results, so they are indirect and non-invasive methods. This is very important, because in opposite to liver biopsy, they may be repeated in regular periods of time without a harm for a patient.

For the purpose of presented research we acquired medical data records from 103 real patients of the Branch of the Gastroenterology and Hepatology of the Independent Public Central Hospital of the Silesian Medical University. Table 1 presents number of examined patients for each fibrosis stage (F0..F4) and Table 2 presents characteristics of acquired medical records.

Table 1 Number of patients with given fibrosis stage [n (%)]

*F*0	*F*1	*F*2	*F*3	*F*4
2 (2 %)	34 (33 %)	5 (5 %)	16 (15 %)	46 (45 %)

Table 2 Blood test results characteristics [mean (std. deviation)]

HB (g/l)	14 (1.80)
RBC ($10^6/\mu l$)	4 (0.69)
WBC ($10^3/\mu l$)	6 (2.35)
PLT ($10^3/\mu l$)	161 (73.08)
PT (sec.)	13 (9.99)
PTP (%)	91 (17.23)
APTT (sec.)	37 (7.29)
INR	1 (0.16)
ASPT (IU/l)	69 (55.14)
ALAT (IU/l)	77 (65.43)
ALP (IU/l)	105 (57.70)
BIL (mg/dl)	2 (2.42)
GGTP (IU/l)	94 (101.31)
KREA (mg/dl)	1 (0.24)
GLU (mg/dl)	93 (17.44)
Na (mmol/l)	138 (3.28)
K (mmol/l)	5 (5.74)
Fe (mmol/l)	92 (63.51)
CRP (IU/l)	5 (28.77)
TG (mg/dl)	107 (53.76)
CHO (mg/dl)	191 (53.23)
Ur. acid (mg/dl)	6 (1.35)
TP (g/dl)	7 (0.84)
TIBC	316 (95.02)
Neutr ($10^3/\mu l$)	3.42 (1.35)
Lymph ($10^3/\mu l$)	2.05 (0.55)
Mono ($10^3/\mu l$)	0.58 (0.19)
Eos ($10^3/\mu l$)	0.17 (0.13)
Baso ($10^3/\mu l$)	0.03 (0.02)
Albu (%)	59.3 (7.04)
Glb. 1(%)	2.9 (1.31)
Glb. 2 (%)	8.8 (2.57)
Glb. (%)	10.8 (1.63)
Glb. (%)	18.4 (6.89)

3 Neural Network Ensemble

The introduced method of classifier ensemble design consists of three main steps:

- Building the pool of individual classifiers.
- Pruning the acquired pool by discarding redundant predictors.
- Using a sophisticated trained fuser to deliver the final output of the ensemble.

3.1 Creating the Pool of Classifiers

One of the most important steps in the ensemble design is preparation of the individual classifiers that are used as base models for the committee [16]. Such models should be complementary to each other, exhibiting at the same time high accuracy and high diversity. By fulfilling this requirements the ensemble classifier may outperform any single model from the pool. Additionally ensemble methods allow, in a natural way, to exploit information coming from different sources—and that is why we have decided to use this approach for our application.

It is a common knowledge that there is no single optimal approach for feature selection task and results obtained on the basis of different methods may differ significantly. Therefore instead of selecting a single best feature selection method we use several of them to reduce the dimensionality of the feature space. Then, on each of their individual output, a neural networkclassifier is build. Therefore for L used feature selection methods we construct a pool of L individual classifiers:

$$\Pi^{\Psi} = \{\Psi_1, \Psi_2, ..., \Psi_L\}. \qquad (1)$$

This way we use all the reduced feature spaces in hope that they will be complementary to each other and provide a valuable contribution to the ensemble.

3.2 Ensemble Pruning

In the MCS design it is assumed that not all of the L models in Π^{Ψ} should be used as ensemble members. There are several different ways in the literature on how to select valuable members to the committee. Among them diversity measures are considered to be one of the most popular [3, 11]. They are based on the idea that ideal ensemble consists of classifiers of high individual accuracy and high diversity. Classifiers with low accuracy but high diversity will produce output of low quality, while adding similar members to the committee will only increase the computational complexity of the model.

Among diversity measures there are two major types: pairwise and non-pairwise. The former ones shows how two classifiers differ from each other, while the latter ones measure the diversity of the whole ensemble. Those two groups have different advantages and weaknesses and it is up to the researcher to select them according to his experience.

For measuring the diversity of whole ensemble we used the *entropy measure*. The highest diversity among classifiers for a particular object $x_j \in X$ is equal to the L/2 of the votes in x_j with the same value (0 or 1) and the other L [L/2] with the alternative value. Denote by $l(x_j)$ the number of classifiers that correctly recognize given sample. With this we can describe entropy-based diversity as:

$$E = \frac{1}{N}\sum_{j=1}^{N}\frac{1}{L-[L/2]}min\{l(z_j), L-l(z_j)\}. \quad (2)$$

E varies between 0 and 1, where 0 indicates no difference and 1 indicates the highest possible diversity.

An exhaustive search is performed to find the pruned pool of K classifiers exhibiting the highest diversity. Pruning ensemble for the proposed method is necessary as it is very likely that some of the used feature selection methods return similar feature subsets, thus leading to creation of similar classifiers. Using diversity-based pruning this situation can be dealt with, as redundant models that have no contribution to the ensemble are eliminated.

3.3 Fusion of Individual Classifiers

Classifier fusion algorithms can make decisions on the basis of class labels given by individual classifiers or they can construct new discriminant functions on the basis of individual classifier support functions. The first group includes voting algorithms [8], while the second group is based on discriminant analysis. The main form of discriminants is the posterior probability typically associated with probabilistic pattern recognition models, although outputs of neural networks or other functions whose values are used to establish the decision of the classifier (so called support functions) are also employed. The design of improved fusion classification models, especially trained fusers, is the focus of current research [19].

Assume that we have K classifiers $\Psi^{(1)}$, $\Psi^{(2)}$, ..., $\Psi^{(K)}$ in a pool after the pruning procedure. For a given object $x \in \mathcal{X}$, each individual classifier decides for class $i \in \mathcal{M} = \{1, ..., M\}$ based on the values of discriminants. Let $F^{(l)}(i, x)$ denote a function that is assigned to class i for a given value of x, and that is used by the l-th classifier $\Psi^{(l)}$. The combined classifier Ψ uses the decision rule:

$$\Psi(x) = i \quad if \quad \hat{F}(i, x) = \max_{k \in M} \hat{F}(k, x), \quad (3)$$

where

$$\hat{F}(i, x) = \sum_{l=1}^{K} w^{(l)} F^{(l)}(i, x) \quad and \quad \sum_{i=1}^{K} w^{(l)} = 1. \quad (4)$$

The weights can be set depending on the classifier and class number: weight $w^{(l)}(i)$ is assigned to the l-th classifier and the i-th class, and given classifier weights assigned to different classes may differ [20].

The used type of trained fuser we employ is a neural fuser can be implemented as a one-layer perceptron [23]. The values of support functions given by each of the base classifiers serve as input, while the output is the weighted support for each of the classes. One perceptron fuser is constructed for each of the classes under consideration. The perceptron may be trained with any standard procedure used in neural network learning; the input weights established during the learning process are then the weights assigned to each of the base classifiers. This method is a quite fast and it exploits the advantage of well-developed training algorithm for searching the solution space.

4 Experimental Results

4.1 Set-up

Eight different feature selection algorithms were used, namely: ReliefF [22], Fast Correlation Based Filter [21], Genetic Wrapper [5], Simulated Annealing Wrapper [5], Forward Selection [5], Backward Selection [5], Quick Branch and Bound [5] and Las Vegas Incremental [5]. Therefore the pool consisted of eight different neural networks.

Neural network architecture was as follows: the number of neurons in the input layer was equal to the number of selected features, the number of output neurons was equal to the number of classes and the number of hidden neurons was equal to the half of the sum of number of neurons from the former layers, as suggested in. Quickprop algorithm was used as a training procedure.

Genetic Wrapper used population equal to 50 with 200 iterations, probability of cross-over equal to 0.7 and probability of mutation 0.3.

As reference methods we have selected most popular ensembles—Bagging, Boosting, Random Forest and Random Subspace—as they were used in our previous work [10]—in there one may also find the details of used parameters for these ensemble classifiers. Additionally we have compared our method with the single best classifier from the pool (i.e. built on the basis of the most effective feature selection algorithm), all classifiers from the pool (i.e. without the pruning procedure) and with simple majority voting (i.e. without the trained fuser). By this we can establish the influence of three steps in our proposed ensemble on the final accuracy.

The combined 5x2 CV F test [1] was carried out to asses the statistical significance of obtained results.

All experiments were carried out in the R environment [15] and computer implementations of the classification methods used were taken from dedicated packages built into the above mentioned software. This ensured that results achieved the best possible efficiency and that performance was not diminished by a bad implementation.

Table 3 Accuracy of the investigated methods [%]

NNE[1]	*Bagg*[2]	*Boost*[3]	*RandS*[4]	*RandF*[5]	*SB*[6]	*All*[7]	*MV*[8]
90.12	80.50	84.92	88.54	87.02	83.12	88.54	86.45
2,3,4,5,6,7,8	–	2,6	2,3,5,6,8	2,3,6	2	2,3,6,7	2,3,6

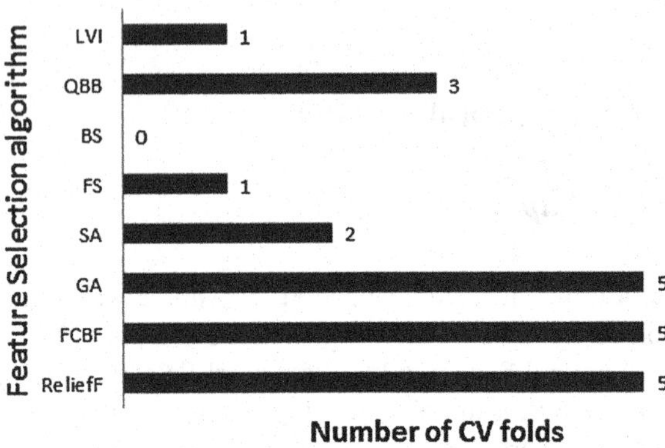

Fig. 1 Number of CV folds in which a neural network based on a given feature selection method was selected to the ensemble

4.2 Results

Results are presented in the Table 3, with small numbers under accuracy indicating from which classifiers this method is significantly better. *NNE* stands for the proposed method, *Bagg* for Bagging, *Boost* for Boosting, *RandS* for Random Subspace, *RandF* for Random Forest, *SB* for Single Best model, *All* for not pruned ensemble and *MV* for ensemble with simple majority voting scheme.

How many times a neural network classifier was selected as the ensemble member for 5x2 CV is presented in Fig. 1.

4.3 Results Discussion

The proposed neural network ensemble, based on feature selection methods, outperforms all the previously used MCS for this problem. Interesting conclusions arise from the analysis of the differences between our model and three simplified versions of it. The weakest results are returned by single best model approach, which highlights the usefulness of utilizing more than one classifier to fully exploit the outputs of feature selection methods. Second biggest accuracy boost lies in the used fuser—trained fusion of individual classifiers allows to derive an optimal linear combination of them. Finally the pruning step has the smallest but still statistically significant impact on the ensemble design. This may be due to the fact that the trained fuser search procedure worked better with the reduced pool of classifiers.

As the ensemble based on all three steps was statistically better than models with one of the steps removed one may conclude that all of them have an impact on the quality of the proposed MCS and should not be discarded.

5 Conclusions

The presented paper shows that, despite some problems (like the fact that it is not easy to get blood test results of patients with diagnosed chronic hepatitis, infected with HCV that have no other medical conditions and are not under any medical therapy, or that blood test results which were available for research were inconsistent, i.e. some patients have one set of blood tests, while other patients have a set of other blood tests) it is possible to reach similar or even lower error level than commercial tests. Applying the proposed MCS based on a neural networks coupled with different feature selection strategies, ensemble pruning and trained fusion lead to high recognition accuracy, that outperformed esteemed off-the-shelf ensemble classification methods. Additionally we proved that each of the three steps embedded in the proposed committee design has an important impact on the quality of the final prediction and thus should not be omitted. The proposed method will be used in practice by several leading Polish hospitals.

Our future works will concentrate on the problem of imbalanced class distribution among biopsy patients and possible presence of class label noise in the data.

References

1. Alpaydin, E.: Combined 5 x 2 cv f test for comparing supervised classification learning algorithms. Neural Comput. **11**(8), 1885–1892 (1999)
2. Bedossa, P., Poynard, T.: An algorithm for the grading of activity in chronic hepatitis c. the metavir cooperative study group. Hepatology **24**, 289–293 (1996)
3. Bi, Y.: The impact of diversity on the accuracy of evidential classifier ensembles. Int. J. Approximate Reasoning **53**(4), 584–607 (2012)
4. BioPredictive. http://www.biopredictive.com/intl/physician/fibrotest-for-hcv/view?set_language=en
5. Guyon, I., Gunn, S., Nikravesh, M., Zadeh L. (eds.): Feature Extraction, Foundations and Applications. Springer, Heidelberg (2006)
6. Hansen, L.K., Salamon, P.: Neural network ensembles. IEEE Trans. Pattern Anal. Mach. Intell. **12**(10), 993–1001 (1990)
7. Ishak, K., Baptista, A., Bianchi, L., Callea, F., De Groote, J., Gudat, F., Denk, H., Desmet, V., Korb, G., MacSween, R.N., et al.: Histological grading and staging of chronic hepatitis. Hepatology **22**, 696–699 (1995)
8. Kittler, J., Alkoot, F.M.: Sum versus vote fusion in multiple classifier systems. IEEE Trans. Pattern Anal. Mach. Intell. **25**(1), 110–115 (2003)

9. Knodell, R.G., Ishak, K.G., Black, W.C., Chen, T.S., Craig, R., Kaplowitz, N., Kiernan, T.W., Wollman, J.: Formulation and application of a numerical scoring system for assessing histological activity in asymptomatic chronic active hepatitis. Hepatology **1**, 431–435 (1981)
10. Krawczyk, B., Woźniak, M., Orczyk, T., Porwik, P., Musialik, J., Błońska-Fajfrowska, B.: Classification techniques for non-invasive recognition of liver fibrosis stage. J. Med. Inform. Technol. **20**, 121–127 (2012)
11. Krawczyk, B., Woźniak, M.: Combining diverse one-class classifiers. In: Corchado, E., Snasel, V., Abraham, A., Wozniak, M., Grana, M., Cho, S-B. (eds.) Hybrid Artificial Intelligent Systems, volume 7209 of Lecture Notes in Computer Science, pp. 590–601. Springer, Berlin (2012)
12. Siemens Medical. http://www.medical.siemens.com/webapp/wcs/stores/servlet/PSGenericDisplay~q_catalogId~e_-111~a_langId~e_-111~a_pageId~e_103713~a_storeId~e_10001.htm
13. Orczyk, T., Pałys, M., Porwik, P., Musialik, J., Błońska-Fajfrowska, B.: Simple and non-invasive liver fibrosis stage prediction method. J. Med. Inform. Technol. **17**, 227–232 (2011)
14. Rokach, L.: Pattern Classification Using Ensemble Methods. Series in Machine Perception and Artificial Intelligence. World Scientific Publishing, Singapore (2010)
15. R Development Core Team: R: A Language and Environment for Statistical Computing. R Foundation for Statistical Computing, Vienna, Austria (2008)
16. Ting, Kai, Wells, Jonathan, Tan, Swee, Teng, Shyh, Webb, Geoffrey: Feature-subspace aggregating: ensembles for stable and unstable learners. Mach. Learn. **82**, 375–397 (2011)
17. Wai, C.T., Greenson, J.K., Fontana, R.J., Kalbfleisch, J.D., Marrero, J.A., Conjeevaram, H.S., Lok, A.S.: A simple noninvasive index can predict both significant fibrosis and cirrhosis in patients with chronic hepatitis c. Hepatology **38**, 518–526 (2003)
18. Wolpert, DH.: The supervised learning no-free-lunch theorems. In: Proceedings of the 6th Online World Conference on Soft Computing in Industrial Applications, pp. 25–42 (2001)
19. Michal Wozniak. Experiments with trained and untrained fusers. In Emilio Corchado, Juan Corchado, and Ajith Abraham, editors, Innovations in Hybrid Intelligent Systems, volume 44 of Advances in Soft Computing, pages 144–150. Springer Berlin / Heidelberg, 2007.
20. Wozniak, Michal, Zmyslony, Marcin: Combining classifiers using trained fuser—analytical and experimental results. Neural Netw. World **13**(7), 925–934 (2010)
21. Yu, L., Liu, H.: Feature selection for high-dimensional data: a fast correlation-based filter solution. In: Proceedings of the Twentieth International Conference on Machine Learning, vol. 2, pp. 856–863 (2003)
22. Yu, L., Liu, H.: Efficient feature selection via analysis of relevance and redundancy. J. Mach. Learning Res. **5**, 1205–1224 (2004)
23. Zmyslony, M., Krawczyk, B., Wozniak, M.: Combined classifiers with neural fuser for spam detection. In: Herrero, A., Snasel, V., Abraham, A., Zelinka, I., Baruque, B., Quintin, H, Calvo, JL., Sedano, J., Corchado, E. (eds.) International Joint Conference CISIS12-ICEUTE12-SOCO12 Special Sessions, volume 189 of Advances in Intelligent Systems and Computing, pp. 245–252. Springer, Heidelberg (2012)

Cooperative and Non-cooperative Equilibrium Problems with Equilibrium Constraints: Applications in Economics and Transportation

Andrew Koh

Abstract In recent years, a plethora of multi-objective evolutionary algorithms (MOEAs) have been proposed which are able to effectively handle complex multi-objective problems. In this paper, we focus on Equilibrium Problems with Equilibrium Constraints. We show that one interpretation of the game can also be handled by MOEAs and then discuss a simple methodology to map the non-cooperative outcome to the cooperative outcome. We demonstrate our proposed methodology with examples sourced from the economics and transportation systems management literature. In doing so we suggest resulting policy implications which will be of importance to regulatory authorities.

1 Introduction

This focus of this paper is on hierarchical optimization problems. Figure 1 shows the structure of a single-leader follower game/bilevel optimization problem [8] which has attracted attention in the Evolutionary Computation community in recent years.[1] In this paper we study a generalized variant of such problems known as Equilibrium Problems with Equilibrium Constraints (EPECs) as illustrated in

The author is grateful for financial support by the Engineering and Physical Sciences Research Council of the UK under Grant EP/H021345/1.

[1] E.g. a Special Session on Bilevel Optimization was convened at the 2012 IEEE Congress on Evolutionary Computation (CEC) (June 10–15) in Brisbane, Australia.

A. Koh (✉)
Institute for Transport Studies, University of Leeds, Leeds, LS2 9JT, Leeds, UK
e-mail: a.koh@its.leeds.ac.uk
URL: http://www.its.leeds.ac.uk

V. Snášel et al. (eds.), *Soft Computing in Industrial Applications*,
Advances in Intelligent Systems and Computing 223, DOI: 10.1007/978-3-319-00930-8_3,

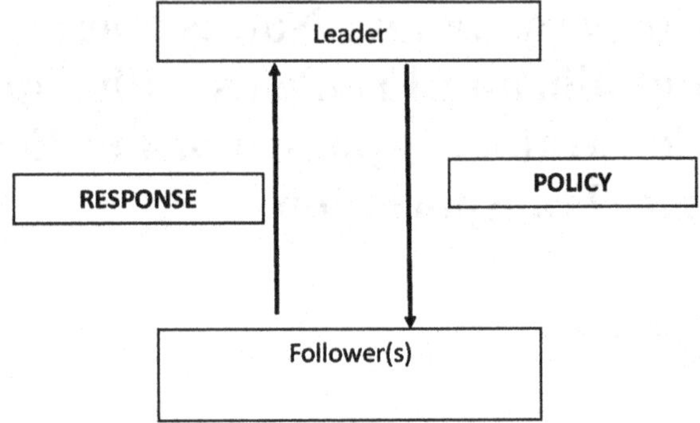

Fig. 1 Bilevel program or mathematical problem with equilibrium constraint (MPEC)

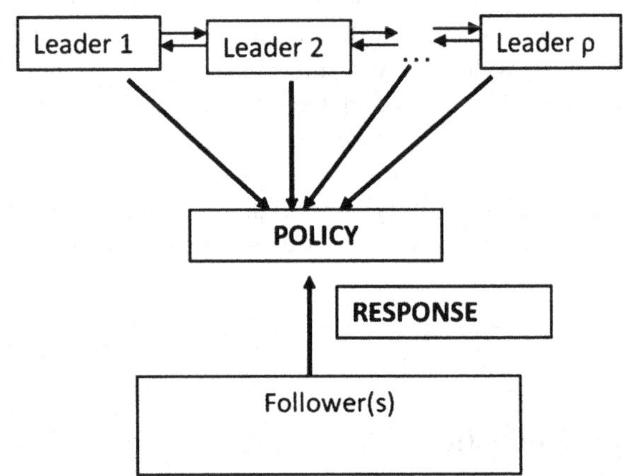

Fig. 2 Equilibrium problem with equilibrium constraint (EPEC)

Fig. 2. Both are hierarchical games where the followers take the leader's variables as given and their responses are subsequently imposed as a nonlinear binding constraint on the actions of the leader(s). The difference is that EPECs are characterized by the presence of multiple leaders.

In this multi-leader generalization of the classic Stackelberg [18] game, researchers have conjectured that there could be two possible behaviors of the leaders [11, 15]. At one extreme, they could cooperate and such a postulate leads naturally to a Multi-Objective EPEC (hereinafter termed MOPEC) [20]. At the other extreme, these leaders could act engage in a non-cooperative Nash [14] game amongst themselves thereby resulting in a Non Cooperative EPEC (NCEPEC).

Under either postulate of leader behavior, we argue that meta-heuristics offer a powerful solution methodology for EPECs that are usually tackled using tools of generalized calculus [11]. For MOPECs, population based MultiObjective Evolutionary Algorithms (MOEAs) are particularly suitable due to their inherent ability to identify multiple Pareto Optimal solutions in a single run [2]. For the latter class of NCEPECs, a Differential Evolution [16] based algorithm exploiting a concept from [10] has been proposed in [9]. In this paper we suggest a methodology that maps the Non Cooperative outcome to the Cooperative outcome by modification of the algorithm proposed in [9].

The rest of this paper is structured thus. We give an overview of notation used in this paper in the next section. In Sect. 3 fundamental notions of Multi-Objective optimization are reviewed before an evolutionary algorithm for solving MO problems is given. Section 4 outlines a solution algorithm for NCEPECs. Section 5 discusses a simple method to map the Non-Cooperative outcome to the Cooperative outcome. Section 6 illustrates the concepts with numerical examples. Section 7 summarizes and provides directions for further research.

2 Notation

In this paper we consider ρ-person games. Focusing on the leaders, each game is defined by a tuple $\{N, X_i, U_i\}$ where N is the set of leaders $\{1, 2, \ldots, \rho\}$, X_i is the strategy/action space for leader $i, i \in N$ and U_i is the payoff function (or reward), $U_i : \mathbb{R}^\rho \to \mathbb{R}^1$, that a leader gets by playing an action/strategy, dependent on the actions which all others take. The collective action of all leaders, often referred to as a strategy profile, is denoted by $\boldsymbol{x} = [x_1, \ldots, x_i, \ldots, x_\rho]^\mathsf{T}$. It is convenient to write x_{-i} when referring to the strategies of every leader *excluding* the leader i i.e. $x_{-i} = [x_1, \ldots, x_{i-1}, x_{i+1}, \ldots, x_\rho]^\mathsf{T}$. With a slight abuse of notation, we also write $\boldsymbol{x} = [x_i, x_{-i}]^\mathsf{T}$. Note that $[x_i, x_{-i}]$ does not mean that the components of $\boldsymbol{x}$ are reordered, so that strategies of leader i becomes the first block. Unless otherwise specified, all vectors are assumed to be column vectors.

The response of the followers that affects the actions of the leaders is assumed to take the form of a Variational Inequality (VI) constraint that defines equilibrium in some parametric system. Following [12] we assume that the solution of this VI exists and is unique for a given a vector of the leaders' strategies.

3 Cooperative EPECs (MOPECS)

A generic MOPEC is shown in Eq. 1. Except for the variational inequality constraint, this problem takes the form of a generic multi objective optimization problem conventionally handled by MOEAs [1, 2].

$$\text{Program MOPEC} \quad \left\{ \max_{\boldsymbol{x} \in X} (U_1(\boldsymbol{x}, \boldsymbol{y}), \ldots, U_\rho(\boldsymbol{x}, \boldsymbol{y}))^\mathsf{T} \right. \tag{1a}$$

where for given $\boldsymbol{x}$, $\boldsymbol{y}$ is the unique solution of the Variational Inequality in 1b:

$$L(\boldsymbol{x}, \boldsymbol{y})^\mathsf{T}(\boldsymbol{y} - \boldsymbol{y}^*) \geq 0, \forall \boldsymbol{y} \in \Upsilon(\boldsymbol{x}) \tag{1b}$$

MOEAs apply stochastic operators to a parent population to evolve a fitter child population to solve multi-objective problems. During the selection phase, a comparison is made between a chromosome $\boldsymbol{a}$ from the parent population and a chromosome $\boldsymbol{b}$ from the child population on the basis of fitness and the weaker of the two is discarded. Since one of the tasks of an MOEA is to identify the entire Pareto front [2], fitness is assigned based on Pareto Domination: $\boldsymbol{a}$ Pareto Dominates $\boldsymbol{b}$ if $\boldsymbol{a}$ is no worse off than $\boldsymbol{b}$ in all objectives *and* $\boldsymbol{a}$ is strictly better than $\boldsymbol{b}$ in at least one objective ([2], Definition 2.5, pp. 28).

Algorithm 1 Multi-Objective Self Adaptive Differential Evolution (MOSADE) [5]

1. Evaluate initial population P of $|P|$ random individuals.
2. Set archive $\mathcal{A}$ to $\emptyset$
3. While stopping criterion not met, do:
   For each individual $P_i, i \in \{1, \ldots, |P|\}$ repeat:
   (a) Use DE to create candidate C_i from parent P_i.
   (b) Evaluate C_i by solving lower level VI 1b
   (c) If P_i dominates C_i, discard C_i else go to Step 4
4. Compare C_i with each member of $\mathcal{A}$,
   (a) if maximum size of $\mathcal{A}$ reached, choose between C_i or each member of $\mathcal{A}$ depending on which occupies the less crowded region of function space
   (b) if C_i dominates any $\mathcal{A}$, remove the member of $\mathcal{A}$ so dominated, accept C_i into $\mathcal{A}$
   (c) if C_i is dominated by any $\mathcal{A}$, reject C_i
5. Update DE control parameters as described in [5]

Algorithm 1 outlines the Multi-Objective Self Adaptive Differential Evolution (MOSADE) Algorithm [5] that was used to generate the Pareto Fronts for the MOPECs to be described in Sect. 6. MOSADE uses an archive to store solutions as they are discovered during the search process. To evaluate the candidate, it is necessary to solve the lower level VI problem in Eq. 1b to maintain the leader-follower paradigm implicit in such hierarchical optimization problems [8].

4 Non Cooperative EPECs (NCEPECS)

In the NCEPEC each leader i treats his competitor's strategic variables as exogenous when maximizing his payoff as in Eq. 2.

$$\forall i \in N, \quad \text{Player } i \text{ solves: } \left\{ \max_{x \in X} U_i(\boldsymbol{x}, \boldsymbol{y}) \right. \tag{2}$$

where for given $\boldsymbol{x}$, $\boldsymbol{y}$ is the unique solution of the Variational Inequality in 1b.

It can be shown that the solution of Eq. 2, if one exists, is a Nash Equilibrium (NE) which is obtained when the condition in Eq. 3 is satisfied [14].

$$U_i(x_i^*,\ x_{-i}^*) \geq U_i(x_i,\ x_{-i}^*)\ \forall x_i \in X_i\ , \forall i \in N \tag{3}$$

Traditional approaches for locating NE are based on fixed point algorithms (e.g. non-linear Gauss-Siedel [4]) or through resolution of a Complementarity Problem formulation [6]. If players can benefit (i.e. increase payoff) from deviating from their current action, then that action cannot be a NE. By counting the number of players that can potentially benefit from deviating, we can compare two chromosomes (representing two strategy profiles, $(\{\boldsymbol{a}, \boldsymbol{b}\} \in X)$) to determine which is closer to a NE and thus deemed "fitter". We say that ***a*** ***Nash Dominates*** ***b*** if there are *fewer* players that benefit from unilaterally deviating (to $\boldsymbol{b}$) when playing $\boldsymbol{a}$ compared to deviating when playing $\boldsymbol{b}$. Based on this principle an evolutionary algorithm for NCEPECS called Nash Domination Evolutionary Multiplayer Optimization (NDEMO) was proposed in [9] as summarized in Algorithm 2. From the proof in [10], the convergence of NDEMO to the NE, if one exists, is theoretically assured for any arbitary NCEPEC.

Algorithm 2 Nash Domination Evolutionary Multiplayer Optimization (NDEMO) [9]

1. Evaluate initial population P of $|P|$ random individuals.
2. While stopping criterion not met, do:
   For each individual $P_i, i \in \{1, \ldots, |P|\}$ repeat:
   (a) Use DE to create candidate C_i from parent P_i.
   (b) Evaluate C_i by solving lower level VI 1b
   (c) If C_i *Nash Dominates* P_i, C_i replaces P_i.
   Else discard C_i.

5 Mapping Non Cooperative to Cooperative Solution

We can map the non-cooperative to the cooperative solution by modifying the objective function of leader i such that he takes into account a proportion $(\alpha, 0 \leq \alpha \leq 1)$ of the payoff of all other leaders when optimizing his own payoff U_i as shown in Eq. 4. It is not hard to see that with $\alpha = 0$ we recover the objective function in Eq. 2.

$$\forall i \in N,\ \text{Player } i \text{ solves: } \left\{ \max_{x \in X} (U_i(\boldsymbol{x},\ \boldsymbol{y}) + \sum_{j=1, j \neq i}^{N} \alpha U_j(\boldsymbol{x},\ \boldsymbol{y})) \right. \tag{4}$$

where for given $\boldsymbol{x}$, $\boldsymbol{y}$ is the unique solution of the Variational Inequality in 1b.

Table 1 The two solutions reported in [12] and indicated on Fig. 3 with ☆

	Solution 1	Solution 2
Profit of Leader 1	840.86	978.89
Profit of Leader 2	485.63	410.97

6 Numerical Examples

In this section, we present two numerical examples that aim to demonstrate the mapping of the NCEPEC to the MOPEC solution.

6.1 Example 1: Competition between Producers

We first consider a 5 player model from [12] using data found in [4] and [13]. The case where 2 of the 5 players emerged as the leaders was discussed in [12] who reported two possible solutions of the resulting MOPEC.[2]

The leaders' optimization problem is subject to the followers' response manifesting in a VI that is imposed as a constraint on the actions of the leaders. In practical implementation, the PATH solver [3] is used to resolve the lower-level VI for a given tuple of the leaders' strategies $\boldsymbol{x}$ to obtain the responses of the lower-level followers $\boldsymbol{y}$. [9] shows that the solutions reported in [12] are only two out of all possible Pareto non dominated solutions found by MOSADE (see Fig. 3; Table 1).

In [9], we also computed the case for when these two leaders engaged in non-cooperative behavior resulting in a NCEPEC. As reported in [9] the NCEPEC solution results in production levels of 97.70 units for Leader 1 and 42.14 units for Leader 2 with corresponding profits of 950.56 and 414.72. In profit space, this point is indicated as $\times$ on Fig. 3. The arrows indicate that the NCEPEC solution is not Pareto Optimal since any leader can be made better off (i.e. increase profit) without making the other worse off. Note that while the solution with $\alpha = 1$ lies on the Pareto Front (and hence Pareto Optimal) this is *different* from Solution 2 reported in [12] and shown in Fig. 4 with ☆.

In order to map the non-cooperative outcome to the cooperative outcome, NDEMO (Algorithm 2) was applied each time fixing α, in steps of 0.2, between 0 and 1. The results are indicated on Fig. 4 with detailed results in Table 2. Notice that the NCEPEC solution reported in [9] is obtained with $\alpha = 0$. It is clear that as α increases, the profits accruing to the leaders tend towards the Pareto Front. We term the path, from the NCEPEC solution ($\alpha = 0$) to the Pareto Front ($\alpha = 1$) the "collusion path".

---

[2] Recall that this is the case where we assumed that the leaders cooperated.

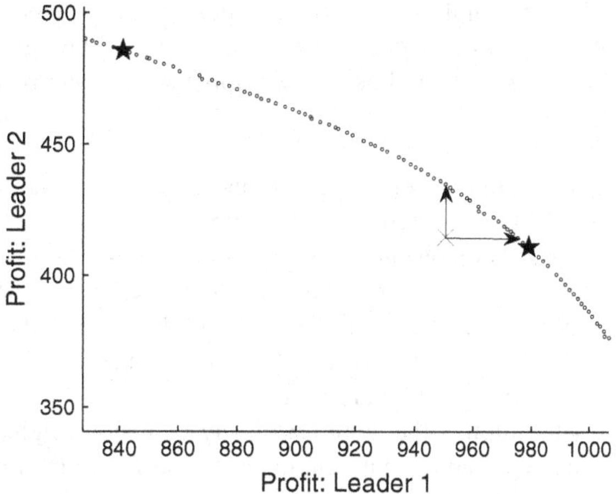

Fig. 3 Example 1: Pareto front generated by MOSADE alongside solutions in [12] indicated by ☆, NCEPEC solution indicated by ×

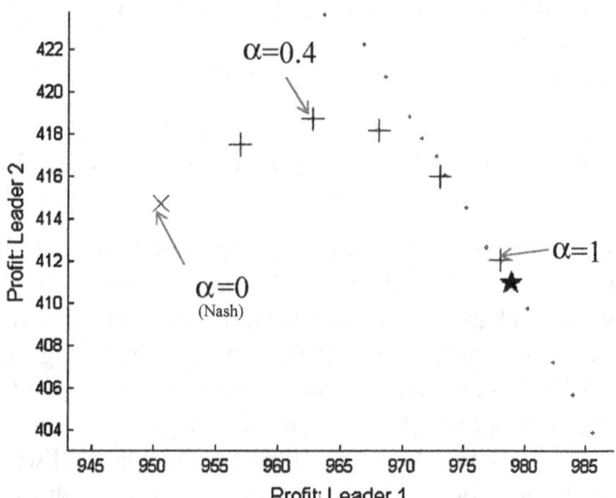

Fig. 4 Example 1 : (zoomed) Pareto front (Cooperative EPEC) and "collusion path" mapping NCEPEC ($\alpha = 0$) to Pareto Front ($\alpha = 1$)

Table 2 Example 1: Production quantities and profits for leaders as α increases

	Leader 1		Leader 2	
α	Quantity	Profit	Quantity	Profit
0 (NCEPEC)	97.70	950.56	42.14	414.73
0.2	95.81	956.95	40.58	417.53
0.4	94.04	962.76	39.01	418.70
0.6	92.40	968.08	37.43	418.19
0.8	90.90	973.07	35.84	416.01
1 (Pareto Front)	89.56	977.93	34.23	412.06

This example shows that it is possible for leaders to signal to each other their intention to cooperate and maximize profit (through the α parameter) and hence engage in "tacit collusion". In doing so, a leader can reduce output resulting in increased total profit. Doing so would send signals to the other leader(s) to indicate their willingness to collude.

Notice in Fig. 4 that the "collusion path" does not lie on a straight line between the NCEPEC solution and the Pareto Front. While Leader 1's profit continues to rise as α increases and thus will have more to gain from collusion, this is not true for Leader 2. In particular, the "collusion path" provides maximum profit for Leader 2 at $\alpha = 0.4$ (c.f. Table 2) but decreases beyond that. In fact the profit for Leader 2 at $\alpha = 1$ is lower than that obtained under NCEPEC even though that solution lies on the Pareto Front.

This implies the stability of any collusion might be undermined. This could eventually lead to an all out quantity war for which Leader 1 could be made worse off if conditions reverted back to the non-cooperative situation. Leader 1 could potentially compensate Leader 2 (e.g. by allocating Leader 2 a share of the profits gained) in such a way that Leader 2 would still be incentivised to cooperate. Regulators clearly need to be aware of such behavior when enforcing anti-trust legislation.

6.2 Example 2: Competition between Authorities

A situation in which two city transportation authorities were the leaders at the upper level was studied in [21]. In this context the strategic variable was the toll price to charge on traffic using road(s) on the network upon which each city exercised jurisdiction with the aim of maximising individual city welfare.

City welfare is a function of the traffic flows due to the routing of traffic on the road network which is in turn influenced by the toll levels charged. Traffic routing must satisfy Wardrop's equilibrium principle [19] which states that traffic arranges on the network such that at an equilibrium, the cost of all used routes/paths connecting any individual Origin-Destination pair is equalized. Wardrop's equilibrium principle can be expressed as a VI [17]. The usual way of obtaining the traffic flows, once the tolls are input, is through traffic assignment [7].

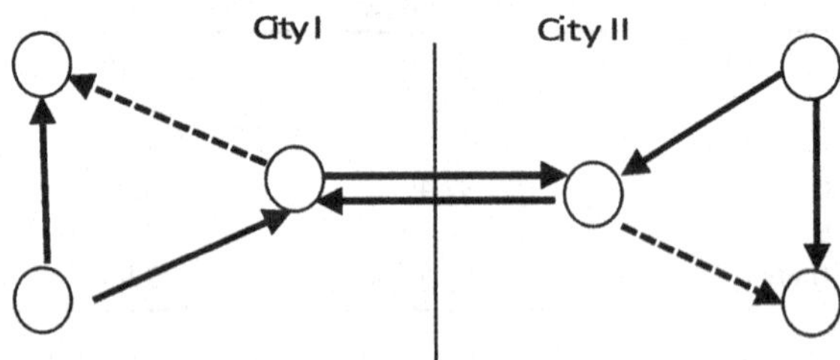

Fig. 5 Directed network [21] for example 2 with the *line down* the middle demarcating authority jurisdiction. *Dashed arcs* are *tolled arcs* in each city

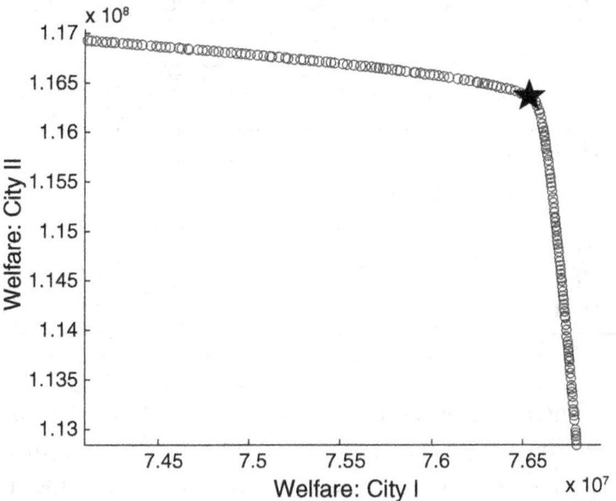

Fig. 6 Example 2: Pareto front (Cooperative EPEC) Cities I and II, Cooperative solution from [21] indicated with ☆

Among several different governance models studied in [21], two cases of most relevance to this work are as follows:

1. the authorities engage in a Nash game by setting tolls with each maximising individual city welfare subject to Wardrop's equilibrium expressed as a VI i.e. a NCEPEC.
2. the authorities cooperate to set tolls to maximise both cities' welfare simultaneously subject to Wardrop's equilibrium expressed as a VI i.e. a MOPEC.

The network is shown in Fig. 5 where dashed arcs are subject to tolls in each Authority respectively. The Pareto Front for the MOPEC where the cities cooperate to maximize welfare is shown in Fig. 6. The single solution reported in [21] also lies on this Pareto Front (indicated by ☆ on Fig. 6).

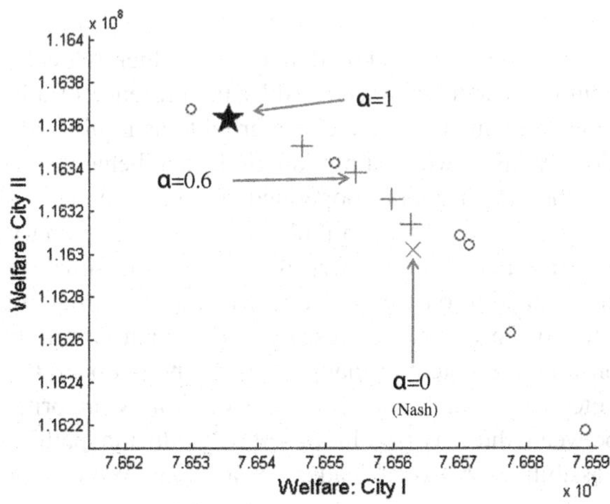

Fig. 7 Example 2: (zoomed) Pareto front (Cooperative EPEC) and "collusion path" mapping NCEPEC ($\times, \alpha = 0$) to Pareto front ($\alpha = 1$)

Table 3 Example 2: Tolls and welfare for leaders as α increases

	City I		City II	
α	Toll (secs)	Welfare (10,000 secs)	Toll (secs)	Welfare (10,000 secs)
0.00 (NCEPEC)	4943	7656.28	4957	11630.22
0.20	4847	7656.25	4931	11631.40
0.40	4748	7655.97	4902	11632.59
0.60	4648	7655.45	4871	11633.80
0.80	4544	7654.65	4838	11635.03
1.00 (Pareto Front)	4439	7653.56	4802	11636.27

Figure 7 shows the result (indicated by $\times$) when $\alpha = 0$ i.e. the NCEPEC solution where the leaders played a Nash game instead. This point is not Pareto Optimal since one city can increase welfare without making the other worse off.

The "collusion path" mapping the NCEPEC to the Pareto Front and Table 3 shows that the welfare for City I marginally decreases as α increases (and the opposite for City II). Hence whether the cooperative solution is sustainable (because City II benefits but City I marginally loses out) is an issue that warrants further research. We also notice that the tolls fall as α rises, again as an indication of signalling behavior to the other authority.

We notice that City I's welfare (c.f. Table 3) continuously decreases as we move from the NCEPEC solution to the MOPEC solution (and eventually lower than the NCEPEC outcome at $\alpha = 1$). Such a situation implies that City II might have to compensate City I so that the latter would be incentivised to cooperate. Again, this opens up a plethora of further research possibilities studying the policy implications in such situations e.g. stability of agreements.

7 Conclusions

In this paper, we studied a class of hierarchical optimization problems with multiple leaders characterized by the presence of a binding variational inequality. This problem is collectively referred to as Equilibrium Problems with Equilibrium Constraints. Two assumption of leader behavior were discussed depending on whether the leaders cooperated to optimize their objectives or otherwise. We showed that advances in multi-objective evolutionary algorithms could be used to generate Pareto Fronts that represented the situation in which the leaders cooperate. In addition we have already proposed an algorithm for the non cooperative situation in our earlier research. The main contribution of this paper is to demonstrate the potential mapping of the non-cooperative solution to the cooperative outcome through the use of evolutionary algorithms. We term the mapping between these two solutions the "collusion path" since it paths the collusion possibilities between leaders in a game. With numerical examples drawn from both the economics and transportation systems management literature, we demonstrated the role that this path plays in assisting policy makers in developing anti-trust legislation.

In terms of policy research, further work should be undertaken to understand the "collusion path" as this will affect the incentives for cooperative action. With regard to evolutionary algorithms, though we have theoretical assurance of the convergence of NDEMO to a NE for arbitary NCEPEC, each run will still take some time. Hence investigating methodologies to speed up convergence of the NDEMO algorithm for the non-cooperative case would be continue to be a useful area of research.

References

1. Coello-Coello, C., Lamont, G.: Applications of Multi-objective Evolutionary Algorithms. World Scientific, Singapore (2004)
2. Deb, K.: Multi-objective Optimization Using Evolutionary Algorithms. John Wiley, Chichester (2001)
3. Ferris, M., Munson, T.: Complementarity problems in GAMS and the PATH solver. J. Econ. Dyn. Control **24**(2), 165–188 (2000)
4. Harker, P.T.: A variational inequality approach for the determination of oligopolistic market equilibrium. Math. Program. **30**(1), 105–111 (1984)
5. Huang, V.L., Qin, A.K., Suganthan, P.N., Tasgetiren, M.F.: Multi-objective optimization based on self-adaptive differential evolution algorithm. In: Proceedings of IEEE CEC, pp. 3601–3608. IEEE Press, Piscataway, New Jersey (2007)
6. Karamardian, S.: Generalized complementarity problems. J. Optimiz. Theory App. **8**(3), 161–168 (1971)
7. Koh, A., Watling D.: Traffic assignment modelling. In: Button, K., Vega, H., Nijkamp, P. (eds.) A Dictionary of Transport Analysis, PP. 418–420. Edward Elgar, Cheltenham (2010)
8. Koh, A.: Solving transportation bi-level programs with differential evolution. Proceedings of the IEEE Congress on Evolutionary Computation, pp. 2243–2250. IEEE Press, Piscataway, New Jersey (2007)
9. Koh, A.: An evolutionary algorithm based on Nash dominance for equilibrium problems with equilibrium constraints. Appl. Soft Comput. **12**(1), 161–173 (2012)
10. Lung, R.I., Dumitrescu, D.: Computing Nash equilibria by means of evolutionary computation. Int. J. Comput. Commun. **III** (Suppl. Issue–ICCCC 2008), 364–368 (2008)
11. Mordukhovich, B.S.: Variational Analysis and Generalized Differentiation I: Basic Theory. Springer, Berlin (2006)
12. Mordukhovich, B.S., Outrata, J.V., Červinka, M.: Equilibrium problems with complementarity constraints: case study with applications to oligopolistic markets. Optimization **56**(4), 479–494 (2007)
13. Murphy, F.H., Sherali, H.D., Soyster, A.L.: A mathematical programming approach for determining oligopolistic market equilibrium. Math. Prog. **24**(1), 92–106 (1982)
14. Nash, J.: Non-cooperative games. Ann. Math. **54**(2), 286–295 (1951)
15. Outrata, J.V.: A note on a class of equilibrium problems with equilibrium constraints. Kybernetika **40**(5), 585–594 (2004)
16. Price, K., Storn, R., Lampinen, J.: Differential Evolution: A Practical Approach to Global Optimization. Springer, Berlin (2005)
17. Smith, M.J.: The existence, uniqueness and stability of traffic equilibria. Transp. Res. B-Meth. **13**(4), 295–304 (1979)

18. von Stackelberg, H.H.: The Theory of the Market Economy. William Hodge, London (1952)
19. Wardrop, J.G.: Some theoretical aspects of road traffic research. In Proceedings of the Institute of Civil Engineers, vol. 1, issue no. 36, pp. 325–378 (1952)
20. Ye, J.J., Zhu, Q.J.: Multiobjective optimization problem with variational inequality constraints. Math. Prog. **96A**(1), 139–160 (2003)
21. Zhang, X.N., Zhang, H.M., Huang, H.J., Sun, L.J., Tang, T.Q.: Competitive, cooperative and Stackelberg congestion pricing for multiple regions in transportation networks. Transportmetrica **7**(4), 297–320 (2011)

Statistical Genetic Programming: The Role of Diversity

Maryam Amir Haeri, Mohammad Mehdi Ebadzadeh and Gianluigi Folino

Abstract In this chapter, a new GP-based algorithm is proposed. The algorithm, named SGP (Statistical GP), exploits statistical information, i.e. mean, variance and correlation-based operators, in order to improve the GP performance. SGP incorporates new genetic operators, i.e. Correlation Based Mutation, Correlation Based Crossover, and Variance Based Editing, to drive the search process towards fitter and shorter solutions. Furthermore, this work investigates the correlation between diversity and fitness in SGP, both in terms of phenotypic and genotypic diversity. First experiments conducted on four symbolic regression problems illustrate the goodness of the approach and permits to verify the different behavior of SGP in comparison with standard GP from the point of view of the diversity and its correlation with the fitness.

1 Introduction

Maintaining diversity in the genetic programming is important, because it helps to prevent the GP process from a premature convergence. The lack of diversity may lead to convergence towards local optima or towards a not optimal behavior in dynamic environments. Therefore, experimental analysis of diversity can give us a better perspective about the population transition and the search process in GP.

M. Amir Haeri (✉) · M. M. Ebadzadeh
Department of Computer Engineering and Information Technology,
Amirkabir University of Technology, Tehran, Iran
e-mail: haeri@aut.ac.ir

M. M. Ebadzadeh
e-mail: ebadzadeh@aut.ac.ir

G. Folino
ICAR-CNR, Rende, Italy
e-mail: folino@icar.cnr.it

V. Snášel et al. (eds.), *Soft Computing in Industrial Applications*,
Advances in Intelligent Systems and Computing 223, DOI: 10.1007/978-3-319-00930-8_4,

According to this, diversity in genetic programming is studied by many researchers working in the GP field. Some of them tried to define appropriate phenotypic or genotypic diversity measures. Rosca [9, 10] suggested a phenotypic measure based on the number of different fitness values in the population. Analogously, Langdon [7] defined genotypic diversity as the number of different structures in the population. Some of the genotypic diversity measures have been defined on the basis of the edit distance between structures in the GP population [2, 3].

Folino et al. [5] analyzed the effectiveness of parallel genetic programming models in maintaining diversity in a population, i.e. island and cellular GP, using phenotypic and genotypic entropy. Their study confirms that the considered parallel models help to promote diversity but the authors conclude no relation between diversity measures and goodness of the fitness can be obtained. Jackson [6] investigated the effects of mutation operator on enhancing the diversity in GP population. He reported that the role of mutation operator in enhancing the diversity depends on the nature of the problem. In three of his test problems mutation did not have a significant effect on any diversity measures, while in one case, mutation operator had a strong influence on improving the structural diversity.

Burke et al. [1] analyzed different types of diversity measures and investigated the importance of these measures and their correlation with the fitness in genetic programming. Their results demonstrate that there is a correlation between fitness and diversity. In particular, a positive correlation between the phenotypic diversity and the fitness and a negative correlation between the genotypic diversity and the fitness were observed in many problems. However, they concluded that this correlation must not be interpreted as a factor of causality, i.e. "...higher diversity does not necessarily cause better performance, but better performance is seen with higher diversity." Finally, in regression problems, they discovered the weakest values of correlation and that is one of the reasons why we decided to explore more deeply the behavior in terms of diversity in this kind of problems.

In recent years, both analyzing diversity and correlation and improving genetic programming has become the focus of many researchers. Among the desired properties, a GP-based algorithm should reduce the code growth and efficiently explore the huge search space of real hard problems considered.

To this aim, a new GP algorithm, named Statistical Genetic Programming (SGP) is introduced in this chapter. The novelty of the method is based on the exploitation of statistical information obtained in the structure of the individuals and in the building of new powerful genetic operators. SGP introduces three new operators, Correlation Based Crossover, Correlation Based Mutation and Variance Based Editing. The effect of these three operators is to decrease the rate of the code growth, while maintaining efficacy in exploring the search space. SGP is particularly apt to cope with symbolic regression problems; however we would like to remark that the algorithm can be also used for other kinds of problems, if the function associated to a node can be computed as a function of the input variables. It will be clearer in the next section. To study the behavior of the search process in SGP, and in regression problems in particular, the population diversity and its correlation with the fitness is analyzed, using phenotypic and genotypic measure of diversity.

The rest of the chapter is organized as follows: In the Sect. 2, Statistical Genetic Programming is introduced. Section 3 presents the diversity measures used in this chapter. Section 4 is devoted to the description of the test problems and to the experimental results. Section 5 concludes the chapter.

2 Statistical Genetic Programming

In this section, a new GP algorithm named *Statistical Genetic Programming* (SGP) is introduced. The SGP utilizes statistical information to improve the performance of the standard GP. Before introducing the operators of SGP, firstly, it should be clarified what we mean by the statistical information of a GP tree.

2.1 Statistical Information of a GP Tree

Statistical information in a GP tree can be exploited in order to drive the evolutionary process in the case in which each node in the GP tree is a function of the input variables, i.e. in symbolic regression problems.

The SGP algorithm computes, for each node of all its subtrees, the following values: $E[g_i] = \frac{1}{M}\sum_{j=1}^{M} g_i(X_j)$, $E[g_i^2] = \frac{1}{M}\sum_{j=1}^{M} g_i^2(X_j)$ and $E[g_i{\cdot}y] = \frac{1}{M}\sum_{j=1}^{M} y_j g_i(X_j)$, where $g_i(X_j)$ is the function of node i. $X_j = (x_{j1}, x_{j2}, \ldots, x_{jn})$ is the vector of input variables and n is the number of variables. M is the number of training data and $y_j = f(X_j)$ is the value of the (to be estimated, in the following named regression) function f at the point X_j. In order to compute these values the mean (m) and variance (σ^2) of each node and the correlation coefficient (ρ) of each node with f can be computed as follows:

$$m = E[g_i] \tag{1}$$

$$\sigma^2 = E[g_i^2] - E[g_i]^2 \tag{2}$$

$$\rho = \frac{E[g_i{\cdot}y] - E[g_i]E[y]}{\sigma_{g_i}\sigma_y} \tag{3}$$

An example of a GP tree and its statistical information is shown in Fig. 1. Suppose that we have a symbolic regression problem with regression data (fitness cases) as shown in Fig. 1, and let the depicted tree represent an individual of the GP population. Each node of the tree implies some function; for instance, the function of node n_6 is $g_6 = 0.6 * x_1$. The "tree function" is the function implied by the root of the tree. Based on the regression data, one can compute statistical information—mean or variance—for each node of the tree (i.e. the function implied by it). For instance, for node n_4 ($g_4 = x_2 + x_1 * 0.6$) and the given regression data, the mean of output of g_4 is equal to $E[g_4] = 0.68$. All relevant data are tabulated in Fig. 1. Another useful statistical information is the correlation coefficient of the outputs of each node function with the desired output values of f (Regression function). This measure can indicate the relation between the function of each node and desired function, and shows how much a subtree is effective in constructing the desired function.

Regression Data		
x_1	x_2	$f(x_1,x_2)$
0.1	0.3	1.38
0.4	0.5	1.93
0.2	0.2	1.42
0.8	0.6	3.42
0	0.9	1.65

Node	Function (g_i)
n_1	$g_1(x_1,x_2)=x_1(x_2+0.6x_1)+0.3$
n_2	$g_2(x_1,x_2)=x_1(x_2+0.6x_1)$
n_3	$g_3(x_1,x_2)=x_1$
n_4	$g_4(x_1,x_2)=x_2+0.6x_1$
n_5	$g_5(x_1,x_2)=x_2$
n_6	$g_6(x_1,x_2)=0.6x_1$
n_7	$g_8(x_1,x_2)=0.6$
n_8	$g_8(x_1,x_2)=x_1$
n_9	$g_8(x_1,x_2)=0.3$

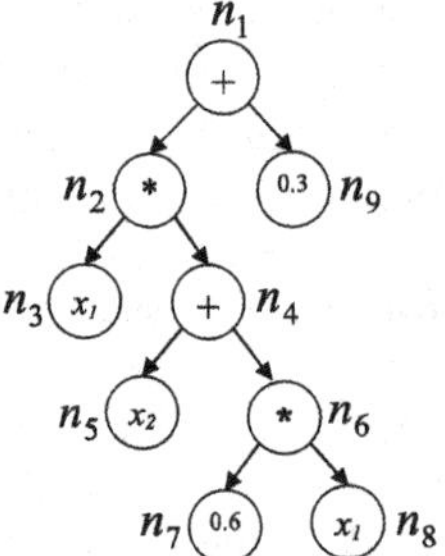

x_1	x_2	$g_1(x_1,x_2)$	$g_2(x_1,x_2)$	$g_3(x_1,x_2)$	$g_4(x_1,x_2)$	$g_5(x_1,x_2)$	$g_6(x_1,x_2)$	$g_7(x_1,x_2)$	$g_8(x_1,x_2)$	$g_9(x_1,x_2)$
0.1	0.3	0.336	0.036	0.1	0.36	0.3	0.06	0.6	0.1	0.3
0.4	0.5	0.596	0.296	0.4	0.74	0.5	0.24	0.6	0.4	0.3
0.2	0.2	0.364	0.064	0.2	0.32	0.2	0.12	0.6	0.2	0.3
0.8	0.6	1.164	0.864	0.8	1.08	0.6	0.48	0.6	0.8	0.3
0	0.9	0.3	0	0	0.9	0.9	0	0.6	0	0.3

Node	Statistics		
	$m = E[g_i]$	$\sigma^2 = E[g_i^2] - E[g_i]^2$	$\rho = \frac{E[g_i f] - E[g_i]E[f]}{\sigma_{g_i} \sigma_f}$
n_1	0.552	0.130496	0.980357
n_2	0.252	0.130496	0.980357
n_3	0.3	0.1	0.926148
n_4	0.68	0.111	0.793819
n_5	0.5	0.075	0.324068
n_6	0.18	0.036	0.926148
n_7	0.6	0	0
n_8	0.3	0.1	0.926148
n_9	0.3	0	0

Fig. 1 An example of a GP tree and its statistical information

Although SGP computes some additional information during the evolution, these computations do not load considerable overheads and they are not very time consuming. Because most of the statistical information is reusable, and those need updating can be computed simultaneity and in parallel with updating the fitnesses.

In practice, SGP uses statistical information of the population to drive the search process. SGP has three operators that use this information: (1) Correlation Based Crossover, Correlation Based Mutation and Variance Based Editing, described in detail in the next subsections.

2.2 Correlation Based Crossover

In the standard crossover, two individuals are selected using a particular selection method and, from each of the parent trees, a subtree is randomly selected and swapped with the subtree of the other parent.

In correlation based crossover (CB crossover), for each parent, the subtrees that are more correlated to the regression function f (i.e. the ones having the maximum value of the correlation coefficient between the subtree root and the regression function f, using the absolute value) have more chance to be selected as swapping subtree. As in tournament selection, the subtrees of each parent compete with each other based on their absolute value of the correlation coefficient with f. The winner subtree of each parent is replaced with the winner of the other parent. The tournament size is proportional to the tree size. On the basis of experimental tries, the tournament size was set from 10 to 20 % of the tree size.

Using this kind of crossover, the nodes, which are more correlated to f, have more chance to be selected as crossover points; so it is more likely that the crossover points are located in the most effective parts of the parent trees. Therefore, the probability of neutral crossover, i.e. is a crossover that results in generating offspring that is not different from its parents, is decreased. Furthermore, more effective subtrees are selected as a swapping genetic material and this should lead to the relocation of valuable subtrees in the population and increase the probability of constructive crossover (crossover generating an offspring that is fitter than its parents).

2.3 Correlation Based Mutation

In the standard mutation, after that an individual is selected, one of its subtrees, randomly selected is replaced by a new random subtree. In CB mutation, the subtrees of the selected individual that are less correlated to the regression function f are more likely to be chosen as the point of mutation. In practice, the probability of choosing each node for mutation is inversely proportional to its absolute correlation. The subtree corresponding to the chosen node is replaced by a random subtree. Unlike the standard mutation, CB mutation selects the mutant subtree *non-uniformly* at random. If a subtree has less correlation (considering the absolute value) with f, it has less influence in constructing the solution tree. Thus, changing this subtree may be productive.

2.4 Variance Based Editing

One of the problems of the GP is code bloat, i.e. producing code which is slower and larger, without a significant improvement in terms of fitness. More precisely, code bloat is a considerable increase in the average code size of the population with no significant change of the fitness. In this work, we use a method based on the editing of the tree in order to perform bloat reduction. In practice, variance and mean of each node are used to edit the trees. Every subtree of each GP individual whose variance of its root is zero is replaced with the mean of its root.

Most of the subtrees of GP trees are introns or just for constructing a numeric constant. The variance of these subtrees is equal to zero. Thus, this editing operator can restrict the code growth significantly.

3 Diversity in Genetic Programming

One of the objectives of this chapter is to try to understand the correlation between the performance of our algorithm and some diversity measures, i.e. the phenotypic and genotypic diversity. This section presents the diversity measures which are used in this chapter.

Phenotypic diversity is related to different fitness values in the population. In this chapter phenotypic entropy is utilized as a phenotypic diversity measure. The phenotypic entropy of the population P can be calculated as follows [9]:

$$H_p(P) = -\sum_{j=1}^{N} p_j \log(p_j)$$

where p_j is the portion ($\frac{n_j}{N}$) of the population P that have fitness j and N is the number of different fitness values in the population P.

As in our case, fitness is a continuous quantity, in order to discretize the fitness values, we used an adaptive procedure, in which the ranges are determined *on the fly*, while the fitness values become known gradually. In practice, for each generation, the first fitness value computed becomes the representative for the first range. Subsequently, we compute the following quantity for each fitness range i:

$$\delta_i = \left| \frac{\text{new fitness value} - \text{avg. fitness in the range } i}{\text{avg. fitness in the range } i} \right|,$$

and if it is less than a predefined threshold τ, we put the new fitness value into that range (in case of ties, the i having minimum δ_i wins). Otherwise, if no such i is found, a new range is created. In the experiments, τ is set to 0.02.

In order to measure the genotypic diversity, the genotypic entropy is used in the chapter. Genotypic diversity is related to the different structures in the population. A tree distance measure is needed to keep into account the different structures. We use the tree edit distance measure, as defined by Ekárt and Németh [4].

The distance between two trees T_1 and T_2 can be computed as follows:

$$dist(T_1, T_2) = \begin{cases} d(a,b) & \text{if neither } T_1 \text{ nor } T_2 \text{ has any children} \\ d(a,b) + K \times \sum_{l=1}^{m} dist(s_l, t_l) & \text{otherwise} \end{cases},$$

where a and b are the roots of T_1 and T_2. T_1 and T_2 have m possible subtrees s and t. The parameter K is set to $1/2$. $d(a,b)$ is 0 if the nodes a and b are equal, 1 if they are different. The edit distance is calculated for each individual against the best individual in the run so far (note that it is different from the best individual in the current population).

As in the case of the phenotypic entropy, genotypic entropy is computed as follows:

$$H_g e(P) = -\sum_{j=1}^{N} ge_j \log(ge_j)$$

where ge_j is the portion of the population that has a given distance from the best individual in the run so far.

4 Experimental Results and Discussion

This section is devoted to assessing the performance of SGP and the effects of the new genetic operators. Specifically, we aim to understand the effect of the new operators on the diversity in the population, using the measure of genotypic and phenotypic diversity, introduced in the previous section.

4.1 Test Problems and GP Parameter Settings

Four real valued symbolic regression problems were chosen in order to perform an experimental evaluation. The benchmark functions were selected from [8, 11]. The benchmark problems are illustrated in Table 1. Each experiment were performed over 30 runs.

Table 1 Test problems

Benchmark number	Benchmark function	Function formula	Domain	Number of instances
Benchmark1	$f_1(x_1,x_2)$	$\frac{(x_1-3)^4+(x_2-3)^3+(x_2-3)}{(x_2-2)^4+10}$	$x_1,x_2 \in [-6,6]$	50
Benchmark2	$f_2(x_1,x_2)$	$x_1x_2 + \sin((x_1-1)(x_2+1))$	$x_1,x_2 \in [-3,3]$	20
Benchmark3	$f_3(x_1,x_2)$	$6\sin(x_1)\cos(x_2)$	$x_1,x_2 \in [-3,3]$	20
Benchmark4	$f_4(x)$	$x^4+x^3+x^2+x$	$x \in [-1,1]$	20

Table 2 GP settings

Population size	100
Function set	$F = \{+,-,\times,\div\}$
Fitness function	Mean squared error (MSE)
Initial population method	Ramped half and half
Selection method	Tournament selection
Tournament size	4
Crossover rate	90%
Mutation rate	5%
Maximum tree depth in initial population	6
Maximum tree depth	17
Maximum generation	200
Number of runs	30 independent runs for each test

The function set is the set $\mathrm{F} = \{+,-,\times,\div\}$. Note that the $\div$ represents *protected division*. The terminal set consists of random numbers, and of the function variables. Standard GP parameters are used, as shown in Table 2. The fitness function is the Mean Squared Error (MSE).

4.2 Accuracy Evaluation

In order to compare the accuracy of standard GP and SGP, we run GP and SGP for 200 generations using a population of 100 individuals on the above described benchmarks.

Figure 2a shows the result of the comparison. According to the figure, SGP performs better than the standard GP in terms of accuracy. Probably, lower probability of neutral crossover, higher constructive crossover rate and more effective mutation lead SGP to explore the GP search space more properly. Furthermore, variance based editing removes introns and decreases the computational cost by making the tree shorter. This can be seen in Fig. 2b, in which standard and statistical GP are compared in terms of average size of trees in the overall population.

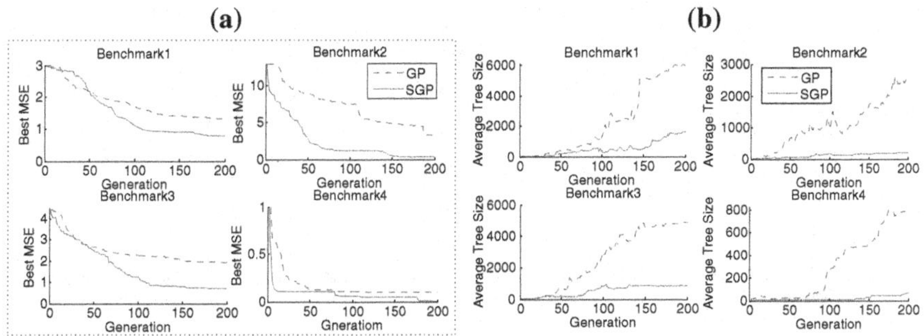

Fig. 2 **a** Comparison of SGP and GP accuracy, **b** Comparison of SGP and GP bloat control

4.3 Diversity and SGP

In this subsection, we want to investigate the correlation between diversity and fitness in SGP, using both the phenotypic and genotypic diversity as defined in the diversity section, and the Spearman correlation, defined later in this section.

Figure 3a shows the phenotypic entropy for SGP in comparison with the standard GP. It can be seen that SGP has a higher phenotypic diversity than the standard GP, probably because the CB crossover operator increases the rate of constructive crossover. In addition to that, CB mutation decreases the rate of ineffective mutation and increases the rate of constructive mutation. Thus, in SGP, the probability of generating offspring, which are better than its parents is higher than in standard GP. Furthermore, VB editing is effective in eliminating the introns and this could help to significantly decrease the rate of neutral genetic operations.

In a sub-optimal tree, higher nodes are more correlated to the regression function f. Hence, in CB crossover the higher nodes of the trees have more chance to be selected as a swapping subtrees. In the measure of genotypic diversity the higher nodes of trees have more influence, because the coefficient K is less than 1 (here is 0.5) . Therefore, as can be seen in Fig. 3b, in SGP the genotypic diversity is higher than in GP.

A second set of experiments aims to answer to the hard question whether populations with higher phenotypic or genotypic diversity can obtain a better solution. In practice, we want to investigate the correlation between the fitness and these measures of diversity.

Similar to [1], Spearman correlation is adopted in order to determinate if a relation between fitness and diversity exists. The Spearman correlation can be defined as $1 - \frac{6\sum_{i=1}^{N} d_i^2}{N^3-N}$, where d_i is the difference between the rank of the best fitness and the rank of the diversity of population i. The population that have a better fitness (less MSE) have a greater fitness rank. Similarly, the diversity rank is higher for the population having higher diversity.

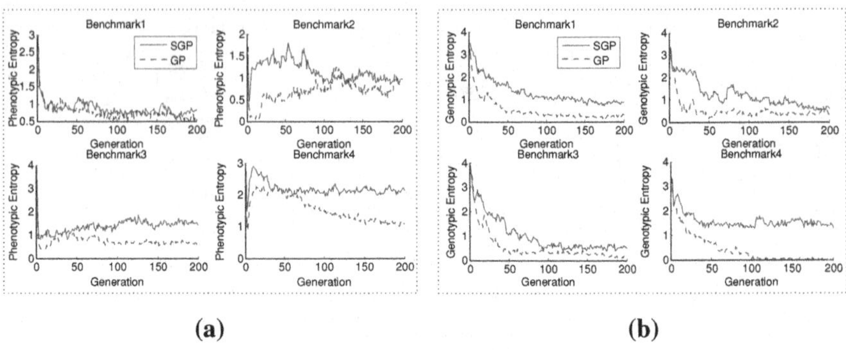

Fig. 3 Phenotypic and genotypic entropy in GP and SGP populations

Figure 4 illustrates the Spearman correlation between the fitness and the phenotypic entropy and the correlation between the fitness and the genotypic entropy. Each point in the graphs depicts the correlation between 30 populations, collected from 30 independent runs in the different phases of the evolutive process.

As can be seen in Fig. 4a, for all benchmarks, at the beginning, as the population is randomly created there is no positive (or negative) correlation between fitness and diversity. Afterwards, a positive correlation can be found both for SGP and GP, then the correlation decreases and no significant correlation can be found. This is probably due, in accordance with the results found in the chapter of Burke, to the presence of many local optima.

In the case of genotypic diversity (see Fig. 4b) in SGP, in very early generations the correlation between the fitness and genotypic entropy is positive. After these early generations, the correlation becomes lower and close to zero and afterward becomes negative. In the experimental results of Burke et al., a similar behavior have been evidenced.

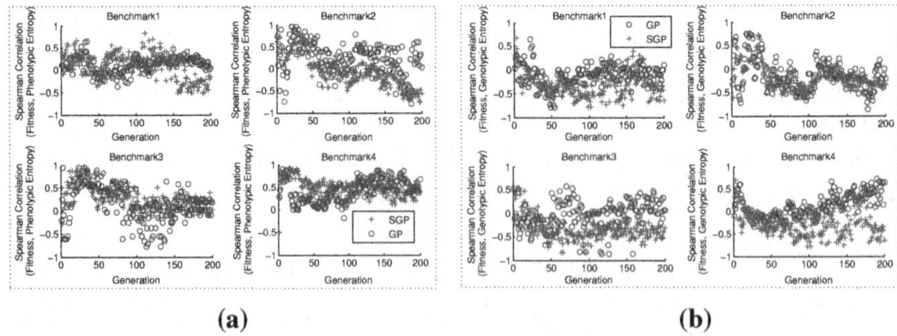

Fig. 4 **a** Correlation between fitness and phenotypic diversity, **b** Correlation between fitness and genotypic diversity

Our investigation results are similar to those of [1]. There is a positive correlation between the phenotypic entropy and fitness and a negative correlation between genotypic diversity and fitness.

It should be considered that, the correlation coefficient represents the association between fitness and diversity, not the causality. This means that, for example, higher phenotypic diversity is not necessarily the cause of better fitness. However, better performance is observed with higher phenotypic diversity. Burke et al. [1] expressed that crossover and selection methods have very important roles in constructing the structures of GP population. Any simple implementation difference may change the diversity of the population. Therefore, care must be taken when inferring causality from diversity.

5 Conclusions

This chapter proposed a new GP paradigm, Statistical Genetic Programming, exploiting the statistical information of the population in order to improve the accuracy of GP, mainly for symbolic regression problems. Experiments conducted on four symbolic regression problems confirm the improvement obtained using the new paradigm. A diversity analysis, based on genotypic and phenotypic diversity measures and on the study of correlation coefficients obtains results comparable with other classical model of GP, with the exception of the capacity of SGP to maintain a higher genotypic diversity. Future works will be conducted in order to try to understand better the relation between the performance of SGP and diversity and to study the different contributions of the three operators introduced.

References

1. Burke, E., Gustafson, S., Kendall, G.: Diversity in genetic programming: an analysis of measures and correlation with fitness. Trans. Evol. Comp. **8**(1), 47–62 (2004)
2. de Jong, E., Watson, R., Pollack, J.: Reducing bloat and promoting diversity using multi-objective methods (2001)
3. Ekárt, A., Németh, S.: A metric for genetic programs and fitness sharing. In: Genetic Programming, pp. 259–270. Springer, Berlin (2000)
4. Ekárt, A., Nemeth, S.: Maintaining the diversity of genetic programs. In: Genetic Programming, pp. 122–135 (2002)
5. Folino, G., Pizzuti, C., Spezzano, G., Vanneschi, L., Tomassini, M.: Diversity analysis in cellular and multipopulation genetic programming. In: CEC'03, vol. 1, pp. 305–311. IEEE (2003)
6. Jackson, D.: Mutation as a diversity enhancing mechanism in genetic programming. In: Proceedings of GECCO '11, pp. 1371–1378. ACM, New York, (2011)
7. Langdon, W.: Genetic programming and data structures: genetic programming+ data structures. Springer, New York (1998)

8. Pennachin, C.L., Looks, M., de Vasconcelos, J.a.A.: Robust symbolic regression with affine arithmetic. In: Proceedings of GECCO '10, pp. 917–924. ACM, New York (2010)
9. Rosca, J.: Entropy-driven adaptive representation. In: Proceedings of the Workshop on Genetic Programming: From Theory to Real-World Applications, vol. 9, pp. 23–32. Tahoe City (1995a)
10. Rosca, J.: Genetic programming: exploratory power and the discovery of functions. In: Proceedings of Evolutionary Programming IV, pp. 719–736. Citeseer (1995b)
11. Vladislavleva, E.J., Smits, G.F., Den Hertog, D.: Order of nonlinearity as a complexity measure for models generated by symbolic regression via pareto genetic programming. Trans. Evol. Comp. **13**, 333–349 (2009)

Breast MRI Tumour Segmentation Using Modified Automatic Seeded Region Growing Based on Particle Swarm Optimization Image Clustering

Ali Qusay Al-Faris, Umi Kalthum Ngah, Nor Ashidi Mat Isa and Ibrahim Lutfi Shuaib

Abstract In this paper, a segmentation system with a modified automatic Seeded Region Growing (SRG) based on Particle Swarm Optimization (PSO) image clustering will be presented. The paper is focused on Magnetic Resonance Imaging (MRI) breast tumour segmentation. The PSO clusters' intensities are involved in the proposed algorithms of the automated SRG initial seed and threshold value selection. Prior to that, some pre-processing methodologies are involved. And breast skin is detected and deleted using the integration of two algorithms, i.e. Level Set Active Contour and Morphological Thinning. The system is applied and tested on the RIDER breast MRI dataset, and the results are evaluated and presented in comparison to the Ground Truths of the dataset. The results show higher performance compared to the previous segmentation approaches that have been tested on the same dataset.

A. Q. Al-Faris (✉) · U. K. Ngah · N. A. M. Isa
Imaging and Computational Intelligence Research Group (ICI),
Universiti Sains Malaysia, Penang, Malaysia
e-mail: alialfaris2009@gmail.com

U. K. Ngah
e-mail: eeumi@eng.usm.my

N. A. M. Isa
e-mail: ashidi@eng.usm.my

I. L. Shuaib
Advanced Medical and Dental Institute (AMDI), Universiti Sains Malaysia,
Penang, Malaysia
e-mail: ibrahim@amdi.usm.edu.my

V. Snášel et al. (eds.), *Soft Computing in Industrial Applications*,
Advances in Intelligent Systems and Computing 223, DOI: 10.1007/978-3-319-00930-8_5,

1 Introduction

Breast cancer today is the leading cause of death amongst cancer patients inflicting women around the world. To date, 1.38 million new breast cancer cases have been diagnosed, which is 23 % of total new cancer cases in the world. 458,400 breast cancer death cases have been recorded making up 14 % of the total cancer deaths in 2008 alone [1]. Several of the commonly used medical screening techniques used for breast screening are mammography, ultrasound and MRI. While the produced images by mammogram demonstrates the contrast between soft tissue and bone density, MRI on the other hand, produces clear and crisp images, which provides a better contrast between different kinds of soft tissue. For that reason, MRI is used for breast screening i.e. to explore the small details between breast tissues. Although this is valuable information, the presented data still needs to be interpreted by the radiologist [2]. For this purpose, image processing methods are used to assist the radiologists in improving the quality of these medical images and in detecting tumour masses.

2 Related Work

Several techniques have been developed and also evolved for medical image segmentation. These includes methods such as Particle Swarm Optimization (PSO) [3], Genetic Algorithm [4] and Artificial Fuzzy Logic [5–7]. The supervised, unsupervised and semi-supervised methods are explored in Azmi et al's study [8]. In their comparison study on MRI Breast RIDER dataset, the researchers found that the supervised segmentation methods such as; K-Nearest Neighbors (KNN), Support Vector Machine (SVM) and Bayesian and the semi-supervised method such as self training and improved self-training (IMPST) lead to high accuracy. However labeled data is needed. Hence, the process becomes difficult, expensive, and involves a lot of time. On the other hand, unsupervised methods such as; Fuzzy C-means (FCM) need no prior knowledge. However, the performance is low [8]. The Seeded Region Growing algorithm which was proposed by Adams and Bischof [9] is widely used in the medical images today because it effectively segments different types of images [10]. In Meinel's study on MRI breast segmentation [11] the SRG was also used. The experiments on breast tumour segmentation returned robust results. However, this approach needed an initial threshold value to be specified by the user. This is used to find the anticipated locations of the tumors. The SRG Feature Extraction algorithm has been proposed on cervical cancer screening. This algorithm extracted four cervical cells features; size of nucleus, cytoplasm, grey level of nucleus and cytoplasm. The data extracted using SRGFE algorithm gave high correlation value when compared with data extracted manually. Still, the user needs to determine the region of interest to select the initial seed pixel. The user also needs to determine the threshold value [12]. Wu et al. proposed texture feature-based automated SRG algorithm on abdominal organ segmentation [13]. The advantage of this algorithm is that it allows minimum user intervention. This is helpful for batch work. However, this approach does have drawbacks. Texture feature based methods all have the assumption that the region should have texture homogeneity. For organs with complex texture, this approach may not work well. Shan et al. proposed an automatic seed point selection algorithm for SRG on ultrasound breast images [14]. The algorithm needs no prior information or training process. Both the homogeneous texture features and spatial features of the breast tumors are taken into account. However, some cases failed because of the shadowing effects of areas having similar intensity as the tumor and right below the tumor. In this study, a proposed segmentation approach with automated features for MRI breast tumor segmentation is presented using modified automatic SRG method based on PSO image clustering.

3 Proposed Approach

The proposed approach starts with a pre-processing phase which is followed by the detection and deletion of the breast skin. Then automatic seed and threshold selection processes are presented based on PSO image clustering before the SRG algorithm is applied. The whole perspective is shown as in Fig. 1.

3.1 Image Pre-Processing

The pre-processing phase is the first process that is executed. The image is split into two sub-images; the right breast image and the left breast image. This process is used only if the MRI breast image is Axial (i.e. the image is taken from the perspective of the patient from head to toe). This process is skipped if the image is Sagittal (i.e. the image is taken according to the lateral view). The splitting process can be done by finding the middle of the X-coordinate of the image and splitting the image vertically from that point. The median filter is then applied in order to enhance the images' resolution and to reduce the presence of the salt and pepper noise while the boundaries and features are kept intact [15].

3.2 Breast Skin Detection and Deletion

The purpose of this process is to delete the breast skin area which has similar intensity range compared to the tumor area's intensity range. This process is also necessary in order to facilitate a better automatic seed selection for the tumor

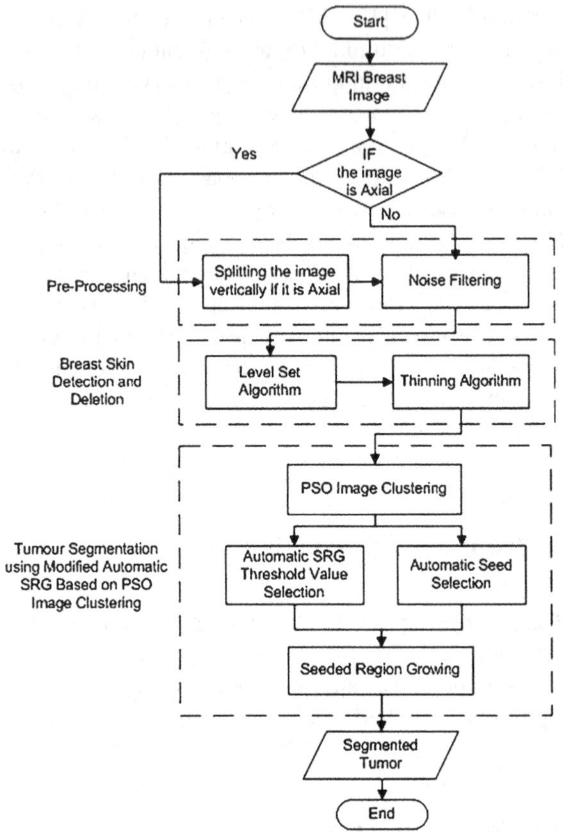

Fig. 1 Methodology flowchart

segmentation in the next stage. To delete the breast skin, an integration of Level Set Active Contour algorithm [16] with Morphological Thinning Algorithm is used in this phase. The Level Set Active Contour (Chunming's algorithm) is used to detect the breast skin border; the algorithm is dynamic curves that move toward the mass border. An external energy moves the zero level curves toward the mass border using the edge indicator function g that is defined in (1)

$$g = \frac{1}{1 + |\nabla G_\sigma * I|^2} \tag{1}$$

where I is the image, G_σ is the Gaussian kernel with a standard deviation σ. By changing the σ parameter value and the number of the iterations, In order to delete the detected breast skin border, the Morphological Thinning Algorithm [17] is used. The thinning level depends on the number of iterations. Whenever the number of iterations is increased, there would be more shrinking of the image's border. The Morphological Thinning Algorithm only accepts a binary version of the image. Therefore, the resultant image after the Chunming's algorithm would be converted to binary image. Furthermore, after applying the thinning, the binary image is reconverted into its original grey scale representation.

3.3 A Modified Automatic SRG Based On PSO Image Clustering

In this study, the SRG algorithm for tumour segmentation is chosen because it is fast, simple and robust [18]. While the chosen image clustering method is PSO-based, because it produces better results compared with other clustering methods such as K-means, Fuzzy C-means, K-Harmonic means and Genetic Algorithms [19–21].

3.3.1 Seeded Region Growing

SRG [9] starts with an initial seed pixel and tries to compare their neighborhood pixels with the seed according to some attributes, such as the intensity or texture. It then merges them if they are similar enough. The eight neighbor pixels will be tested according to the intensity next sorted into the segmented pixel list if the tested neighbor pixel has similar intensity. Subsequently, the eight neighbors of the new pixel will be tested and sorted too. The process then continues in the same manner. Fig. 2 shows the initial seed pixel and the eight neighbors. SRG has two variable factors which are usually selected manually. The first factor is determining the initial seed pixel that the SRG can start growing. The second factor is the threshold value for measuring the difference between the pixel and their neighbors. In this work, an automated version of the seed selection algorithm and SRG threshold based on the PSO image clustering are presented.

3.3.2 Particle Swarm Optimization (PSO) Image Clustering

Applying the PSO image clustering would be organizing the image into groups whose members are having similar intensity range. Therefore, each cluster represents different intensity range of image. Various versions of image clustering based on PSO have been proposed in [19–21]. The method used in this chapter proposed in [19] and described in [21] as below:

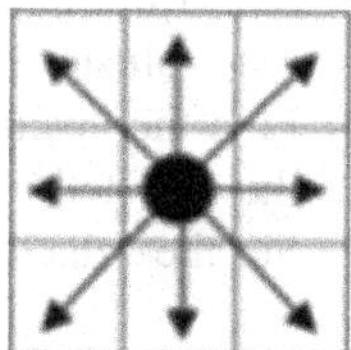

Fig. 2 The initial seed pixel and their eight neighbor pixels

N_c is the number of clusters to be formed, Z_p is the p-th pixel, C_j is the subset of pixel vectors that form cluster j, $x_i = (m_{i1}, \ldots m_{ij}, \ldots m_{iNc})$ where refers to the j-th cluster centroid vector of the i-th particle.

1. Initialize each particle to contain N_c randomly selected cluster means.
2. FOR t = 1 to t_{max} (maximum number of iterations)

FOR each particle i

FOR each pixel Z_p

Calculate $d(Z_p, M_{ij})$ for all clusters

Assign Z_p to c_{ij} where

$d(Z_p, m_{ij}) = min_{\forall} c = 1, \ldots, N_c\{d(Z_p, M_{ic})\}$

$d(Z_p, m_{ij})$ represents the Euclidean distance between the p-th pixel zp and the centroid of j-th cluster of particle i.

Calculate the fitness function $f(x_i(t), z)$ where Z is a matrix representing the assignment of pixels to clusters of particle i.

Update the personal best and the global best positions.

Update the cluster centroids using Eqs. (2) and (3).

$$v_i(t+1) = wv_i(t) + c_1r_1(t)(y_i(t) - x_i(t)) + c_2r_2(t)(y_g(t) - x_i(t)) \quad (2)$$

$$x_i(t+1) = x_i(t) + V_i(t+1) \quad (3)$$

where x_i is the current position of the particle, v_i is the current velocity of the particle, y^i, is the best position that particle has achieved so far, y_g is the location of overall best value, w is the inertia weight,c_1 and $c_2$are the acceleration constants, $_1(t)$ and $_2(t)$ are random numbers generated in the range between 0 and 1. 3. Segment the image using the optimal number of clusters and the optimal clusters centroids given by the best global particle.

3.3.3 The Proposed Automatic SRG Initial Seed Selection.

After applying the PSO image clustering, the clusters' intensities would be ranked in ascending order. Subsequently, the cluster which has the highest intensity would be chosen. Then the centre of the chosen cluster region is selected as the initial seed. The steps taken for this process are as follows:

1. Apply PSO Image clustering on the MRI breast image.
2. Rank the PSO clusters according to their intensity values in ascending order.
3. Select the regions with the highest clusters' intensity values and eliminate the other cluster regions.
4. Find the position (x, y coordinates) of the center pixel of the maximum area in the selected regions.
5. Set the selected position in step 4 as the position of the initial seed.

3.3.4 The Proposed Automatic SRG Threshold Value Selection.

The importance of this process is because of the ranges of the grayscale representations for the tumour and the other parts of the breast are not consistent from one image to another. Therefore, the proposed method has the capability of changing the SRG threshold value according to the respective image's gray scale distribution. The method is based upon finding the optimum estimated value from the PSO clusters' intensities mean values. The average for clusters' intensities except the highest cluster's intensity (which contains the tumor region) has to be calculated first using Eq. 4.

$$CAvg = \frac{\sum_{i=0}^{N_{c-1}} CI_i}{N_c - 1} \tag{4}$$

where i is the PSO clusters counter, is the maximum clusters number, CI is the cluster intensity and CAvg is the clusters' intensity average.

The CAvg is used to examine the optimum SRG threshold value as described in the pseudo code:

SET Zero to (IntensitySum), Set DefaultValue (the default value is value between CAvg-255)

FOR i=1 to N_c Do

IntensitySum= IntensitySum+CI (i)

SRGThreshold=CAvg - (IntensitySum/i)

IF (SRGThreshold< CAvg) THEN

SET SRGThreshold as the chosen threshold value

Break

ELSE

SRGThreshold=DefaultValue

ENDIF

END FOR

4 Experimental Results and Discussion

The methodology explained earlier is applied and tested on the RIDER Breast MRI dataset which is downloaded from the National Biomedical Imaging Archive (NBIA) [22]. This website belongs to the U.S. National Cancer Institute. The dataset includes breast MRI images for five patients. All images are Axial 288 × 288 pixels. The dataset also include Ground Truth (GT) segmentation, which have been identified manually by a radiologist. Three sequences with their GT are selected for each patient to be used in the experiments as test images. GT is used as a benchmark for performance evaluation of segmentation methods in our experiments. Figure 3 illustrates one RIDER MRI breast image after applying the different processes of the approach.

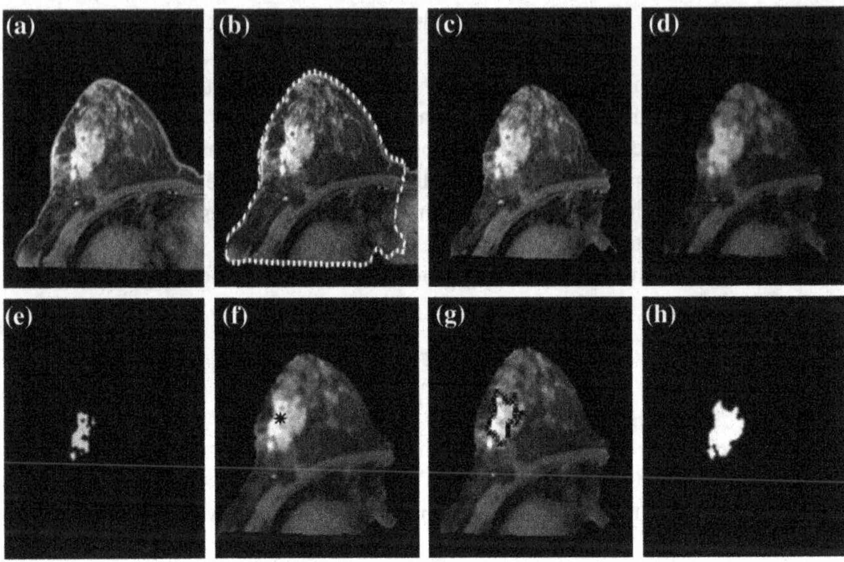

Fig. 3 The tumor segmentation process applied on RIDER MRI image **a** MRI image after splitting process. **b** Breast skin detected by the Level Set Algorithm (marked in white dotted line). **c** Breast skin deleted by the Thinning Algorithm. **d** The resultant images after applying the PSO image clustering. **e** The highest PSO clusters' intensity region after other regions are eliminated. **f** Initial seed selected automatically (marked in black star) as the center of the region selected in (**e**). **g** SRG using the automatic threshold value is applied (marked in black dotted line). **h** The segmented tumor area

Different measures are used in this study in order to evaluate the segmentation accuracy. The number of pixels of (R_s) and (R_t) have to be found first, where (R_s) represents the segmented region by the proposed approach, while (R_t) represents the ground truth regions segmented by the experts. The evaluation measures used in this study are; True Positive Fraction (TPF) (also called Sensitivity), True Negative Fraction (TNF) [23–26], Relative Overlap (RO) (also called segmentation precision) and Misclassification Rate (MCR) which have been used before for brain segmentation [27] and in breast segmentation [8, 28]. The calculations are made using the Eqs. 5– 8.

$$TPF = \frac{R_s \bigcap R_t}{R_t} \tag{5}$$

$$TNF = 1 - \frac{R_s - R_t}{R_t} \tag{6}$$

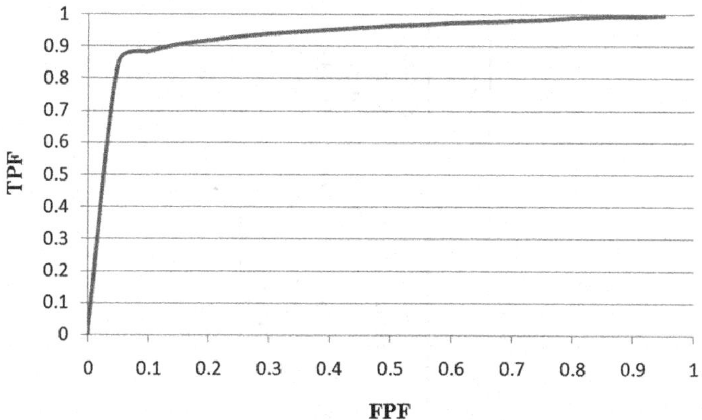

Fig. 4 The ROC curve for all tested RIDER dataset images

$$RO = \frac{R_s \bigcap R_t}{R_s \bigcup R_t} \tag{7}$$

$$MCR = 1 - \frac{R_s \bigcap R_t}{R_t} \tag{8}$$

The results of the proposed approach are compared with the results of the previous works involving five different segmentation approaches. The previous work's results have been stated in the comparison study of [8]. The approaches are K-Nearest Neighbors (KNN), Support Vector Machine (SVM), Bayesian, Fuzzy C-means (FCM) and Improved Self-Training (IMPST). The tested data are the same dataset which is (Breast MRI RIDER dataset). The results of the five approaches and the proposed approach are stated in Table 1.

The results of the proposed approach show improved performance compared with the results of the previous approaches. The proposed approach not only provides improved performance, it also facilitates automated selection of the suspected regions without the need to manually select the region of interest (ROI) as is necessary in the previous approaches. Table 1 show the results of the RO mean (0.704) and TNF mean (0.851) of the proposed approach are improved compared with the previous approaches. Nevertheless, the TPF mean (0.792) and MCS mean (0.209) of the proposed approach are superior to the K.N.N, SVM, Bayesian and IMPST approaches. However, the values are less in accuracy when compared with one approach (the FCM) because it uses manually drawn ROI to select the suspected window before starting the segmentation process. Fig. 3 shows the Receiver Operating Characteristic (ROC) curve which is used to illustrate the True Positive Fraction compared with the False Positive Fraction. From the curve, it can be observed that the Area under the Curve (AUC) is 0.95. The high AUC indicates improved segmentation performance.

Table 1 Segmentation results for the proposed approach and other approaches (K.N.N, SVM, Bayesian, FCM and IMPST)

The method	KNN		SVM		Bayesian		FCM		IMPST		The proposed approach	
Statistic	Mean	Standard deviation	Mean	Standard deviation	Mean	Standard deviation	Mean	Standard deviation	Mean	Standard deviation	Mean	Standard deviation
RO	0.595	0.121	0.6	0.187	0.555	0.174	0.587	0.250	0.677	0.167	**0.704**	**0.143**
MSC	0.27	0.87	0.25	0.81	0.24	0.81	0.18	0.73	0.21	0.81	**0.209**	**0.133**
TPF	0.73	0.13	0.75	0.19	0.76	0.19	0.82	0.27	0.79	0.19	**0.792**	**0.133**
TNF	0.75	0.15	0.71	0.21	0.59	0.25	0.42	0.55	0.83	0.15	**0.851**	**0.237**

5 Conclusion

A modified automatic Seeded Region Growing based on PSO image clustering system for MRI breast tumour segmentation has been presented in this paper. The modification has been made by proposing two automatic approaches for selecting the SRG variable factors which are usually selected manually. The first approach selects the position of the initial seed pixel; along with the second approach which determines the SRG threshold value for measuring the difference between the pixel and their neighbors. Both approaches are based on the clusters' intensities of the PSO image clustering. Prior to that, some necessary pre-processing processes are made such as; splitting the axial images, noise reduction and deletion of the breast skin using the integration of Level Set Active Contour and Morphological Thinning algorithms. The study is then supported by the results of applying the methodology on the RIDER breast MRI dataset. The evaluation has been made using the ground truth of the dataset as a benchmark. The results are then compared with previous works, which are also based on the same dataset. Not only is the performance significantly improved; the proposed approach also avoided the need for manual selection of the suspected region window, seed pixel and threshold value processes. These processes are replaced with automated methods. The methods are also generic for any grayscale representation of the breast MRI images.

References

1. Jemal, A., et al.: Global cancer statistics. CA A Cancer J. Clin. **61**(2):69–90 (2011)
2. Gardiner, I.: CAD improves breast MRI workflow: increasing throughput while maintaining accuracy in breast MRI reads requires powerful workflow tools. In: Imaging Technology News (2010)
3. Ibrahim, S., Khalid, N.E.A., Manaf, M.: Empirical study of brain segmentation using particle swarm optimization. In: International Conference on Information Retrieval and Knowledge Management, CAMP10, p. 235–239. Shah Alam, Selangor (2010)
4. Ganesan, R., Radhakrishnan, S.: Segmentation of computed tomography brain images using genetic algorithm. Int. J. Soft Comput. **4**(4), 157–161 (2009)
5. Hussain, R., et al.: Fuzzy clustering based malignant areas detection in noisy breast magnetic resonant (MR) images. Int. J. Acad. Res. **3**(2) (2011)
6. Kannan, S., Sathya, A., Ramathilagam, S.: Effective fuzzy clustering techniques for segmentation of breast MRI. Soft Comput. Fusion Found. Methodol. Appl. **15**(3), 483–491 (2011)
7. Noor, N.M., et al.: Adaptive neuro-fuzzy inference system for brain abnormality segmentation. In: 2010 IEEE Control and System Graduate Research Colloquium, ICSGRC 2010 (2010)
8. Azmi, R., et al.: IMPST: a new interactive self-training approach to segmentation suspicious lesions in breast MRI. J. Med. Signals Sens. **1**(2), 138–148 (2011)
9. Adams, R., Bischof, L.: Seeded region growing. IEEE Trans. Pattern Anal. Machine Intell. **16**, 641–647 (1994)

10. Khalid, N.E.A., et al.: Seed-based region growing study for brain abnormalities segmentation. In: International Symposium on Information Technology 2010 (ITSim 2010), p. 856–860. Kuala Lumpur, (2010)
11. Meinel, L.A.: Development of computer-aided diagnostic system for breast MRI lesion classification. Dissertation, in Biomedical Engineering, University of Iowa: Iowa (2005)
12. Mat-Isa, N.A., Mashor, M.Y., Othman, N.H.: Seeded region growing features extraction algorithm; its potential use in improving screening for cervical cancer. Int. J. Comput. Internet Manag. **13**(1) (2005)
13. Wu, J., et al.: Texture feature based automated seeded region growing in abdominal MRI segmentation. In: BioMedical Engineering and Informatics. BMEI 2008. Sanya (2008)
14. Shan, J., Cheng, H.D., Wang, Y.: A novel automatic seed point selection algorithm for breast ultrasound images. In: 19th International Conference on Pattern Recognition, ICPR (2008)
15. Chun-yu, N., Shu-fen, L., Ming, Q.: Research on removing noise in medical image based on median filter method. In: IEEE International Symposium in IT in Medicine & Education, ITIME '09. I.C. Publications, Editor 2009, 384–388 (2009)
16. Li, C., et al.: Level set evolution without re-initialization: a new variational formulation. In: IEEE International Conference on Computer Vision and Pattern Recognition (CVPR), San Diego (2005)
17. Lam, L., Lee, S.W., Suen, C.Y.: Thinning methodologies-a comprehensive survey. IEEE Trans. Pattern Anal. Mach. Intell. **14**(9), 879 (1992)
18. Ibrahim, S., et al.: Particle swarm optimization vs seed-based region growing: brain abnormalities segmentation. Int. J. Artif. Intell. **7**(1), 174–188 (2011)
19. Omran, M.G.H.: A PSO-based clustering algorithm with application to unsupervised image classification. University of Pretoria etd (2005)
20. Ouadfel, S., Batouche, M., Taleb-Ahmed, A.: A modified particle swarm optimization algorithm for automatic image clustering. In: International Symposium on Modelling and Implementation of, Complex Systems, MISC'2010, (2010)
21. Wong, M.T., He, X., Yeh, W.-C.: Image clustering using particle swarm, optimization. (2011)
22. "RIDER Breast MRI", National Biomedical Imaging Archive (NBIA), U.o. Michigan, Editor 2007, U.S. National Cancer Institute.
23. Chalana, V., Kim, Y.: A methodology for evaluation of boundary detection algorithms on medical images. IEEE Trans. Med. Imag. **16**, 642–652 (1997)
24. Fenster, A., Chiu, B.: Evaluation of segmentation algorithms for medical imaging. In: 27th Annual Conference on IEEE Engineering in Medicine and Biology. Shanghai, China (2005)
25. Metz, C.E.: ROC methodology in radiologic imaging. Invest. Radiol. **21**, 720–733 (1986)
26. McNeil, B.J., Hanley, J.A.: Statistical approaches to analysis of receiver operating characteristic ROC curves. Med. Decis. Making **14**, 137–150 (1984)
27. Song, T., et al.: A hybrid tissue segmentation approach for brain MR images. Med. Biol. Eng. Comput. **44**, 242 (2006)
28. Ertas, Gökhan, et al.: Breast MR segmentation and lesion detection with cellular neural networks and 3D template matching. Comput. Biol. Med. **38**, 116–126 (2008)

Differential Evolution and Tabu Search to Find Multiple Solutions of Multimodal Optimization Problems

Erick R. F. A. Schneider and Renato A. Krohling

Abstract Many real life optimization problems are multimodal with multiple optima. Evolutionary Algorithms (EA) have successfully been used to solve these problems, but they have the disadvantage since that they converge to only one optimum, even though there are many optima. We proposed a hybrid algorithm combining differential evolution (DE) with tabu search (TS) to find multiple solutions of these problems. The proposed algorithm was tested on optimization problems with multiple optima and the results compared with those provided by the Particle Swarm Optimization (PSO) algorithm.

1 Introduction

Evolutionary Algorithms (EA) and Particle Swarm Optimization (PSO) techniques are effective and robust techniques for solving optimization problems [4]. Typically, these algorithms converge to a single final solution. However, many real--world problems are multimodal in nature and may have many satisfactory solutions [5]. Niching methods have been developed in recent decades to find solution of multimodal optimization problems with multiple optima. These niching methods play an important role when incorporated into evolutionary algorithms to promote the diversity of the population and maintain multiple solutions within a stable population.

E. R. F. A. Schneider (✉)
Graduate Program in Computer Science (PPGI), Federal University of Espírito Santo, Vitória 29060-970, Brazil
e-mail: erickrfas@gmail.com.br

R. A. Krohling
Department of Production Engineering and Graduate Program in Computer Science (PPGI), Federal University of Espírito Santo, Vitoria 29060-970, Brazil
e-mail: krohling.renato@gmail.com
URL: http://www.inf.ufes.br/~rkrohling

V. Snášel et al. (eds.), *Soft Computing in Industrial Applications*,
Advances in Intelligent Systems and Computing 223, DOI: 10.1007/978-3-319-00930-8_6,

Several niching methods have been proposed in recent years, and the most relevant include [5]: fitness sharing, derating, restricted tournament selection, crowding, deterministic crowding, clustering, clearing, parallelization, speciation, among others. Most of the methods listed above present difficulties for solving multimodal optimization problems with multiple local or global optima because they need to specify the parameters of niching, which are difficult to tune and they generally are dependent on the problem to be optimized.

An interesting approach recently proposed in [5], which does not require specification of any niching parameters uses a PSO with ring topology. In this case, the "local memory" of the individual particles of the PSO is able to maintain the best positions found so far, while the particles explore the search space. In the niching PSO without niching parameters proposed in [5] it is shown that large populations using PSO with ring topology is capable of forming stable niche and able to find multiple local and global optima. The promising results suggest that this method presents good results without requiring parameters to tune.

Based on the method developed in [5], motivated us to extend this method to Differential Evolution (DE). In fact, DE with ring topology [2, 3] were applied to multimodal problems with promising results to find a single solution of optimization problems. However, for problems with multiple optima, as far as we know, DE with ring topology has not been used. The DE version proposed in [2] called DEGL has a good capacity of exploration / exploitation of the search space.

In case of PSO with ring topology, the personal best positions *pbest* of all particles in the population are used to form the memory of the swarm and retain the best solutions found so far in the population, and the positions of the particles act to explore the search space. In this work, inspired by the PSO with ring topology, the DEGL is used to explore the search space, but as DEGL has no memory, this motivated us to combine the algorithm with Tabu Search (TS) to fulfill the function of memory. In this context, the hybrid method, which consists of a genetic algorithm (GA) embedded with a niching method and Tabu search has been proposed in [6]. The main disadvantage of that method is that is necessary to specify the niching parameters. So, to overcome this problem, we propose the DEGL with ring topology combined with Tabu search. As in [6], the method proposed here also consists of two stages. The first stage uses DEGL to explore the search space. In the second stage, the TS takes the initial solutions provided by DEGL in the first stage and performs a local search aiming to improve the population of solutions of the DEGL. At the end of the algorithm run, a population of local/global solutions is found.

2 Differential Evolution

In the following, the standard Differential Evolution and the Differential Evolution with ring topology are described.

2.1 Standard Differential Evolution

Differential Evolution (DE) is an optimization method introduced in [8]. Similar to other Evolutionary algorithms (EAs), it is based on the idea of evolution of populations of possible candidate solutions, which undergoes the operations of mutation, crossover and selection [9]. The candidate solutions of the optimization problem in DE are represented by vectors. The components of the vectors are the variables of the optimization problem and the set of vectors make up the population. Unless stated otherwise in our study, we are considering minimization problems. Therefore, the higher the value of the fitness, the smaller the values of the objective function.

Consider a population of NP individuals in an N-dimensional search space. The individuals in the population are initialized according to:

$$x_{i,j} = x_{j,min} + U(0,1)_{i,j}(x_{j,max} - x_{j,min}) \tag{1}$$

where j is the index of the j th component of the i th individual of the population, $x_{j,max}$ and $x_{j,min}$ are the upper and lower bounds of the j-th component of the N-dimensional vector $\vec{X}$, respectively, and U (0,1) is a random number drawn from a uniform distribution.

By means of the mutation operator, a new vector $\vec{V}_i$ is generated by the following equation:

$$\vec{V}_i = \vec{X}_{r1,i} + F(\vec{X}_{r2,i} - \vec{X}_{r3,i}) \tag{2}$$

where F is a scaling factor, and $r_1, r_2, r_3 \in [1, NP]$, such that $r_1 \neq r_2 \neq r_3$.

In the crossover operation, a new vector $\vec{U}_i$ is generated according to:

$$u_{i,j,G} = \begin{cases} v_{i,j,G} & \text{if } \text{U }(0,1)_{i,j} \leq C_r \text{ or } j = j_{rand} \\ x_{i,j,G} & \text{otherwise} \end{cases} \tag{3}$$

where C_r is the crossover rate and j_{rand} is a random component of each individual to ensure that at least one component of the vector $\vec{V}_i$ is part of the new vector.

In the selection, the individual is chosen according to:

$$\vec{X}_{i,G+1} = \begin{cases} \vec{U}_{i,G} & \text{if } f(\vec{U}_{i,G}) \leq f(\vec{X}_{i,G}) \\ \vec{X}_{i,G} & \text{otherwise} \end{cases} \tag{4}$$

where $f(.)$ is the objective function.

2.2 Differential Evolution with Ring Topology

One disadvantage of standard DE is the premature convergence to local optima in multimodal functions, losing the diversity of the population. To increase the diversity of the population, it is used a neighborhood with ring topology as shown in Fig. 1. So, the mutation operator is implemented through the creation of two vectors before the generation of the vector $\vec{V}_i$.

First, the vector is calculated as:

$$\vec{L}_{i,G} = \vec{X}_{i,G} + \alpha(\vec{X}_{n_{besti}} - \vec{X}_{i,G}) + \beta(\vec{X}_{p,G} - \vec{X}_{q,G}) \tag{5}$$

where n_{best_i} indicates the best vector in the neighborhood of $\vec{X}_{i,G}$. p, q are indices in the neighborhood chosen such that $p \neq q \neq i$ and $p, q \in [i-k, i+k]$,and where k is the length of the neighborhood.

Next, the vector $\vec{g}_i$ is calculated as:

$$\vec{g}_{i,G} = \vec{X}_{i,G} + \alpha(\vec{X}_{gbest} - \vec{X}_{i,G}) + \beta(\vec{X}_{r1,G} - \vec{X}_{r2,G}) \tag{6}$$

where *gbest* indicates the best of the entire population, r_1 and r_2 are such that $r_1 \neq r2 \neq i$, and $r_1, r_2 \in [1, NP]$.

Then, one calculates the vector $\vec{V}_i$ as follows:

$$\vec{v}_{i,G} = w\vec{g}_{i,G} + (1-w)\vec{L}_{i,G} \tag{7}$$

where w is a weighting factor. Small values of w increase the diversity of the population. An illustration of the ring topology is shown in Fig. 1, where the length of the neighborhood is $k = 2$ and the number of individuals is NP. The DE algorithm pseudo-code of DE with ring topology is shown in Fig. 2.

3 Proposed Method

In recent years, several approaches combining EA with local search methods have been proposed [1, 5, 6, 7, 10]. In order to solve the optimization problem with multiple optima we propose a hybrid algorithm based on DE and TS. In our approach, the DE algorithm is used for global search in order to increase the

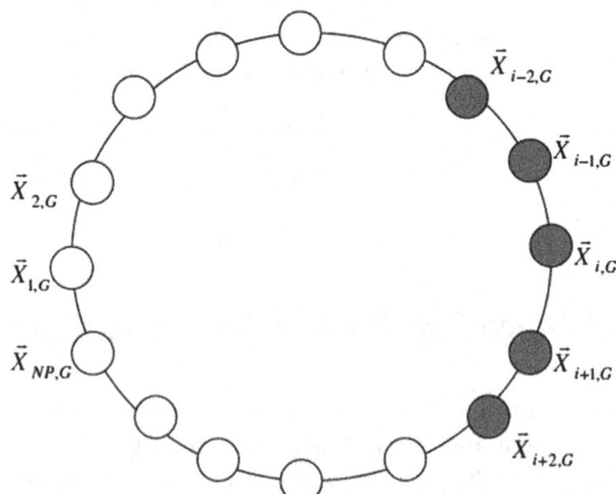

Fig. 1 Illustration of the ring topology

```
Input: Population size NP
for i= 1 to NP
    for j = 1 to N
        x_{i,j} = x_{j,min} + U(0,1)_{i,j} (x_{j,max} - x_{j,min})
    end for
end for
while termination condition not met
    for i=1 to NP
        choose p and q such that p ≠ q ≠ i from [i - k, i + k]
        L_{i,G} = X_{i,G} + α(X_{n_best i} - X_{i,G}) + β(X_{p,G} - X_{q,G})
        choose r1 and r2 such that r1 ≠ r2 ≠ i from [1,NP]
        g_{i,G} = X_{i,G} + α(X_{gbest} - X_{i,G}) + β(X_{r1,G} - X_{r2,G})
        V_{i,G} = w g_{i,G} + (1 - w) L_{i,G}
        j_rand = U(1, N)
        for j = 1 to N
            if U(0,1)_{i,j} ≤ C_r or j = j_rand
                u_{i,j} = v_{i,j}
            else
                u_{i,j} = x_{i,j}
            end if
        end for
        if f(U_i) ≤ f(X_i)
            X_i = U_i
        end if
    end for
end while
```

Fig. 2 Pseudo-code of the DE algorithm with ring topology

diversity of the population. Then, the TS algorithm is used for local search to refine the search around promising solutions. The pseudo-code of TS algorithm is presented in Fig. 3. In order to generate the neighborhood two parameters are used: the number of neighbors (nn) and the step size Δ. So, the neighbors are generated according to:

```
Input: population size NP, step size Δ and number of neighbors nn
for i= 1 to NP
    s = V⃗_i
    best = s
    T = ∅
    while stopping condition not reached
        s' = best neighbor of s that is not tabu
        if f(best) > f(s')
            best = s'
        end if
        s = s'
        update tabu list T with s'
    end while
    return best
end for
```

Fig. 3 Pseudo-code of Tabu Search algorithm

$$S'i,j = s_{i,j} + U(-1,1)_{i,j}\Delta \tag{8}$$

where U(−1,1) is a random number drawn from a uniform distribution.

To verify if the best neighbor is tabu, it is necessary to perform the comparison:

$$(S'j \geq t_{i,j} - \varepsilon) \quad or \quad (S'j \leq t_{i,j} + \varepsilon) \tag{9}$$

where S'_j is the j-th component of the best neighbor and $t_{i,j}$ is the j-th component of the i-th element of the tabu list and ε is a parameter of the algorithm.

The tabu list has a fixed length (L). So, elements are inserted into the head of the list, so that the last element is deleted when an element is inserted.

4 Simulation Results

4.1 Experimental Settings

The DE algorithm with ring topology has been tested on four benchmark functions as described in Table 1. Figure 4 illustrates the multimodality of the benchmarks. The algorithm parameters used in the experiments are presented in Table 2.

Table 1 Benchmarks functions (Li, 2010)

Function	Range
$F_1(x) = \sin^6 5\pi x$	$0 \leq x \leq 1$
$F_2(x) = e^{-2\log 2\left(\frac{x-0.1}{0.8}\right)} \sin^6 5\pi x$	$0 \leq x \leq 1$
$F_3(x) = \sin^6 5\pi\left(x^{\frac{3}{4}} - 0.05\right)$	$0 \leq x \leq 1$
$F_4(x) = e^{-2\log 2\left(\frac{x-0.08}{0.854}\right)} \sin^6 5\pi\left(x^{\frac{3}{4}} - 0.05\right)$	$0 \leq x \leq 1$

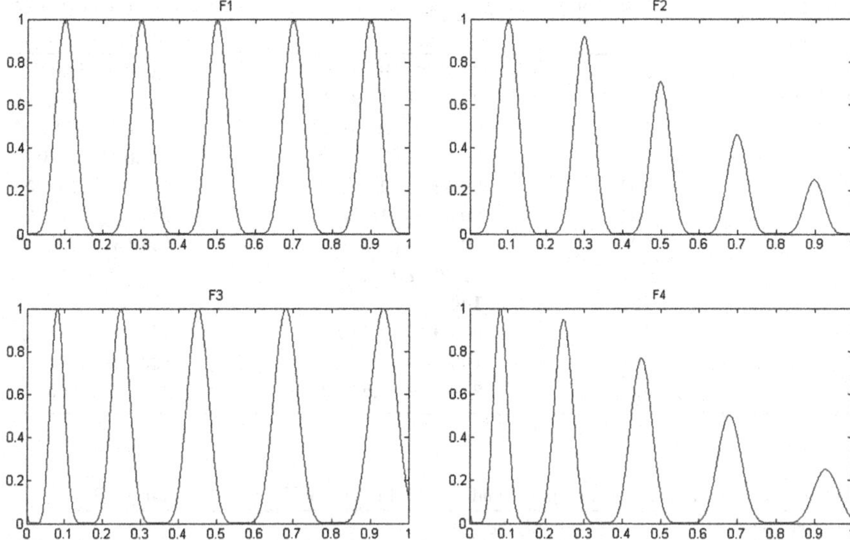

Fig. 4 Plot of the benchmarks functions used in the experiments

Table 2 Algorithm parameters used in DE and DE+TS

c_r	0.9
α	0.8
β	0.8
W	0.1
L	10
Δ	0.1
nn	10
ε	0.01

The algorithm DE with ring topology has been tested on four different population sizes. Next, we tested the performance of the proposed method combining DE with ring topology and Tabu Search. In order to compare our results, we have also carried out the experiments for the benchmarks with the state of art PSO with ring topologies:

- PSO_r3: ring topology with two neighbors for each individual.
- PSO_r3_lhc: similar to the first, but without overlapping neighborhoods.
- PSO_r2: ring topology with one neighbor for each individual.
- PSO_r2_lhc: similar to the third, but without overlapping neighborhoods.

The experiments are run for 2000 iterations. For the proposed algorithm DE+TS after 2000 iterations of DE more 100 iterations are performed for TS. The results are presented in Tables 3 and 4.

Table 3 Success rate in terms of mean and standard deviation for the algorithms DE and DE+TS

Function	NP	DE	DE + TS
F1	51	94.40 % (10.61%)	**98.00 %(6.00%)**
F2	51	**80.00 % (13.86%)**	71.20% (13.95%)
F3	51	92.40 % (10.50%)	**98.40 % (5.43%)**
F4	51	81.20 % (14.09%)	**82.40 % (5.43%)**
F1	102	**99.60 % (2.80%)**	**99.60 % (2.80%)**
F2	102	**90.80 % (10.74%)**	78.00 % (10.77%)
F3	102	99.60 % (2.80%)	**100 %**
F4	102	**91,20 % (10,70%)**	90.00 % (10.00%)
F2	201	**100 %**	**100 %**
F3	201	**91.60 % (9.87%)**	87.60 % (9.71%)
F4	201	**100 %**	**100 %**
F5	201	**97.60 % (6.50%)**	96.40 % (7.68%)

Table 4 Success rate in terms of mean and standard deviation for the algorithm PSO

Function	NP	r3	r3_lhc	r2	r2_lhc
F1	51	98.00 % (7.21%)	**99.20 % (3.92%)**	**99.20 % (3.92%)**	**99.20 % (3.92%)**
F2	51	20.00 % (0%)	**80.40 % (12.96%)**	20.00 % (0%)	77.20 % (17.44%)
F3	51	92.80 % (11.14%)	**96.80 % (7.33%)**	96.40 % (7.68%)	97.60 % (6.50%)
F4	51	20.00 % (0%)	**91.20 % (11.43%)**	20.00 % (0%)	87.60 % (14.91%)
F1	102	99.60 % (2.80%)	**100 % (0%)**	**100 % (0%)**	**100 % (0%)**
F2	102	20.00 % (0%)	**94.80 % (8.77%)**	20.00 % (0%)	89.60 % (13.41%)
F3	102	98.80 % (4.75%)	**100 % (0%)**	99.60 % (2.80%)	99.60 % (2.80%)
F4	102	21.60 % (11.20%)	**98.40 % (5.43%)**	23.20 % (15.68%)	**98.40 % (5.43%)**
F2	201	**100 % (0%)**	**100 % (0%)**	**100 % (0%)**	**100 % (0%)**
F3	201	21.60 % (11.20%)	**98.40 % (5.43%)**	24.80 % (19.00%)	96.80 % (7.33%)
F4	201	99.60 % (2.80%)	**100 % (0%)**	**100 % (0%)**	**100 % (0%)**
F5	201	26.40 % (21.70%)	99.60 % (2.80%)	31.20 % (27.76%)	**100 % (0%)**

4.2 Results Discussion

From Table 3 and considering a population of 51 individuals, the hybrid algorithm presents better performance than the DE algorithm in 3 out of 4 benchmarks. Increasing the population size, we noticed that the DE algorithm and the DE+TS present very similar performance. The DE algorithm with population size 201 shows a superior performance on 2 out of 4 benchmarks. The cause of this may occur because TS refines the previous found solution (local optima) allowing reach the global ones.

The PSO algorithm with topology r3_lhc provided the best results among the four topologies investigated. Considering a population size of 51 the performance of PSO with topology r3_lhc is better than DE and DE+TS in 3 out of 4 benchmarks. Increasing the population size to 201 the performance of DE and DE+TS is getting closer to PSO r3_lhc. In this case, for two benchmarks the performance is the same, and for the other two benchmarks PSO r3_lhc outperforms DE and DE+TS.

5 Conclusions

In this paper, we propose a DE with ring topology without overlapping neighborhood combined with Tabu Search. The results of the DE+TS have been compared to DE for 3 different population sizes. Increasing the population size, we noticed that DE provided better performance. This may be caused by the fact that DE+TS refines the solution found by DE, resulting in a loss of diversity of the population. We compare our results with a state of the art PSO with 4 different ring topologies (r2, r2_lhc, r3, r3_lhc). The best results using PSO with ring topologies were obtained with the topology r3_lhc. Increasing the population size, we notice that the performance of the proposed algorithm DEGL+TS is similar to PSO r3_lhc. The next step of this research is to investigate the use of other local search methods, e.g., Nelder-Mead, or Hooke and Jeves for a larger suite of benchmark functions.

References

1. Chelouah, R., Siarry, P.: A hybrid method combining continuous tabu search and Nelder-Mead simplex algorithms for the global optimization of multiminima functions. Eur. J. Oper. Res. **161**, 636–654 (2005)
2. Das, S., Abraham, A., Chakraborty, U.K., Konar, A.: Differential evolution using a neighborhood based mutation operator. IEEE Trans. Evol. Comput. **13**, 526–553 (2009)
3. Dorronsoro, B., Bouvry, P.: Improving classical and decentralized differential evolution with new mutation operator and population topologies. IEEE Trans. Evol. Comput. **15**, 67–98 (2011)
4. Eberhart, R.C., Shi, Y.: Computational intelligence—concepts to implementations. Morgan Kaufmann, Burlington (2007)
5. Li, X.: Niching without niching parameters: particle swarm optimization using a ring topology. IEEE Trans. Evol. Comput. **14**(1), 150–169 (2010)
6. Li, Z., Li, H., Chen, Y., Sallam, A.: A hybrid strategy based on niche genetic algorithm and tabu search and its convergence property, pp. 328–332. In: IEEE Fifth International Conference on Bio-Inspired computing: Theories and Applications (BIC-TA) (2010)
7. Mashinchi, M.H., Orgun, M.A., Pedrycz, W.: Hybrid optimization with improved tabu search. Appl. Soft Comput. **11**, 1993–2006 (2011)
8. Storn, R., Price, K.V.: Differential evolution—a simple and efficient adaptive scheme for global optimization over continuous spaces. Technical Report TR-95-012. ICSI (1995)
9. Storn, R., Price, K.: Differential evolution—a simple and efficient heuristic for global optimization over continuous spaces. J. Global Optim. **11**, 341–359 (1997)
10. Wei, L., Zhao, M.: A niche hybrid genetic algorithm for global optimization of continuous multimodal functions. Appl. Math. Comput. **160**(3), 649–661 (2005)

A New RBFNDDA-KNN Network and Its Application to Medical Pattern Classification

Shing Chiang Tan, Chee Peng Lim, Robert F. Harrison and R. Lee Kennedy

Abstract In this paper, a new variant of the Radial Basis Function Network with the Dynamic Decay Adjustment algorithm (i.e., RBFNDDA) is introduced for undertaking pattern classification problems with noisy data. The RBFNDDA network is integrated with the k-nearest neighbours algorithm to form the proposed RBFNDDA-KNN model. Given a set of labelled data samples, the RBFNDDA network undergoes a constructive learning algorithm that exhibits a greedy insertion behaviour. As a result, many prototypes (hidden neurons) that represent small (with respect to a threshold) clusters of labelled data are introduced in the hidden layer. This results in a large network size. Such small prototypes can be caused by noisy data, or they can be valid representatives of small clusters of labelled data. The KNN algorithm is used to identify small prototypes that exist in the vicinity (with respect to a distance metric) of the majority of large prototypes from different classes. These small prototypes are treated as noise, and are, therefore, pruned from the network. To evaluate the effectiveness of RBFNDDA-KNN, a series of experiments using pattern classification problems in the medical domain is conducted. Benchmark and real medical data sets are experimented, and the results are compared, analysed, and discussed. The outcomes show that RBFNDDA-KNN is able to learn information with a compact network structure and to produce fast and accurate classification results.

S. C. Tan (✉)
Faculty of Information Science and Technology, Multimedia University, Melaka, Malaysia
e-mail: sctan@mmu.edu.my

C. P. Lim
Centre for Intelligent Systems Research, Deakin University, Burwood, Australia

R. F. Harrison
Department of Automatic Control and Systems Engineering, University of Sheffield, Sheffield, UK

R. L. Kennedy
School of Medicine, Deakin University, Burwood, Australia

V. Snášel et al. (eds.), *Soft Computing in Industrial Applications*,
Advances in Intelligent Systems and Computing 223, DOI: 10.1007/978-3-319-00930-8_7,

1 Introduction

During the past few decades, many different soft-computing techniques such as artificial neural networks [1–3], decision trees [4–6], and fuzzy system [7–9] have been successfully applied as intelligent data processing tools to various domains, e.g., biomedicine [1, 4, 7], manufacturing processes [2, 5, 8], and power systems [3, 6, 9]. This line of research continues to attract the attention of the soft-computing community to further develop more efficient models for learning information from databases and for dealing with various complex problems encountered in different domains. Learning from databases is a challenging task. Normally, real-world data samples are complicated to analyse, as they usually contain noise. Indeed, noisy data denote a group of unrepresentative samples in the database [10]. On the other hand, the database can also contain some small (with respect to a threshold) clusters of labelled data that contain important information for supervised learning. As an example, in medical diagnosis, some patients may develop some distinctive reactions or symptoms pertaining to a treatment, e.g. a severe chest pain [11] can be caused by a novel occurrence of drug allergy, and it should be captured as an important case for further study. Hence, it is useful to have a learning model that is able to first identify small prototypes (i.e. representatives of small clusters of labelled data) in the vicinity (with respect to a distance metric) of large ones. The learning model needs to estimate whether the small prototypes are noise, or they are valid representatives of small clusters of labelled data (which carry important information). The former needs to be pruned, while the latter needs to be preserved in the network. This is the main motivation of this study.

The main purpose of this study is to devise a new variant of the Radial Basis Function Network with Dynamic Decay Adjustment algorithm (i.e., RBFNDDA) [12] model for pattern classification with noisy data. Specifically, the k-nearest neighbour algorithm [13, 14] is integrated with RBFNDDA to form the proposed RBFNDDA-KNN model to achieve this purpose. The original RBFNDDA network has a fast constructive supervised learning process, and is able to produce good classification rates. One of the salient features of RBFNDDA is that its structure grows incrementally during the training process, whereby new hidden neurons are created to include new training instances in form of prototypes. However, RBFNDDA is sensitive to noise. Its constructive learning algorithm manifests a greedy insertion behaviour, which can lead to an over-sized network that contains a lot of small prototypes (hidden neurons). The proposed RBFNDDA-KNN network is designed with the aim to prune small noisy prototypes while preserving other small but useful prototypes in its structure. To demonstrate the effectiveness of the proposed RBFNDDA-KNN network, a number of case studies in medical pattern classification are conducted. Both benchmark and real medical data sets are used in the experiments. The results are compared and discussed.

The organisation of this paper is as follows. In Sect. 2, the background of RBFNDDA-based networks is first presented. Then, the proposed RBFNDDA-KNN network is explained in detail. In Sect. 3, an empirical study using a number of benchmark data sets from the UCI machine-learning repository [15] and a real medical data set from a UK hospital is described. The results are compared and analysed. In Sect. 4, a summary of the paper and suggestions for further work is included.

2 Classification Methods

In this section, the dynamics of RBFNDDA and KNN are explained. This is followed by a description of the proposed RBFNDDA-KNN network for selectively pruning small prototypes for undertaking pattern classification problems.

2.1 Radial Basis Function Neural Network with Dynamic Decay Adjustment

RBFNDDA inherits the salient features of the Probabilistic Neural Network (PNN) [16] and the Restricted Coulomb Energy Network (RCEN) [17, 18]. It exploits the probabilistic characteristics of a static PNN and the ability of adjusting the prototype (hidden neuron) width of the RCEN, and embeds them into a structure that can grow incrementally during training. Figure 1 shows the RBFNDDA network structure.

RBFNDDA requires only two user-defined thresholds (i.e., θ^+ and θ^-) for adjusting the prototype width during the network training phase. These two user-defined thresholds are shown in Fig. 2. The thresholds separate a prototype from its neighbours (prototypes) of different classes. For each training pattern, threshold θ^+ sets the minimum correct-classification probability for the correct class, whereas threshold θ^- sets the highest probability allowable for an incorrect class. This corresponds to an area of conflict whereby neither matching nor conflicting training patterns can reside, as shown in Fig. 2. In other words, thresholds θ^+ and θ^- control the size of the overlapping region of each prototype in RBFNDDA. According to [10, 12], $\theta^+ = 0.40$ and $\theta^- = 0.20$ are suitable to be used as the default settings, and they are good enough to give satisfactory classification results.

RBFNDDA employs the DDA algorithm to construct its structure. During training, new prototypes are inserted into the network to learn new information from the incoming data samples. All committed prototypes are formed during training. Figure 3 illustrates the learning procedure of RBFNDDA in a single epoch.

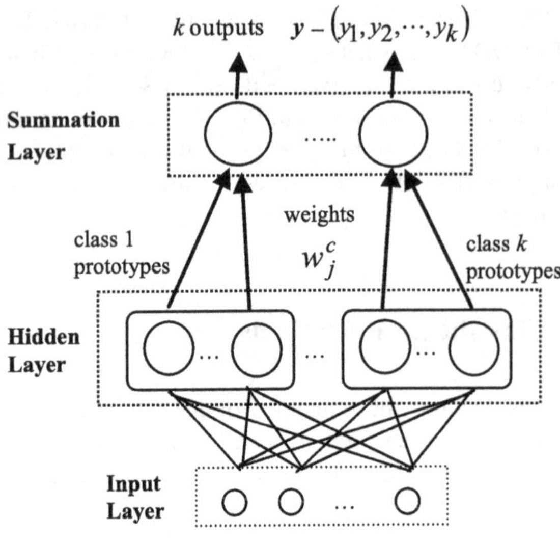

Fig. 1 The RBFNDDA structure

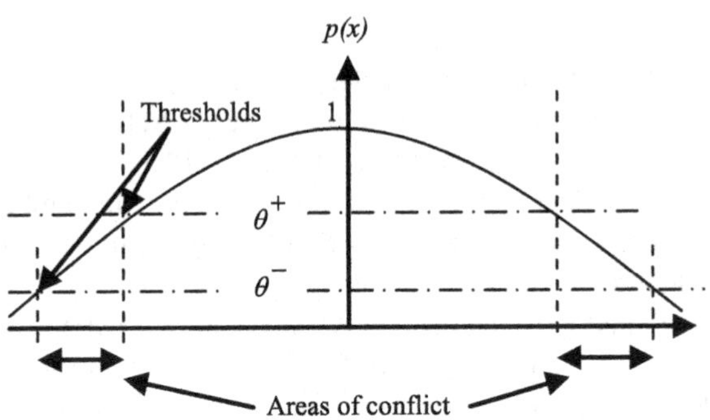

Fig. 2 The two user-defined thresholds of a radial basis hidden node

The RBFNDDA training procedure can be summarised as follows. First, all the weights of the prototypes are initialized to zero (in order to prevent accumulation of duplicated information pertaining to the training patterns). On presentation of a new training sample, if the sample is correctly classified by a few existing prototypes, the weight of the largest prototype is increased. However, if the training sample is incorrectly classified, a new prototype is introduced to include the sample in the network. In this case, the new prototype has the sample as its centre, with its weight set to one. Its initial width is then set in such a way to avoid overlapping with the neighbouring prototypes of different classes. The next step is to perform width shrinking for all prototypes of different classes, if their activations caused by the training sample are above θ^-. Details of the RBFNDDA can be found in [10, 12].

Fig. 3 The learning algorithm of RBFNDDA in one epoch [10]

// reset weights
FORALL prototypes p_j^c **DO**
$\quad w_j^c = 0$
ENDFOR
// train one complete epoch
FORALL training patterns (x, c) **DO**:
$\quad$ **IF** $\exists p_j^k : p_j^k(x) \geq \theta^+$ **THEN**
$\qquad w_j^c = w_j^c + 1$
$\quad$ **ELSE**
$\quad$ // introduce a new prototype (hidden neuron)
$\quad m_c = m_c + 1$
$\quad w_{m_c}^c = 1$
$\quad z_{m_c}^c = x$
$\quad$ // adapt radii
$\quad r_{m_c}^c = \min_{\substack{s \neq c \\ 1 \leq j \leq m_s}} \left\{ \sqrt{-\frac{\left\| z_j^s - z_{m_c}^c \right\|^2}{\ln \theta^-}} \right\}$
$\quad$ **END**
// shrink radii of conflicting prototypes (hidden neurons)
$\quad$ **FORALL** $s \neq c,\ 1 \leq i \leq m_s$ **DO**
$\qquad r_i^s = \min \left\{ r_i^s, \sqrt{-\frac{\left\| z_j^s - z_{m_c}^c \right\|^2}{\ln \theta^-}} \right\}$
$\quad$ **END**
END

2.2 The k-Nearest Neighbour (KNN) Algorithm

The KNN algorithm [13, 14] is a simple instance-based learning method. It classifies a new instance (training sample) based on the k closest instances in the data space. A distance metric, e.g., the Euclidean distance, is used to measure the closeness between the new sample and the existing ones. Using a single neighbour classification scheme, the new sample is classified to the class of the closest sample. If more than one (i.e., k-) nearest neighbours from the existing samples is considered for classification, then the new sample is assigned to the majority class of its closest k neighbours.

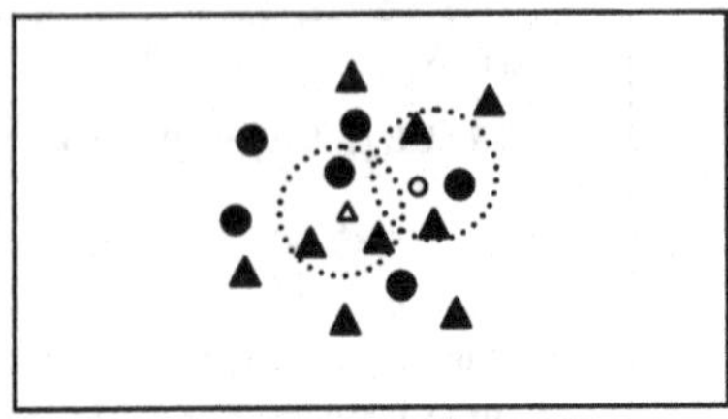

Fig. 4 The use of the KNN algorithm to identify and remove noisy prototypes (hidden neurons). Filled elements represent "large" prototypes (larger than v) and hollow elements represent "small" prototypes (smaller than v) from two classes (C1 = circle and C2 = triangles). With $k = 3$, the small C2 prototype (*hollow triangle*) is preserved while the small C1 prototype (*hollow circle*) is removed from the RBFNDDA-KNN network

2.3 The RBFNDDA-KNN Network

RBFNDDA-KNN is an extension of RBFNDDA. Its learning process encompasses an instance-based KNN algorithm to deal with noisy information. In RBFNDDA-KNN, a collection of prototypes is formed after a training epoch. These prototypes capture information from all training samples. Some of the prototypes are committed to cover only a small number of training samples. The small prototypes are caused either by noisy data or valid representatives of small clusters of labelled data. The KNN algorithm helps avoid establishing a large number of small prototypes in the network, especially in the presence of noisy samples. In each training epoch, such small prototypes, P_j^c, $c \in \{1, 2, \ldots, C\}$, $j \in \{1, 2, \ldots, m_c\}$, are identified if their weights are smaller than a threshold value, v. On the other hand, those prototypes with weights larger than v, i.e., $p_l^c \quad c \in \{1, 2, \ldots, C\}$, $l \in \{1, 2, \ldots, m_c\}$, are deemed useful.

When applying the KNN algorithm, let $N_k(p_l^c)$ be a set of k centres of p_l^c prototypes that are the closest to each centre of p_j^c measured using the standard Euclidean distance metric. The majority class of $N_k(p_l^c)$ is determined according to

$$s = \arg \max_{c \in \{1, 2, \ldots, C\}} N_k(p_l^c) \tag{1}$$

If $c \neq s$, prototype p_j^c is removed from the network because it is considered as a noisy prototype locating close to a group of large prototypes from other classes in the data space. However, if $c = s$, it is retained in the network because both prototypes belong to the same class, except one is a large prototype while another is a small prototype (which may contain specific information) that needs to be preserved. Figure 4 shows a two-dimensional example that uses the KNN algorithm to identify and remove a small noisy prototype from the RBFNDDA-KNN network.

3 Experimental Study

The RBFNDDA-KNN network was first evaluated with three medical data sets from the UCI Machine Learning Repository [15]. Then, a data set comprising real Myocardial Infarction (MI) patient records collected from a hospital in the UK was used. The purpose of the experiment was two-fold: (i) to compare RBFNDDA-KNN with RBFNDDA-based (as in [10]) and Support Vector Machine (SVM)-based (as in [19]) methods using the UCI benchmark data sets; (ii) to demonstrate the applicability of RBFNDDA-KNN using the real MI data set. The results were analysed and discussed.

3.1 Benchmark Medical Problems

Two case studies were conducted. In the first case study, three medical data sets from the UCI machine learning repository were used, i.e. the *Diabetes*, *Cancer*, and *Heart* data sets. Their respective number of data samples, number of attributes, number of classes were 768, 8, 2, 699, 9, 2, 270, 13 and 2. The study aimed to compare the classification performances of RBFNDDA-KNN (from this study), RBFNDDA-T, and original RBFNDDA (both from [10]) based on the same experimental setup as in [10]. Each data set was divided equally into a training set and a test set. For clarity, RBFNDDA-T is a modified version of RBFNDDA that has been enhanced with an online pruning algorithm for removing noise from the network after every training epoch. If a prototype does not contain a sufficient number of training samples (by comparing the number of training samples against a threshold), it is treated as a noise and is kept in a blacklist. The prototype is removed for subsequent training epochs.

In [10], both RBFNDDA and RBFNDDA-T were trained in multiple epochs with $\theta^+ = 0.40$, $\theta^- = 0.20$. In addition, RBFNDDA-T used a threshold setting of $v = 2$ [10]. In this study, RBFNDDA-KNN was trained in multiple epochs with the same setting of $\theta^+ = 0.40$, $\theta^- = 0.20$, and with $k = 10$. The experiment was conducted using a personal computer with a Genuine Intel (R) CPU 2160 at 1.80 GHz, 4 GB of memory. The experiment was repeated eight runs, and the average accuracy rates and number of prototypes were computed. Table 1 shows the overall results. Note that the results of RBNDDDA and RBFNDDA-T are those reported in [10].

Table 1 shows that the accuracy rates of three networks are similar. RBFNDDA-KNN used approximately 50 % fewer number of prototypes than RBFNDDA. By analysing the RBFNDDA-KNN structure, it was noticed that the "small" prototypes, which were located among other prototypes representing the general concept of other class(es) in the data space, could be identified as noise; hence removed. The RBFNDDA-KNN size was larger than that of RBFNDDA-T. The former did not simply remove all small prototypes during training, while the

Table 1 Performance comparison with RBFNDDA-based networks (Acc.-average test accuracy rate; #Prototypes–average number of prototypes (hidden neurons)

Data set	RBFNDDA (from [10])		RBFNDDA-T (from [10])		RBFNDDA-KNN (from this study)	
	Acc. (%)	#Prototypes	Acc. (%)	#Prototypes	Acc. (%)	#Prototypes
Diabetes	74.35	288.5	73.50	65.6	71.14	161.9
Cancer	96.86	70.6	96.90	38.3	96.85	38.7
Heart	79.26	83.6	79.82	32.6	79.90	39.5

latter would remove all prototypes smaller than the specified threshold and put them in a blacklist.

In the second case study, a performance comparison between RBFNDDA-KNN and SVM classifiers (as in [19]) was conducted using the same benchmark problems as in the first case study. However, in accordance with [19], the missing samples were moved in the Cancer data set, resulting in a total of 683 samples. The Diabetes and Heart data sets remained the same. In [19], a series of SVM classifiers, either a single or an ensemble entity, was constructed using the RBF, linear and polynomial kernels. Following the procedure in [19], an experiment with the ten-fold cross-validation method was performed. Table 2 shows the overall results.

RBFNDDA-KNN achieved the best result for the Cancer data set, and moderate results for the Diabetes and Heart data sets. RBFNDDA-KNN produced a higher accuracy rate than all polynomial-based SVM classifiers for the Heart data set. RBFNDDA-KNN performed better than almost all RBF-based SVM classifiers

Table 2 Performance comparison (accruacy in percentages) with SVM-based classifiers (all the SVM results are taken from [19])

Classification method	Data Set		
	Diabetes	Cancer	Heart
MAdaBoostSVM (RBF Kernel)	74.05	95.54	78.30
SingleSVM (RBF Kernel)	74.29	95.26	79
BaggingSVM (RBF Kernel)	77.21	96.57	83.48
Arc-x4SVM (RBF Kernel)	71.43	95.35	78.56
AdaboostSVM (RBF Kernel)	71.95	95.21	78.41
MAdaBoostSVM (Linear Kernel)	76.96	96.72	83.19
SingleSVM (Linear Kernel)	77.17	96.69	83.37
BaggingSVM (Linear Kernel)	77.24	96.78	83.19
Arc-x4SVM (Linear Kernel)	72.92	95.99	79.22
AdaboostSVM (Linear Kernel)	76.86	96.65	82.11
MAdaBoostSVM (Polynomial Kernel)	76.42	95.31	77.70
SingleSVM (Polynomial Kernel)	75.55	94.12	75.41
BaggingSVM (Polynomial Kernel)	76.18	94.85	78.48
Arc-x4SVM (Polynomial Kernel)	72.82	94	76.26
AdaboostSVM (Polynomial Kernel)	75.83	94.37	76.89
RBFNDDA-KNN	74.89	97.07	80.00

(except one) using the Heart and Diabetes data sets. Overall, the results indicate that RBFNDDA-KNN is as competitive as single- and ensemble-based SVM classifiers as reported in [19].

3.2 A Real Medical Problem

In this experiment, a data set comprising 500 patient records collected from the Northern General Hospital, Sheffield, UK, was used to evaluate the applicability of the RBFNDDA-KNN network to real-world medical classification problems. The data set was used for missing data analysis previously [20]. In this study, the task was to categorise the data samples into to two categories: suspected Myocardial Infarction (MI) and non-MI cases. There were about 31 and 69 % MI and non-MI cases, respectively, in the data set. Each data sample comprised a total of 29 features that were found to be significant to MI diagnosis, which included clinical measurements (e.g. ECG readings), physical symptoms (e.g. sweating, vomiting), and other relevant information (e.g. age, smoker/ex-smoker). The features were coded as binary numbers, with a 1/0 indicating presence/absence of symptoms. There were two real-valued features, i.e., age and duration of pain. All features were normalised between 0 and 1. To better evaluate the stability of the performances, the experiment was repeated 100 times with different sequences of training samples. The average results and their 95 % confidence intervals were estimated using the bootstrap method [21]. The bootstrap method was used because of its ability to provide accurate estimates of statistical parameters (e.g. the average results and their confidence intervals) based on a small sample size. In addition, the Data Retention Rate (DRR) [22], i.e. a useful measure of network compactness, was computed, as follows.

$$\text{data retention rate} = \frac{\text{number of hidden neurons (prototypes)}}{\text{number of training samples}} \tag{2}$$

Table 3 summarises the bootstrap results. To further compare the results statistically, the bootstrap hypothesis test was conducted for (i) RBFNDDA versus RBFNDDA-KNN; (ii) RBFNDDA-T versus RBFNDDA-KNN; and (iii) RBFNDDA versus RBFNDDA-T, respectively, with three performance metrics, i.e., accuracy rate, number of prototypes and training time. The significance level was set at 0.05. The null hypothesis stated that the performance metric (e.g., accuracy rate of RBFNDDA versus RBFDDDA-KNN) was the same, whereas the alternative hypothesis claimed that the former was lower than the latter. All p-values from the hypothesis tests are listed in Table 4.

Referring to Tables 3 and 4, the average accuracy rate of RBFNDDA-KNN is statistically higher than those of RBFNDDA-T and RBFNDDA because the p-values are lower than 0.05. The numbers of prototypes for RBFNDDA versus RBFNDDA-KNN and RBFNDDA versus RBFNDDA-T show significant

Table 3 The (boostrap) average results for the RBFNDDA-based networks (the numbers in the square brackets indicate the 95 % confidence intervals with bootstrap)

Classifier	Acc. (%)	#Prototypes	DDR	Time (s)
RBFNDDA	69.69 [69.28 70.12]	119.0 [119.3 125.4]	0.48	2.5 [2.4 2.5]
RBFNDDA-T	69.54 [69.15 69.98]	30.9 [29.9 31.8]	0.12	3.4 [3.3 3.5]
RBFNDDA-KNN	70.27 [69.73 70.84]	69.8 [67.3 72.3]	0.28	2.7 [2.6 2.7]

Table 4 The (bootstrap) p-values for the RBFNDDA-based networks

Hypothesis	Acc.	#Prototypes	Time (s)
RBFNDDA versus RBFNDDA-KNN	0.043	1.000	0.000
RBFNDDA-T versus RBFNDDA-KNN	0.024	0.000	1.000
RBFNDDA versus RBFNDDA-T	0.667	1.000	0.000

difference (with p-value of 1), i.e., they are significantly (at the 95 % confidence interval) fewer than that of RBFNDDA. The empirical results indicated that RBFNDDA-KNN was able to mitigate the influence of unrepresentative data samples and to produce high classification accuracy. Between RBFNDDA-KNN and RBFNDDA-T, the former created significantly more number of prototypes than the latter, as shown by the DRR results and the p-value. However, the computational time of RBFNDDA-KNN was significantly faster than that of RBFNDDA-T, although the former had more than twice the number of prototypes than the latter. As explained in Sect. 3.1, RBFNDDA-T stored a list of unrepresentative samples to check against noise from the training samples in each training epoch. Hence, additional computational time was required. This problem could become severe when a large and noisy training data set was used. In short, RBFNDDA-KNN was able to establish a compact network structure (as compared with RBFNDDA) and produce fast and accurate classification results (as compared with both RBFNDDA and RBFNDDA-T) in this MI case study.

4 Summary

In this paper, a new variant of the RBFNDDA network has been proposed for pattern classification with noisy data. The proposed network, known as RBFNDDA-KNN, is based on integration between RBFNDDA and the KNN algorithm. The proposed RBFNDDA-KNN network is able to identify small noisy prototypes and preserve small useful prototypes that represent small clusters of important labelled data. A series of experiments using both benchmark and real medical data sets has been conducted to evaluate the effectiveness of RBFNDDA-KNN. The results have shown that RBFNDDA-KNN is able to learn and encode information from the database using a smaller number of prototypes than RBFNDDA, and with improved classification results. As compared with

RBFNDDA-T, RBFNDDA-KNN has exhibited a larger network structure. However, RBFNDDA-T does not need to keep a list of unrepresentative samples (as what RBFNDDA-T does to identify noisy data) during the training process. As such, RBFNDDA-KNN is able to produce fast and accurate results in the presence of noisy samples for undertaking pattern classification tasks.

For future work, additional experiments to ascertain the effectiveness of the proposed RBFNDDA-KNN network in various real-world classification domains will be conducted. In addition, rule extraction from RBFNDDA-KNN can be carried out so that it can provide explanation for its predicted outcomes.

References

1. Lee, J., Steele, C.M., Chau, T.: Classification of healthy and abnormal swallows based on accelerometry and nasal airflow signals. AI Med. **52**, 17–25 (2011)
2. Yu, J.-b., Xi., L.-f.: A neural network ensemble-based model for on-line monitoring and diagnosis of out-of-control signals in multivariate manufacturing processes. Expert Syst. Appl. **36**, 909–921 (2009)
3. Barakat, M., Druaux, F., Lefebvre, D., Khalil, M., Mustapha, O.: Self adaptive growing neural network classifier for faults detection and diagnosis. Neurocomputing **74**, 3865–3876 (2011)
4. Son, C.-S., Kim, Y.-N., Kim, H.-S., Park, H.-S., Kim, M.-S.: Decision-making model for early diagnosis of congestive heart failure using rough set and decision tree approaches. J. Biomed. Inform. **45**, 999–1008 (2012)
5. Ciflikli, C., Kahya-Özyirmidokuz, E.: Implementing a data mining solution for enhancing carpet manufacturing productivity. Knowl. Based Syst. **23**, 783–788 (2010)
6. Upendar, J., Gupta, C.P., Singh, G.K.: Statistical decision-tree based fault classification scheme for protection of power transmission lines. Electr. Power Energy Syst. **36**, 1–12 (2012)
7. Alayón, S., Robertson, R., Warfield, S.K., Ruiz-Alzola, J.: A fuzzy system for helping medical diagnosis of malformations of cortical development. J. Biomed. Inform. **40**, 221–235 (2007)
8. Piltan, M., Mehmanchi, E., Ghaderi, S.F.: Proposing a decision-making model using analytical hierarchy process and fuzzy expert system for prioritizing industries in installation of combined heat and power systems. Expert Syst. Appl. **39**, 1124–1133 (2012)
9. Kazemi, M.V., Moradi, M., Kazemi, R.V.: Minimization of powers ripple of direct power controlled DFIG by fuzzy controller and improved discrete space vector modulation. Expert Syst. Appl. **89**, 23–30 (2012)
10. Paetz, J.: Reducing the number of neurons in radial basis function networks with dynamic decay adjustment. Neurocomputing **62**, 79–91 (2004)
11. Davey, P., Lalloo, D.G.: Drug induced chest pain—rare but important. Postgrad Med J. **76**, 420–422 (2000)
12. Berthold, M.R., Diamond, J.: Constructive training of probabilistic neural networks. Neurocomputing **19**, 167–183 (1998)
13. Cover, T.M., Hart, P.E.: Nearest neighbor pattern classification. IEEE Trans. Inf. Theory **13**, 21–27 (1967)
14. Dasarathy, B.V.: Nearest neighbor (NN) norms: NN pattern classification techniques. A generalized knearest neighbor rule, pp. 64–84. IEEE Computer Society Press, Los Alamitos (1991)

15. Asuncion, A., Newman, D.J.: UCI machine learning repository. University of California, School of Information and Computer Science, Irvine. [http://www.ics.uci.edu/~mlearn/MLRepository.html] (2007)
16. Specht, D.F.: Probabilistic neural networks. Neural Netw. **3**, 109–118 (1990)
17. Hudak, M.H.: RCE Classifiers: theory and practice. Cybern. Syst. **23**, 483–515 (1992)
18. Reilly, D.L., Cooper, L.N., Elbaum, C.: A neural model for category learning. Biol. Cybern. **45**, 35–41 (1982)
19. Wang, S.-J., Mathew, A., Chen, Y., Xi, L.-F., Ma, L., Lee, J.: Empirical analysis of support vector machine ensemble classifiers. Expert Syst. Appl. **36**, 6466–6476 (2009)
20. Lim, C.P., Kuan, M.M., Harrison, R.F.: Application of fuzzy ARTMAP and fuzzy c—means clustering to pattern classification with incomplete data. Neural Comput. Appl. **14**, 104–113 (2005)
21. Efron, B.: Bootstrap methods: another look at the jackknife. Ann. Stat. **7**, 1–26 (1979)
22. Lam, W., Keung, C.-K., Liu, D.: Discovering useful concept prototypes for classification based on filtering and abstraction. IEEE Trans. Pattern Anal. Mach. Intell. **24**, 1075–1090 (2002)

An Approach to Fuzzy Modeling of Anti-lock Braking Systems

Radu-Codruţ David, Ramona-Bianca Grad, Radu-Emil Precup, Mircea-Bogdan Rădac, Claudia-Adina Dragoş and Emil M. Petriu

Abstract This chapter proposes an approach to fuzzy modeling of Anti-lock Braking Systems (ABSs). The local state-space models are derived by the linearization of the nonlinear ABS process model at ten operating points. The Takagi-Sugeno (T-S) fuzzy models are obtained by the modal equivalence principle, where the local state-space models are the rule consequents. The optimization problems are defined in order to minimize the objective functions expressed as the squared modeling errors, and the variables of these functions are a part of the parameters of input membership functions. Simulated Annealing algorithms are implemented to solve the optimization problems and to obtain optimal T-S fuzzy models. Real-time experimental results are included to validate the new optimal T-S fuzzy models for ABS laboratory equipment.

R.-C. David (✉) · R.-B. Grad · R.-E. Precup · M.-B. Rădac · C.-A. Dragoş
Department of Automation and Applied Informatics, "Politehnica" University of Timisoara, Bd. V. Parvan 2, ON 300223, Timisoara, Romania
e-mail: davidradu@gmail.com

R.-B. Grad
e-mail: bibi23grad@yahoo.com

R.-E. Precup
e-mail: radu.precup@aut.upt.ro

M.-B. Rădac
e-mail: mircea.radac@aut.upt.ro

C.-A. Dragoş
e-mail: claudia.dragos@aut.upt.ro

E. M. Petriu
School of Electrical Engineering and Computer Science, University of Ottawa, 800 King Edward, Ottawa K1N 6N5, Canada
e-mail: petriu@eecs.uottawa.ca

V. Snášel et al. (eds.), *Soft Computing in Industrial Applications*,
Advances in Intelligent Systems and Computing 223, DOI: 10.1007/978-3-319-00930-8_8,

1 Introduction

The development of fuzzy models for Anti-lock Braking Systems (ABSs) is a challenging problem because of the importance of these nonlinear safety subsystems. The approaches to fuzzy modeling of ABSs aim the slip controller design. The ABS process identification and the robust adaptive control of an active suspension system by hierarchical T-S fuzzy-neural models are discussed in [1]. A T-S fuzzy model of deceleration based on analyzing the braking process and dynamic model of vehicle and wheel is proposed in [2]. A quarter vehicle braking model with four-degrees of freedom subject to irregular excitation from a road surface is offered in [3] and applied to ABS fuzzy control. Two discrete-time dynamic Takagi-Sugeno (T-S) fuzzy models of ABS processes based on the modal equivalence principle are suggested in [4]. Combinations of fuzzy models applied to intelligent ABS controllers based on fuzzy control, neural networks and sliding mode control are given in [5–8].

This chapter offers a simple approach to fuzzy modeling of ABSs. Our approach starts with the derivation of an initial discrete-time T-S fuzzy model of the process by the modal equivalence principle; this fuzzy model is characterized by a set of local linearized state-space models of the process which are placed in the rule consequents. A part of the parameters of the input membership functions (m.f.s) is optimized by a Simulated Annealing (SA) algorithm in order to solve the optimization problems which aim the minimization of the objective functions (o.f.s) expressed as the sum of squared modeling errors, viz. the differences between the process output (the wheel slip) and the T-S fuzzy model output. Other successful applications related to the optimal tuning of fuzzy models by means of SA algorithms are given in [9–14].

Our approach is different to the state-of-the-art [1–8] because it starts with the first-principle mathematical model of the process. This approach is advantageous because of the performance of the optimal T-S fuzzy model is checked by real-time experiments on the ABS laboratory equipment. Although this approach cannot guarantee the reaching of the global minimum of the o.f., this chapter shows that a serious decrease of o.f. is exhibited. Therefore the performance improvement of our T-S fuzzy model is clearly indicated.

The chapter treats the following topics. Section 2 is dedicated to the mathematical modeling of the process and to the design of discrete-time dynamic T-S fuzzy models focused on ABS laboratory equipment. The main aspects concerning the implementation of our SA algorithm are discussed in Sect. 3. Real-time experimental results are offered in Sect. 4 to validate the new optimal T-S fuzzy models. The conclusions are highlighted in Sect. 5.

2 ABS Process Models

The nonlinear state-space model of the ABS laboratory equipment [15] is derived starting with the first-principle equations [15, 16]

$$\begin{aligned} J_1\dot{x}_1 &= F_n r_1 \mu(\lambda) - d_1 x_1 - M_{10} - M_1, \\ J_2\dot{x}_2 &= -F_n r_2 \mu(\lambda) - d_2 x_2 - M_{20}, \end{aligned} \tag{1}$$

where λ is the longitudinal slip (the wheel slip), J_1 and J_2 are the inertia moments of wheels, x_1 and x_2 are the angular velocities, d_1 and d_2 are the friction coefficients in wheels' axes, M_{10} and M_{20} are the static friction torques that oppose the normal rotation, M_1 is the brake torque, r_1 and r_2 are the radii of wheels, F_n is the normal force that the upper wheel pushes upon the lower wheel, $\mu(\lambda)$ is the friction coefficient, and $\dot{x}_1$ and $\dot{x}_2$ are the wheels' angular accelerations. The identification by experiment-based measurements leads to the parameter values [16]: $r_1 = r_2 = 0.99$ m, $F_n = 58.214$ N, $J_1 = 7.53 \cdot 10^{-3}$ kg m^2, $J_2 = 25.6 \cdot 10^{-3}$ kg m^2, $d_1 = 1.1874 \cdot 10^{-4}$ kg m^2/s, $d_2 = 2.1468 \cdot 10^{-4}$ kg m^2/s, $M_{10} = 0.0032$ N m, $M_{20} = 0.0925$ N m. The wheel slip and the nonlinear term $S(\lambda)$ are expressed as

$$\lambda = (r_2 x_2 - r_1 x_1)/(r_2 x_2),\ x_2 \neq 0,\ S(\lambda) = \mu(\lambda)/\{L[\sin \Phi - \mu(\lambda) \cos \Phi]\}, \tag{2}$$

where $L = 0.37$ m is the arm's length which fixes the upper wheel, and $\phi = 65.61^{\circ}$ is the angle between the normal direction in wheels' contact point and L's direction.

The nonlinear state-space equations of the process are

$$\begin{aligned} \dot{x}_1 &= S(\lambda)(c_{11}x_1 + c_{12}) + c_{13}x_1 + c_{14} + (c_{15}S(\lambda) + c_{16})s_1 M_1, \\ \dot{x}_2 &= S(\lambda)(c_{21}x_1 + c_{22}) + c_{23}x_2 + c_{24} + c_{25}S(\lambda)s_1 M_1, \\ \dot{M}_1 &= c_{31}(b(u) - M_1), \end{aligned} \tag{3}$$

where u is the control signal applied and the actuator's nonlinear model is highlighted in the third equation. The expressions of the parameters in (3) are [16]

$$\begin{aligned} &c_{11} = r_1 d_1/J_1, c_{12} = (M_{10} + M_g)r_1/J_1, c_{13} = -d_1/J_1, c_{14} = -M_{10}/J_1, \\ &c_{15} = r_1/J_1, c_{16} = -1/J_1, c_{21} = -r_2 d_1/J_2, c_{22} = -(M_{10} + M_g)r_2/J_2, \\ &c_{23} = -d_2/J_2, c_{24} = -M_{20}/J_2, c_{25} = -r_2/J_2. \end{aligned} \tag{4}$$

The introduction of λ as controlled output in the model (3) is done by the substitution of x_1 from (3). This leads to the state-space equations of the ABS process given in [4, 16], to the state vector $\mathbf{x} = [\lambda \quad x_2 \quad M_1]^T$; T indicates the matrix transposition.

The steps of our modeling approach are:

- the definition of the m.f.s of the input variables λ, x_2 and M_1,
- the derivation of an initial T-S fuzzy model of the process, with the state variables λ, x_2 and M_1 as input variables, and the discrete-time state-space process models with the matrices $\mathbf{A}_{d,i}$, $\mathbf{B}_{d,i}$, $\mathbf{C}_{d,i}$, $i = 1\ldots10$, in the rule consequents,
- the definition of the optimization problem where the vector variable of the o.f. consists of a part of the parameters of the input m.f.s,
- the application of the SA algorithm to obtain the optimal input m.f. parameters which lead to the optimal T-S fuzzy model.

The derivation of the initial T-S fuzzy model starts with setting the largest domains of variation of the state variables in all ABS operating regimes [4, 16]: $0.1 \le \lambda \le 1$, $20 \le x_2 \le 178$ and $0 \le M_1 \le 10$. The fuzzification in the T-S fuzzy model employs the linguistic terms assigned to the input variables and defined as follows. The first input variable, λ, uses five linguistic terms, $LT_{\lambda,j}$, $j = 1\ldots5$, with the triangular m.f.s

$$\begin{aligned} &\mu_{LT_{\lambda,1}} : [0, 0.2] \to [0, 1], \quad \mu_{LT_{\lambda,2}} : [0.1, 0.4] \to [0, 1], \quad \mu_{LT_{\lambda,3}} : [0.2, 0.8] \to [0, 1], \\ &\mu_{TL_{\lambda,4}} : [0.4, 1] \to [0, 1], \quad \mu_{LT_{\lambda,5}} : [0.8, 1.1] \to [0, 1], \end{aligned} \tag{5}$$

having the expressions

$$\mu_{LT_{\lambda,j}}(x) = \begin{cases} 0, & x < a_{\lambda,j}, \\ 1 + (x - b_{\lambda,j})/(b_{\lambda,j} - a_{\lambda,j}), & x \in a_{\lambda,j} \le x < b_{\lambda,j}, \\ 1 - (x - b_{\lambda,j})/(c_{\lambda,j} - b_{\lambda,j}), & x \in b_{\lambda,j} \le x < c_{\lambda,j}, \\ 0, & x \ge c_{\lambda,j}, \end{cases} \quad a_{\lambda,j} < b_{\lambda,j} < c_{\lambda,j}, j = 1...5. \tag{6}$$

The parameters $a_{\lambda,j}$, $j = 1\ldots5$, and $c_{\lambda,j}$, $j = 1\ldots5$, are variable and they belong to the vector variable ρ of the o.f., and the parameters $b_{\lambda,j}$, $j = 1\ldots5$, which stand for the modal values of the m.f.s, are fixed: $b_{\lambda,1} = 0.1$, $b_{\lambda,2} = 0.2$, $b_{\lambda,3} = 0.4$, $b_{\lambda,4} = 0.8$ and $b_{\lambda,5} = 1$. The second input variable, x_2, uses two linguistic terms, $LT_{x_2,j}$, $j = 1\ldots2$, with the triangular m.f.s of type (6), $\mu_{LT_{x_2,1}} : [0, 150] \to [0, 1]$ and $\mu_{LT_{x_2,2}} : [50, 180] \to [0, 1]$. The parameters $a_{x_2,j}$, $j = 1\ldots2$, and $c_{x_2,j}$, $j = 1\ldots2$, are variable and they belong to ρ, and the parameters $b_{x_2,j}$, $j = 1\ldots2$, are fixed: $b_{x_2,1} = 50$ and $b_{x_2,2} = 150$. The third input variable, M_1, uses one linguistic term, $LT_{M_1,1}$, with the triangular m.f. of type (6), $\mu_{LT_{M_1,1}} : [0, 11] \to [0, 1]$. The parameters $a_{M_1,1}$ and $c_{M_1,1}$ are variable and they belong to ρ, and the parameter $b_{M_1,1}$ is fixed: $b_{M_1,1} = 10$. Other m.f. shapes can be used in different industrial applications [17–22].

The complete rule base of the discrete-time dynamic T-S fuzzy model of the process consists of the rules R^i, $i = 1\ldots10$, expressed as:

$$R^1: \text{IF } \lambda \text{ IS } LT_{\lambda,1} \text{ AND } \mathrm{x}_2 \text{ IS } LT_{x_2,1} \text{ AND } \mathrm{M}_1 \text{ IS } LT_{M_1,1} \text{ THEN} \begin{cases} \mathbf{x}_{k+1} = \mathbf{A}_{d,1}\mathbf{x}_k + \mathbf{B}_{d,1}u_k, \\ y_{k,m} = \mathbf{C}_{d,1}\mathbf{x}_k, \end{cases}$$
$$\ldots$$
$$R^{10}: \text{IF } \lambda \text{ IS } LT_{\lambda,5} \text{ AND } \mathrm{x}_2 \text{ IS } LT_{x_2,2} \text{ AND } \mathrm{M}_1 \text{ IS } LT_{M_1,1} \text{ THEN} \begin{cases} \mathbf{x}_{k+1} = \mathbf{A}_{d,10}\mathbf{x}_k + \mathbf{B}_{d,10}u_k, \\ y_{k,m} = \mathbf{C}_{d,10}\mathbf{x}_k, \end{cases} \tag{7}$$

where the discrete-time state-space models in the rule consequents are obtained by the discretization of the continuous-time state-space linearized models at ten operating points. The state-space model matrices in the rule consequents of R^1 and R^{10} are

$$\mathbf{A}_{d,1} = \begin{bmatrix} 0.9441 & -0.0025 & 0.0194 \\ -1.0335 & 1.0012 & -0.0601 \\ 0 & 0 & 0.8157 \end{bmatrix}, \quad \mathbf{B}_{d,1} = \begin{bmatrix} 0.0021 \\ -0.0059 \\ 0.1843 \end{bmatrix}, \quad \mathbf{C}_{d,1} = [1 \ \ 0 \ \ 0],$$
$$\mathbf{A}_{d,10} = \begin{bmatrix} 1.0041 & -0.0003 & 0.0069 \\ -0.3192 & 1 & -0.0519 \\ 0 & 0 & 0.8157 \end{bmatrix}, \quad \mathbf{B}_{d,10} = \begin{bmatrix} 0.0007 \\ -0.0054 \\ 0.1843 \end{bmatrix}, \quad \mathbf{C}_{d,10} = [1 \ \ 0 \ \ 0]. \tag{8}$$

The modal equivalence principle is applied to obtain the initial fuzzy model (7) with the rule consequent parameters given in (8) as the coordinates of the operating points being represented by the modal values of input m.f.s. The sampling period $T_s = 0.01$ s is used in this context, and k in (7) indicates the number of current sampling interval. The SUM and PROD operators are used in the inference engine, and the weighted average method is employed for defuzzification. Different operators and defuzzification lead to modified nonlinear input–output model maps [23–26].

3 Simulated Annealing Algorithm

SA is applied to solve the optimization problem

$$\rho^* = \arg\min_{\rho \in D} J(\rho), \tag{9}$$

where the parameter vector of the fuzzy model (the vector variable of the o.f.) is

$$\rho = [a_{\lambda,1} \ \ c_{\lambda,1} \ \ a_{\lambda,2} \ \ c_{\lambda,2} \ \ a_{\lambda,3} \ \ c_{\lambda,3} \ \ a_{\lambda,4} \ \ c_{\lambda,4} \ \ a_{\lambda,5} \ \ c_{\lambda,5} \\ a_{x_2,1} \ \ c_{x_2,1} \ \ a_{x_2,2} \ \ c_{x_2,2} \ \ a_{M_1,1} \ \ c_{M_1,1}]^T, \tag{10}$$

$J(\rho)$ is the o.f. defined as

$$J(\rho) = \frac{1}{N}\sum_{k=1}^{N}(y_k(\rho) - y_{k,m}(\rho))^2 = \frac{1}{N}\sum_{k=1}^{N}(e_{k,m}(\rho))^2, \quad (11)$$

ρ^* is the optimal parameter vector of the fuzzy model and the solution to the optimization problem (9), $y_k(\rho) = \lambda_k(\rho)$ is the process output at k^{th} sampling interval, $y_{k,m}(\rho)$ is the fuzzy model output, $e_{k,m}(\rho) = y_k(\rho) - y_{k,m}(\rho)$ is the modeling error, N is the length of the time horizon, and D is the feasible domain of (9).

The SA algorithm is adapted from the general SA algorithms proposed in [27, 28] and from the SA suggested in [13] and applied to the optimal tuning of fuzzy controller parameters. The steps of our SA algorithm are

- *Step 1*. Set $\mu = 0$, where μ is the iteration number, $s_{r\max}$ and $r_{r\max}$ (the maximum success and rejection rates defined in [29]) with the initial values $s_r = 0$ and $r_r = 0$, and the minimum temperature $\theta_{\min}$. Choose the initial temperature θ_0.
- *Step 2*. Generate a random initial solution ϕ and compute its fitness value $C(\phi)$, where C is the fitness function.
- *Step 3*. Generate a probable solution ψ by disturbing ϕ, and evaluate the fitness value $C(\psi)$.
- *Step 4*. Compute the difference $\Delta C_{\phi\psi} = C(\phi) - C(\psi)$. If $\Delta C_{\phi\psi} \geq 0$, then accept ψ as the new solution. Otherwise, set the random parameter q_n, $0 \leq q_n \leq 1$, and compute the probability of ψ to be the next solution:

$$p_\psi = \begin{cases} 1 & \text{if } \Delta C_{\phi\psi} > 0, \\ \exp(\Delta C_{\phi\psi}/\theta_\mu) & \text{otherwise.} \end{cases} \quad (12)$$

If $p_\psi > q_n$, then ψ is the new solution.

- *Step 5*. If the new solution is accepted, then update the new solution and C, increment s_r and set $r_r = 0$, where r_r is the rejection rate. Otherwise, increment r_r. If r_r has reached its maximum value, $r_{r\max}$, the algorithm is stopped; otherwise, continue with Step 6.
- *Step 6*. Increment s_r. If s_r has reached its maximum value $s_{r\max}$, go to Step 7; otherwise increment μ. If μ has reached its maximum value $\mu_{\max}$, go to Step 7; otherwise, go to Step 2.
- *Step 7*. Alleviate the temperature according to the temperature decrement rule

$$\theta_{\mu+1} = \alpha_{cs}\theta_\mu,\ \alpha_{cs} = \text{const},\ \alpha_{cs} < 1,\ \alpha_{cs} \approx 1. \quad (13)$$

- *Step 8*. If $\theta_\mu > \theta_{\min}$ then go to Step 3, otherwise the algorithm is stopped indicating that it has reached the solution ψ.

This SA algorithm is mapped onto (9) by means of the following relations between the fitness and objective functions and between the parameter vectors:

$$J(\rho) = C(\psi), \; J(\rho) = C(\phi), \; \rho = \psi, \; \rho = \phi. \tag{14}$$

4 Experimental Results

A part of the results and implementation details of our modeling approach is presented as follows. The maximum success and rejection rates were set to $r_{r\max} = 100$ and $s_{r\max} = 50$. The SA algorithm was stopped after 84 iterations, with the final temperature $\theta_{84} = 0.090235$. The initial temperature and solution were to $\theta_0 = 1$ and [16]:

$$\begin{aligned} \rho = [0 \quad 0.2 \quad 0.1 \quad 0.4 \quad 0.2 \quad 0.8 \quad 0.4 \quad 1 \quad 0.8 \quad 1.1 \\ 0 \quad 150 \quad 50 \quad 180 \quad 0 \quad 11]^T, \end{aligned} \tag{15}$$

respectively, and the final solution is:

$$\begin{aligned} \rho^* = \psi = [0.008615 \quad 0.1614 \quad 0.09901 \quad 0.4067 \quad 0.2295 \quad 0.8471 \quad 0.4019 \quad 0.9578 \\ 0.847 \quad 1.12 \quad 0.02202 \quad 169 \quad 59.85 \quad 180.5 \quad -0.3227 \quad 11.04]^T. \end{aligned} \tag{16}$$

The control signal u was generated in order to cover different ranges of magnitudes. The evolution of the control signal versus time is presented in Fig. 1. Figure 1 gives a total set of 35,000 input–output data points which are separated in

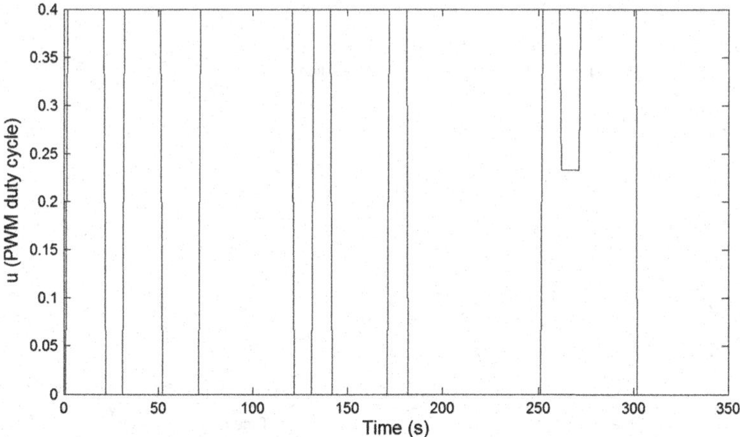

Fig. 1 Control signal u versus time, applied to real-world process and T-S fuzzy model. T-S fuzzy model

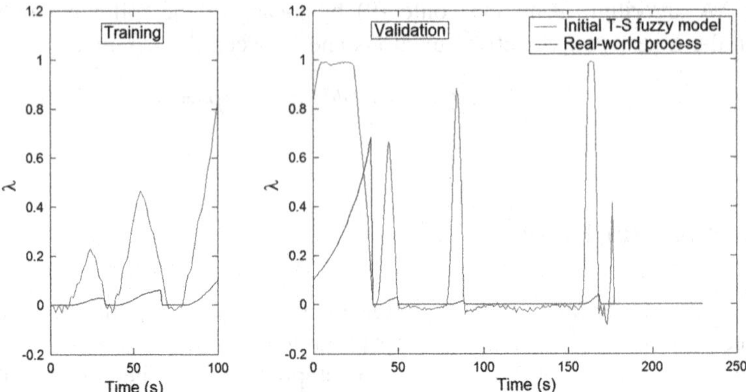

Fig. 2 Real-time experimental results: wheel slip λ versus time for initial T-S fuzzy model and for real-world process

the training data and the validation data set for cross-validation and to assess the performance of the T-S fuzzy models. The first $N = 10000$ data points (corresponding to the time frame from 0 to 100 s) which result from Fig. 1 are the training data set. The rest of $N = 25000$ data points (corresponding to the time frame from 100 to 350 s) which result from Fig. 1 are the validation data set.

The experimental results are presented in Fig. 2 and in Fig. 3 as the outputs of the real-world process (the ABS laboratory equipment), of the T-S fuzzy model before the application of the SA algorithm and of the T-S fuzzy model after the application of the SA algorithm. Figs. 2 and 3 clearly show that the performance of the T-S fuzzy model is strongly improved by the application of our SA algorithm from the point of view the modeling errors. The modeling errors are seriously

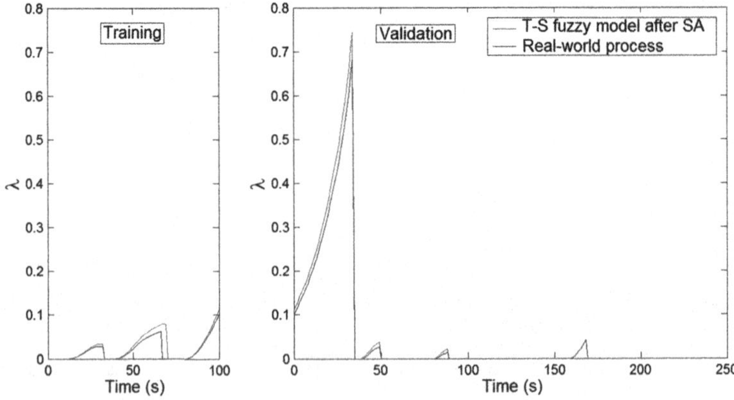

Fig. 3 Real-time experimental results: wheel slip λ versus time for T-S fuzzy model after optimization by SA algorithm and for real-world process

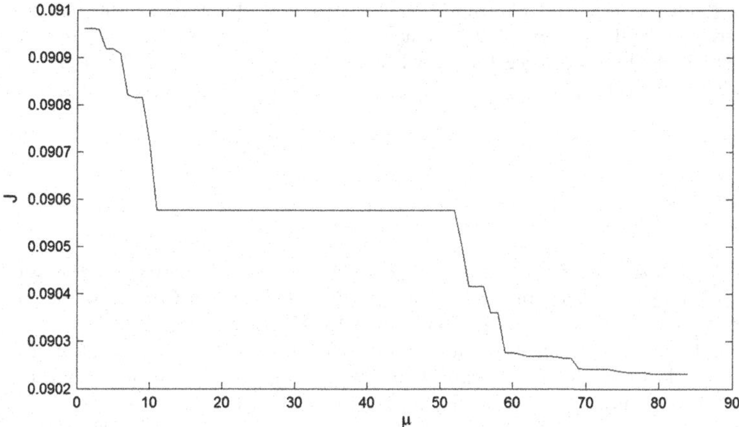

Fig. 4 Objective function J versus iteration number μ for validation data set

alleviated in both training and validation data sets. Fig. 4, which corresponds to the validation data set, shows that the solution to (9) obtained by our SA algorithm ensures a strong decrease of the o.f. Although the minimum of o.f. cannot be guaranteed, Fig. 4 points out that the improvement can continue by increasing the number of iterations.

5 Conclusions

This chapter has proposed an approach to obtain discrete-time T-S fuzzy models dedicated to ABSs. The models result by the application of an SA algorithm which ensures the optimal tuning of the input m.f. parameters of T-S fuzzy models initially obtained by the modal equivalence principle.

Our approach is important because it is relatively simple, it produces models which can be implemented easily, and it can be generalized to a wide category of industrial applications. The limitation of our approach is represented by the random generation of the initial solution of the SA algorithm. However, the results are presented as an average of the best five runs of the SA algorithm, and they convincingly show the performance improvement of the T-S fuzzy model of ABS laboratory equipment.

Future research will deal with the extension of the SA algorithm to the modeling of multi input-multi output nonlinear systems. The inclusion of o.f. gradients in other evolutionary-based algorithms [17–22, 29], the data-driven computation of gradients and the correlation analysis will be considered.

Acknowledgments This work was supported by a grant in the framework of the Partnerships in priority areas—PN II program of the Romanian National Authority for Scientific Research ANCS, CNDI—UEFISCDI, project number PN-II-PT-PCCA-2011-3.2-0732, and by a grant of the NSERC of Canada.

References

1. Zhao, Z., Yu, Z., Sun, Z.: Research on fuzzy road surface identification and logic control for anti-lock braking system. In: Proceedings of IEEE International Conference on Vehicular Electronics and Safety, Shanghai, China, pp. 380–387 (2006)
2. Zheng, T., Ma, F., Zhang, K.: Estimation of reference vehicle speed based on T-S fuzzy model. In: Proceedings of international conference on advanced in control engineering and information science, Dali, China, vol. 15, pp. 188–193 (2011)
3. Wang, W.-Y., Chen, M.-C., Su, S.-F.: Hierarchical T-S fuzzy-neural control of anti-lock braking system and active suspension in a vehicle. Automatica **48**, 1698–1706 (2012)
4. Precup, R.-E., Spătaru, S.V., Rădac, M.-B., Petriu, E.M., Preitl, S., Dragoş, C.-A., David, R.-C.: Experimental results of model-based fuzzy control solutions for a laboratory antilock braking system. In: Hippe, Z.S., Kulikowski, J.L., Mroczek, T., (eds.), Human-Computer Systems Interaction: Backgrounds and Applications 2, Part 2, AICS, vol. 99, pp. 223–234. Springer, New York (2012)
5. Naderi, P., Farhadi, A., Mirsalim, M., Mohammadi, T.: Anti-lock and Anti-slip braking system, using fuzzy logic and sliding mode controllers. In: Proceedings of IEEE Vehicle Power and Propulsion Conference, Lille, France, p. 6 (2010)
6. Topalov, A.V., Oniz, Y., Kayacan, E., Kaynak, O.: Neuro-fuzzy control of antilock braking system using sliding mode incremental learning algorithm. Neurocomputting **74**, 1883–1893 (2011)
7. Bhandari, R., Patil, S., Singh, R.K.: Surface prediction and control algorithms for anti-lock brake system. Trans. Res. Part C Emerg. Technol. **21**, 181–195 (2012)
8. Aleksendrić, D., Jakovljević, Ž., Ćirović, V.: Intelligent control of braking process. Expert Syst. Appl. **39**, 11758–11765 (2012)
9. Garibaldi, J.M., Ifeachor, E.C.: Application of simulated annealing fuzzy model tuning to umbilical cord acid-base interpretation. IEEE Trans. Fuzzy Syst. **7**, 72–84 (1999)
10. Liu, G., Yang, W.: Learning and Tuning of fuzzy membership functions by simulated annealing algorithm. In: Proceedings of 2000 IEEE Asia-Pacific Conference on Circuits and Systems, Tianjin, China, pp. 367–370 (2000)
11. Almaraashi, M., John, R., Coupland S., Hopgood A.: Time series forecasting using a TSK fuzzy system tuned with simulated annealing. In: Proceedings of 2010 IEEE International Conference on Fuzzy Systems, Barcelona, Spain, p. 6 (2010)
12. Yanara, T.A., Akyürek, Z.: Fuzzy model tuning using simulated annealing. Expert Syst. Appl. **38**, 8159–8169 (2011)
13. Precup, R.-E., David, R.-C., Petriu, E.M., Preitl, S., Rădac, M.-B.: Fuzzy control systems with reduced parametric sensitivity based on simulated annealing. IEEE Trans. Industr. Electron. **59**, 3049–3061 (2012)
14. David, R.-C., Dragoş, C.-A., Bulzan, R.-G., Precup, R.-E., Petriu, E.M., Rădac, M.-B.: An approach to fuzzy modeling of magnetic levitation systems. Int. J. Artif. Intell. **9**, 1–18 (2012)
15. Inteco: ABS: The laboratory anti-lock braking system controlled from PC—user manual, Inteco Sp. z o. o., Krakow, Poland (2007)
16. Grad, R.-B.: Biologically inspired optimization algorithm for fuzzy modeling of an anti-lock braking system laboratory equipment. M.Sc. thesis, "Politehnica" University of Timisoara, Timisoara, Romania (2012)

17. Köppen, M.: Light-weight evolutionary computation for complex image-processing applications. In: Proceedings of 6th International Conference on Hybrid Intelligent Systems, Auckland, New Zealand, pp. 3–3 (2006)
18. Deb, K., Gupta, S., Daum, D., Branke, J., Mall, A.K., Padmanabhan, D.: Reliability-based optimization using evolutionary algorithms. IEEE Trans. Evol. Comput. **13**, 1054–1074 (2009)
19. Blažič, S., Matko, D., Škrjanc, I.: Adaptive law with a new leakage term. IET Control Theory Appl. **4**, 1533–1542 (2010)
20. Carrano, E.G., Takahashi, R.H.C., Fonseca, C.M., Neto, O.M.: Non-linear network optimization—an embedding vector space approach. IEEE Trans. Evol. Comput. **14**, 206–226 (2010)
21. Ko, M., Tiwari, A., Mehnen, J.: A review of soft computing applications in supply chain management. Appl. Soft Comput. **10**, 661–674 (2010)
22. Kudelka, M., Horak, Z., Snásel, V., Krömer, P., Platos, J., Abraham, A.: Social and swarm aspects of co-authorship network. Logic J. IGPL **20**, 634–643 (2012)
23. Johanyák, Z.C.: Student evaluation based on fuzzy rule interpolation. Int. J. Artif. Intell. **5**, 37–55 (2010)
24. Vaščák, J., Madarász, L.: Adaptation of fuzzy cognitive maps—a comparison study. Acta Polytech. Hung. **7**, 109–122 (2010)
25. Castillo, O., Melin, P., Garza, A.A., Montiel, O., Sepúlveda, R.: Optimization of interval type-2 fuzzy logic controllers using evolutionary algorithms. Soft. Comput. **15**, 1145–1160 (2011)
26. Schaefer, G., Hu, Q., Zhou, H., Peters, J.F., Hassanien, A.E.: Rough C-means and fuzzy rough C-means for colour quantisation. Fundam. Inf. **119**, 113–120 (2012)
27. Kirkpatrick, S., Gelatt Jr, C.D., Vecchi, M.P.: Optimization by simulated annealing. Science **20**, 671–680 (1983)
28. Geman, S., Geman, D.: Stochastic relaxation, gibbs distribution and the bayesian restoration in images. IEEE Trans. Pattern Anal. Mach. Intell. **6**, 721–741 (1984)
29. Precup, R.-E., David, R.-C., Petriu, E.M., Rădac, M.-B., Preitl, S., Fodor, J.: Evolutionary optimization-based tuning of low-cost fuzzy controllers for servo systems. Knowl. Based Syst. **38**, 74–84 (2013)

An Improved Evolutionary Algorithm to Sequence Operations on an ASRS Warehouse

José A. Oliveira, João Ferreira, Guilherme A. B. Pereira and Luis S. Dias

Abstract This paper describes the hybridization of an evolutionary algorithm with a greedy algorithm to solve a job-shop problem with recirculation. We model a real problem that arises within the domain of loads' dispatch inside an automatic warehouse. The evolutionary algorithm is based on random key representation. It is very easy to implement and allows the use of conventional genetic operators for combinatorial optimization problems. A greedy algorithm is used to generate active schedules. This constructive algorithm reads the chromosome and decides which operation is scheduled next. This option increases the efficiency of the evolutionary algorithm. The algorithm was tested using some instances of the real problem and computational results are presented.

1 Introduction

The efficiency of an automatic storage system depends, among other factors, on the plan for loading operations on trucks. In the Automated Storage and Retrieval System (AS/RS) warehouses, where a large number of truckloads are performed on a daily basis, it is necessary to plan and execute accurately the loading

J. A. Oliveira (✉) · J. Ferreira · G. A. B. Pereira · L. S. Dias
Centre ALGORITMI, University of Minho, Guimarães, Portugal
e-mail: zan@dps.uminho.pt
URL: http://pessoais.dps.uminho.pt/zan

J. Ferreira
e-mail: joao.aoferreira@gmail.com

G. A. B. Pereira
e-mail: gui@dps.uminho.pt

L. S. Dias
e-mail: lsd@dps.uminho.pt

V. Snášel et al. (eds.), *Soft Computing in Industrial Applications*,
Advances in Intelligent Systems and Computing 223, DOI: 10.1007/978-3-319-00930-8_9,

procedures in order to fulfil the delivery deadlines. The issue of AS/RS warehouses has been studied since the 70s, increasing their interest in recent years [1–8]. The problem for sequencing operations in automated warehouses was presented by Oliveira [4, 5], and the problem was modelled as a variant of the job-shop scheduling problem (JSSP). Recently, Figueiredo et al. [8] modelled the problem in which the operations have general processing times and they presented a genetic algorithm to solve the problem.

In this paper we describe an evolutionary algorithm that represents an improvement that represents an improvement to the work presented by Figueiredo et al. [8]. In their work the authors present a genetic algorithm to generate non-delay schedules. The authors support their option on the fact that the solution space is shorter than the solution space of active schedules. In this paper we study the hybridization of a Evolutionary Algorithm (EA) with a greedy algorithm to generate active schedules. Although the search space being larger, we show that this EA gets better results, without an increase of CPU time.

The paper is organized in the following way: next section describes the automatic warehouse type of operations; the third section presents the model adopted, some remarks about its application and also discusses some extensions to the model; the fourth section is dedicated to the characterization of the solution's methodology adopted; the fifth section presents computational results of the developed algorithm; finally the conclusions about the work are discussed.

2 The Storage System

The real scheduling problem reported by Oliveira [5] occurs in a AS/SR type automatic warehouse. The AS/RS is formed by several aisles of pallets racks, with capacity about forty thousand pallets. In each aisle operates a stacker crane (also S/R machine). In the retrieval process the S/R machine moves the pallets from their storage position to the collector at the top of the aisle. Next, several forklift trucks move the pallets and place them inside the trucks. According to this process the real problem was modelled as a job-shop scheduling problem with recirculation [4]—that means a job is processed more than once in the same machine. In this case, a load for one truck could receive more than one pallet from the same aisle and the same S/R machine. For details about the warehouse functioning we address the reader to [4, 5].

In the real problem, there are about a hundred loads to dispatch per day (truck loads—also called "trips"). The strategy adopted to program the trucks' loading consists on defining a partition of the whole load into disjoint subsets of loads, which are processed simultaneously, called batches (blocks). This strategy guarantees that every load of a batch is finished before starting to prepare the loads for the next batch. The batches' dimensioning respects the imposed limits, in order to fulfil the delivery deadlines. The number of docking bays limits the maximum number of loads in a batch. A standard workday can originate plans with 15–20 batches, with 6–13 loads each [8].

Oliveira [5] demonstrated that the time spent to process the batches depends on the number of loads of the batch, and it is calculated based on the nominal values of the system's throughput. It is assumed that the storage system guarantees a nominal flow of pallets' dispatch at a constant rate, and therefore it considers that the duration of the batch is proportional to the total number of batches' pallets. Figueiredo et al. [8] demonstrated that time to process the batch depends also on the sequence for retrieving the pallets, which is to say, on the S/R machines' operations scheduling. Due to precedence constraints between pallets, the order in which the pallets of the batches are retrieved by the S/R machines may lead to different processing times. The problem of establishing the best pallets' retrieving operations sequence in each aisle is also an optimization problem, and the authors modelled it as a Job Shop Scheduling Problem (JSSP) with recirculation.

In Oliveira [5] there is a detailed description of the process to retrieve the pallets—from the aisle to the truck. Part of the movement is performed by forklifts that are controlled by Radio Frequency. In his work, Oliveira [5] assumes identical processing times to transport pallets independently of the location of the aisle and the truck. In this work, a new model considering general processing times is presented, and it takes into account the location of the truck in the docking bays and the aisle where the pallets are retrieved. This model is believed to better representing the real problem, but also turns the associated job shop problem more difficult to solve.

3 Job Shop Scheduling Problem

Some Combinatorial Optimization Problems are very hard to solve and therefore require the use of heuristic procedures. One of them is the Job Shop Scheduling Problem. The use of exact methods to solve the JSSP is limited to small size instances. According to Zhang et al. [9] the Branch and Bound methods do not solve instances larger than 250 operations in a reasonable time. The heuristic methods have become very popular and have gained much success in solving job shop scheduling problems. In the last twenty years a huge number of papers has been published, presenting several metaheuristic methods. From Simulated Annealing to Particle Swarm Optimization, there are several variants of the same method class. EAs are very popular among researchers [2, 4, 5, 8, 10–12]. In 1996, Vaessens et al. [13] stated a goal for the Job Shop Problem: to achieve an average error of less than two percent within 1,000 s total computation time. In this work the authors presented the Genetic Algorithms as the less effective metaheuristic to solve the JSSP. A possibility to increase the efficiency and the effectiveness is an algorithm that includes specific knowledge of the problem.

In the JSSP each job is formed by operations that have to be processed on a set of machines. Each job has a technological definition that determines a specific order to process the job's operations, and it is also necessary to guarantee that there is no overlap, in time, when processing different operations in the same machine;

and also that there is no overlap, in time, when processing operations of the same job. The objective of the JSSP is to finish all jobs as soon as possible, that is, to minimize the makespan. The classical JSSP model considers a set of n jobs, and a set of m machines. Each job consists of a set of m operations (one operation on each machine), among which precedence relations exist and that defines a single processing sequence. Each operation is processed in one machine only, during p time units. The instant when the job is concluded is C. All operations are concluded at C_{max} (called makespan). For the convenience of the representation, operations are numbered consecutively from 1 to N, in which N is the total number of operations. The classic model considers that all the jobs are processed once in every machine, and the total number of operations is $N = n.m$. In a more general model, a job can have a number of operations different from m (number of machines). The case in which a job is processed more than once in the same machine, is called a job shop model with recirculation [4]. The computational and practical significance of JSSP have motivated the attention of researchers for the last several decades.

The solutions (schedules) for the JSSP can be classified in three sets: semi-actives (SA), actives (A) and non-delayed (ND). Typically, these sets have the particularity that A is a proper subset of SA, and ND is a proper subset of A. In relation to the optimal solution of the problem (minimization of C_max), it is known that it is an active schedule but not necessarily a non-delayed schedule. In this work, the solutions obtained from the EA belong to the active set. This choice was taken in order to obtain better solutions.

4 Methodology

In this work we adopted a method based on evolutionary algorithms. This technique's simplicity to model complex problems and its easy integration with other optimization methods were factors that were considered for its choice. Traditionally, genetic algorithms used bit string chromosomes. These chromosomes consisted of only '0s' and '1s'. Modern genetic algorithms more often use problem-specific chromosomes with evidence that the use of real or integer value chromosomes often outperformed bit string chromosomes. The permutation code was adequate to permutation problems. In this kind of representation, the chromosome is a literal of the operations sequence on the machines. For the classical JSSP case, Oliveira [5] presents a chromosome that is composed by m sub-chromosomes, one for each machine, each one composed by n genes, one for each operation. The i gene of the sub-chromosome corresponds to the operation processed in i place in the corresponding machine. The allele identifies the operation's index in the disjunctive graph. Nevertheless, in this work, the random key code presented by Bean [14] is used, attending the easiness to model complex systems with this type of representation, as it is the case to model the loading of trucks in an automatic warehouse.

The important feature of random keys is that all offspring formed by the genetic operators are feasible solutions—this occurs when it is used as a constructive procedure based on the available operations to schedule and the priority is given by the random key allele. Another advantage of the random key representation is the possibility of using the conventional genetic operators. This characteristic allows the use of EA with other optimization problems, adapting only a few routines related to the problem. Equally easy becomes the hybridization with other heuristics when a genetic algorithm with random keys is used. Also, it is possible to hybridize with a MIP solver, once the fractional value of a variable could be used as an allele of the random key representation. A chromosome represents a solution to the problem and it is encoded as a vector of random keys (random numbers). In this work, according to Cheng et al. [10], the problem representation is indeed a mix from priority rule-based representation and random keys representation.

4.1 Constructive Algorithm

The solutions represented by chromosome are decoded by an algorithm, which is based on Giffler and Thompson's algorithm (GTa) [15]. While the GTa can generate all the active plans, the constructive algorithm only generates the plan in agreement with the chromosome. As advantages of this strategy, we pointed out the small dimension of solution space, which includes the optimum solution and the fact that it does not produce impossible or disinteresting solutions from the optimization point of view. On the other hand, since the dimensions between the representation space and the solution space are very different, this option can represent a problem because two chromosomes can represent the same solution. The constructive algorithm has N stages and in each stage an operation is scheduled. To assist the algorithm's presentation, consider the following notation for stage t:

P_t—the partial schedule of the $(t-1)$ scheduled operations;
S_t—the set of operations schedulable at stage t, i.e. all the operations that must precede those in S_t are in P_t;
σ_k—the earliest time that operation o_k in S_t could be started;
ϕ_k—the earliest time that operation o_k in S_t could be finished, that is $\phi_k = \sigma_k + p_k$ and p_k is the processing time of operation o_k;
M^*—the selected machine where $\phi^* = min_{o_k \in S_t}(\phi_k)$;
S^*—the conflict set formed by $o_k \in S_t$ processed in M^* and $\sigma_k < \phi^*$;
o_k^*—the selected operation to be scheduled at stage t.

The constructive algorithm of solutions is presented in Fig. 1 in a similar format to the one used by Cheng et al. [10]. In Step 3 instead of using a priority dispatching rule, the information given by the chromosome is used. If the maximum allele value is equal for two or more operations, one is chosen randomly.

Step 1	Let $t=1$ with P_1 being null. S_t will be the set of all operations with no predecessors.
Step 2	Find $\phi^* = \min_{o_k \in S_t}\{\phi_k\}$ and identify M^*. Form S_t^*.
Step 3	Select operation o_t^* in S_t^* with greatest allele value.
Step 4	Move to next stage by (1) adding o_t^* to P_t, so creating P_{t+1}; (2) removing o_j^* from S_t and creating S_{t+1} by adding to S_t the operation that directly follows o_t^* in its job (unless o_t^* completes its job); (3) incrementing t by 1.
Step 5	If there are any operations left unscheduled $(t < N)$, go to *Step 2*. Otherwise, stop.

Fig. 1 Algorithm 1—Constructive algorithm

4.2 The Evolutionary Algorithm Structure

The EA has a very simple structure and can be represented in Algorithm 2, illustrated in Fig. 2. It begins with population generation and her evaluation. Attending to the fitness of the chromosomes the individuals are selected to be parents. Comparing the fitness of the new elements and of their progenitors the former population is updated.

On building of the offspring population we implemented an elitist strategy. Firstly, we select the better individuals of the current population to be progenitors, (10 %). Then, the remaining parents are chosen by the routines of the roulette wheel (60 %) and tournament (30 %). So, one individual can be selected more than once to be progenitor. Secondly, we apply the genetic operators over this population according to the following partition: we use the crossover on two parents (40 %); the crossover on two parents plus the mutation (30 %); and the mutation (30 %). The Uniform Crossover (UX) is used in this work. This genetic operator uses a new sequence of random numbers and swaps both progenitors' alleles if the random key is greater than a prefixed value. This genetic operator uses a new sequence of random numbers and swaps both progenitors' alleles if the random key is greater than a prefixed value. Fig. 3 illustrates the UX's application on two parents (*prnt*1, *prnt*2), and swaps alleles if the random key is *geq* 0.75.

Fig. 2 Algorithm 2—Evolutionary algorithm

Step 1	**begin**
Step 2	$P \leftarrow$ GenerateInitialPopulation()
Step 3	Evaluate(P)
Step 4	**while** termination conditions not meet **do** (1) P' ← Select_progenitors(P) (2) P'' ← Apply_GeneticOperators(P') (3) Evaluate(P'') (4) $P \leftarrow$ Select($P \cup P''$)
Step 5	**end while**

i	*1*	*2*	*3*	*4*	*5*	*6*	*7*	*8*	*9*	*10*
prnt1	0.89	0.48	0.24	0.03	0.41	0.11	0.24	0.12	0.33	0.30
prnt2	0.83	0.41	0.40	0.04	0.29	0.35	0.38	0.01	0.42	0.32
randkey	0.64	0.72	0.75	0.83	0.26	0.56	0.28	0.31	0.09	0.11
dscndt1	0.89	0.48	0.40	0.04	0.41	0.11	0.24	0.12	0.33	0.30
dscndt2	0.83	0.41	0.24	0.03	0.29	0.35	0.38	0.01	0.42	0.32

Fig. 3 The UX crossover

The genes 3 and 4 are changed and it originates two descendants (*dscndt*1, *dscndt*2). Descendant 1 is similar to parent 1, because it has about 75 % of the genes of this parent.

5 Computational Experiments

Computational experiments were carried out with some representative instances of the real problem. We use the same representative instances of the real problem presented by Figueiredo et al. [8] that were generated randomly and they concern a job-shop problem with recirculation. The dimension of these instances corresponds to the defined maximum dimension of the real problem. Instances jr_13_1, jr_13_2 are constituted by 13 jobs (docking bay number) and 35 operations per job (one load's dimension), which adds up to a total of 455 operations and 11 machines (number of aisles). The 35 operations of the major part of the jobs are processed in five machines. This situation corresponds to the real problem. Usually, the load of a truck (35 pallets) comes from the five aisles closer to the docking bay. The recirculation situation happens when a job is processed more than once in the same machine. In the real problem this corresponds to collecting several pallets from the same aisle. Table 1 presents the results of computational experiences, and summarize the results obtained by Figueiredo et al. [8] with non-delay schedules and also the results for our active schedules. Globally an improvement of 1 % is obtained by active schedules. Furthermore, for some instances an improvement of 2 and 3 % is reached—these instances correspond to greater sizes in terms of number of operations.

The algorithm was implemented in C++ and the code was compiled using GNU Compiler Collection (GCC) version 4.6.1. The tests were run on a computer Intel i3, with 4 MB of RAM, on the Ubuntu 11.10 in Windows7 using VirtualBox. The experiments were performed using two different populations—a small one with 20 individuals, and a larger one with 100 individuals. The third column of Table 1 indicates the number of jobs, and column Op indicates the number of operations.

Table 1 Experimental results

				NonDelay Schedules			Active Schedules		
Name	Pop	J	Op	Best	Aver.	Worst	Best	Aver.	Worst
jr_1_1	20	1	35	83	83	83	83	83	83
	100	1	35	83	83	83	83	83	83
jr_3_1	20	3	105	90	90	90	90	90	90
	100	3	105	90	90	90	90	90	90
jr_4_1	20	4	140	93	93	93	93	93	93
	100	4	140	93	93	93	93	93	93
jr_4_2	20	4	140	9	94.7	96	94	94.2	95
	100	4	140	94	94.4	95	94	94.2	95
jr_6_1	20	6	210	110	111.6	113	109	110.2	112
	100	6	210	109	110.3	112	108	109.3	111
jr_6_2	20	6	210	98	100.07	102	99	99.3	101
	100	6	210	99	100.07	101	98	99.1	100
jr_8_1	20	8	280	148	148.5	150	148	148	148
	100	8	280	147	148.1	149	147	147.7	148
jr_8_2	20	8	280	141	141.8	143	141	141.2	142
	100	8	280	141	141	141	141	141	141
jr_10_1	20	10	350	138	140.4	142	135	137.5	139
	100	10	350	137	139.5	141	134	136.6	138
jr_10_2	20	10	350	149	150.2	152	146	147.5	149
	100	10	350	149	150.4	151	145	146.5	148
jr_13_1	20	13	455	197	201.1	204	198	198.8	200
	100	13	455	197	199.9	202	194	197.7	200
jr_13_2	20	13	455	166	170.4	173	166	167.7	171
	100	13	455	166	168.8	171	164	165.8	167

This table presents the best value obtained from 15 runs of each configuration, and the average value of 15 runs. For all experiments 500 iterations were performed.

The results are promising, while achieving better solutions in early stages of the optimization process. Also the algorithm proved to be robust and consistent, with similar performances in all experiments considered. With larger populations it is possible to achieve better results since the beginning of the optimization process, although requiring extra CPU time. Despite the code is not yet optimized for calculations speed, a run for the largest instance (jr_13_2), performing 500 iterations takes about 85 s with a population size of 100, and about 18 s with a population size of 20. Fig. 4 shows these runs—population size 20 and population size 100 for both types of schedules (non delay/active). Fig. 4 also illustrates the advantage of using a larger population (better solutions since the beginning), although this option takes more CPU time. With a population five times bigger the algorithm returns a better result within just 20 % of the number of iterations. For this particular run, it is possible to verify that the generation of active schedules outperforms the non-delay schedules' generation.

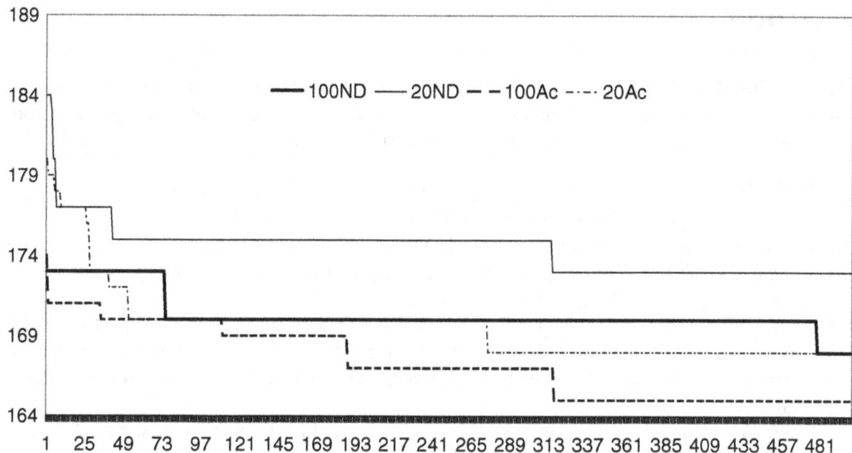

Fig. 4 Results for different population sizes

6 Conclusions

This paper presents a model for scheduling load operations in an automatic warehouse and a Evolutionary Algorithm to solve the JSSP with recirculation. The schedules are built using information given by the genetic algorithm to sequence the operations. The algorithm incorporates a constructive algorithm to build the solution from the chromosome, and it always generates feasible plans, and in particular active schedules. This constructive algorithm allows the inclusion of specific knowledge of the problem and makes the algorithm very efficient. This algorithm easily shows the solution for several daily problems that may occur in a warehouse. The algorithm was tested with success on instances of equal dimension of the real problem. The computation time suggests that it is an efficient algorithm, allowing its integration in a decision support system to evaluate different alternatives in useful time. The main advantage in generating active plans is the fact that all schedules are valid in terms of programming the S/R machines, without generating deadlocks.

As future developments, we do think it would be beneficial to try using both constructive algorithms. In fact, for some instances the optimal solution is an active schedule, and for other instances it is a non-delay schedule. It is feasible to design an hybrid constructive algorithm.

Since the random keys representation has no Lemarkin property it is our intention to develop a variation of random keys representation to avoid this weakness.

Acknowledgments This work was funded by the “Programa Operacional Fatores de Competitividade—COMPETE” and by the FCT—Fundação para a Ciência e Tecnologia in the scope of the project: FCOMP-01-0124-FEDER-022674.

References

1. Rashid, M.M., Kasemi, B., Rahman, M.: New Automated Storage and Retrieval System (ASRS) using wireless communications. In: 2011 4th International Conference on Mechatronics: Integrated Engineering for Industrial and Societal, Development, ICOM'11 (2011)
2. Crdenas, J.J., Garcia, A., Romeral, J.L., Andrade, F.: A genetic algorithm approach to optimization of power peaks in an automated warehouse. In: Proceedings of Industrial Electronics Conference (IECON), pp. 3297–3302 (2009)
3. Hausman, W.H., Schwarz, L.B., Graves, S.C.: Optimal storage assignment in automatic warehousing systems. Manage. Sci. **22**, 629–638 (1976)
4. Oliveira, J.A.: A genetic algorithm with a quasi-local search for the job shop problem with recirculation. In: Abraham, A., de Barts, B., Köppen, M., Nickolay, B. (eds.) Applied Soft Computing Technologies: The Challenge of Complexity, pp. 221–234. Springer, Heidelberg (2006)
5. Oliveira, J.A.: Scheduling the truckload operations in automatic warehouses. Eur. J. Oper. Res. **179**, 723–735 (2007)
6. Ko, M., Tiwari, A., Mehnen, J.: A review of soft computing applications in supply chain management. Appl. Soft Comput. **10**, 661–674 (2010)
7. Poon, T.C., Choy, K.L., Cheng, C.K., Lao, S.I., Lam, H.Y.: Effective selection and allocation of material handling equipment for stochastic production material demand problems using genetic algorithm. Expert Syst. Appl. **38**, 12497–12505 (2011)
8. Figueiredo, J., Oliveira, J.A., Dias, L., Pereira, G.: A genetic algorithm for the job shop on an ASRS warehouse. Computational science and its applications ICCSA 2012. Lect. Notes Comput. Sci. **7335**, 133–146 (2012)
9. Zhang, C., Li, P., Guan, Z., Rao, Y.: A tabu search algorithm with a new neighborhood structure for the job shop scheduling problem. Comput. Oper. Res. **53**, 313–320 (2007)
10. Cheng, R., Gen, M., Tsujimura, Y.: A tutorial survey of job-shop scheduling problems using genetic algorithms—I. representation. Comput. Ind. Eng. **30**, 983–997 (1996)
11. Park, B.J., Choi, H.R., Kim, H.S.: A hybrid genetic algorithm for the job shop scheduling problems. Comput. Ind. Eng. **45**, 597–613 (2003)
12. Oliveira, J.A., Dias, L., Pereira, G.: Solving the job shop problem with a random keys genetic algorithm with instance parameters. In: Proceedings of 2nd International Conference on Engineering Optimization (EngOpt 2010), Lisbon, Portugal, (CDRom) (2010)
13. Vaessens, R., Aarts, E., Lenstra, J.K.: Job shop scheduling by local search. INFORMS J. Comput. **8**, 302–317 (1996)
14. Bean, J.C.: Genetics and random keys for sequencing and optimization. ORSA J. Comput. **6**, 154–160 (1994)
15. Giffler, B., Thompson, G.L.: Algorithms for solving production scheduling problems. Oper. Res. **8**, 487–503 (1960)

Fuzzy Reliability Analysis of Washing Unit in a Paper Plant Using Soft-Computing Based Hybridized Techniques

Komal and S. P. Sharma

Abstract The present study deals with the fuzzy reliability analysis of washing unit in a paper plant utilizing available uncertain data which reflects their components' failure and repair pattern. Paper computes different reliability parameters of the system in the form of fuzzy membership functions. Two soft-computing based hybridized techniques namely Genetic Algorithms Based Lambda-Tau (GABLT) and Neural Network and Genetic Algorithms Based Lambda-Tau (NGABLT) along with traditional Fuzzy Lambda-Tau (FLT) technique are used to evaluate the fuzzy reliability parameters of the system. In FLT, ordinary fuzzy arithmetic is utilized while in GABLT and NGABLT ordinary arithmetic and nonlinear programming approach are used. The computed results, as obtained by these techniques, are compared. Crisp and defuzzified results are also computed. Based on results some important suggestions are given for future course of action in maintenance planning.

1 Introduction

The important performance measure for repairable system are system reliability and availability. When system reliability is low, efforts are desired to improve it by reducing the failure rate or increasing the repair rate for each subsystem/component. To this effect the knowledge of system (or components) failure/repair

Komal (✉)
Department of Mathematics, University of Petroleum and Energy Studies (UPES), Dehradun, Uttarakhand 248007, India
e-mail: karyadma.iitr@gmail.com

S. P. Sharma
Department of Mathematics, Indian Institute of Technology Roorkee (IITR), Roorkee, Uttarakhand 247667, India
e-mail: sspprfma@iitr.ernet.in

V. Snášel et al. (eds.), *Soft Computing in Industrial Applications*,
Advances in Intelligent Systems and Computing 223, DOI: 10.1007/978-3-319-00930-8_10,

behavior is customary in order to plan and adapt suitable maintenance strategies. Reliability analysts analyze the system reliability with the help of various qualitative and quantitative techniques such as reliability block diagram (RBD), fault tree analysis (FTA), event tree analysis (ETA), markov models (MM), Petri-nets (PN), failure mode and effect analysis (FMEA), Baysian approach etc [1–5]. These techniques generally require knowledge of precise numerical probabilities and functional component dependencies, information which are sometimes relatively difficult to obtain in any large-scale system. In view of these problems, selection of the appropriate method depends upon the complexity of the system and measures used to analyse system reliability. After selecting appropriate method or technique, system behavior is analysed by using collected or available historical data. But, data either collected or historical are often inaccurate, imprecise, vague and collected under different operating and environmental conditions. The causes may be age, adverse operating conditions and the vagaries of manufacturing processes which affect each part/unit of the system differently, and thus the issue is subject to uncertainty [6]. Considering these facts, many researchers pay attention on these issues and analyzed various industrial systems behavior in terms of reliability/ availability using traditional markovian approach without quantifying uncertainties involved in the data [7]. Knezevic and Odoom [3] introduced the concept of FLT utilizing quantified data for analyzing the behavior of a general repairable system. In their approach, PN is used to model the system while fuzzy set theory is used to quantify the uncertain, vague and imprecise data. Fuzzy arithmetic is used for computation of different reliability indices of a general repairable system [8]. Rajeev [5] analysed reliability of different subsystems in a paper mill using FLT and FMEA. Komal et al. [1] established a new technique based on nonlinear programming approach and named it as GABLT for evaluating fuzzy reliability indices of repairable industrial systems. All the above discussed approaches are limited only for those systems whose components? functional dependencies are precisely known. So, these approaches cannot be applied to evaluate those systems reliability indices for which components functional dependencies are partially known. To overcome the problem, Sharma et al. [2] extended GABLT technique by coupling it with ANN and named it as NGABLT technique. The major benefit of using ANN is that it can be used effectively in the situations where output pattern is precisely known (supervised) and where output pattern is partially known (unsupervised). Present study is based on the systems whose functional dependencies are precisely known.

The objective of the chapter is to analyze the fuzzy reliability of washing system in a paper plant using available information and uncertain data through two soft-computing based hybridized techniques GABLT and NGABLT along with traditional FLT technique.

2 A Brief Overview of FLT, GABLT and NGABLT Techniques

The basic assumptions used in these techniques are given as:

- component failures and repair rates are statistically independent, constant, very small and obey exponential distribution function;
- the product of the failure rate and repair time is small (less than 0.1);
- after repairs, the repaired component is considered as good as new;
- system structure is precisely known.

2.1 FLT Technique

Fuzzy Lambda-Tau methodology is a traditional method for analyzing system fuzzy reliability [3]. The methodology is based on qualitative modeling using PN and quantitative modeling using Lambda-Tau method (Table 1) of solution with basic events (AND-gates and OR-gates) represented by triangular fuzzy numbers. This approach is limited as the number of components of the system increases or system structure becomes more complex, the computed reliability indices (Table 2) in the form of fuzzy membership function have wide spread due to various fuzzy arithmetic operations used in the computations [8]. It means these indices have high range of uncertainty and cannot give exact idea about the system reliability and consequently its performance. Thus this approach is not suitable for large and complex repairable industrial systems' reliability analysis when data is imprecise and represented by fuzzy numbers.

2.2 GABLT Technique

To analyze the complex industrial system reliability stochastically up to a desired degree of accuracy, GABLT technique has been used in this chapter [1]. GABLT utilizes ordinary arithmetic and mathematical programming approach instead of

Table 1 Basic expressions of lambda-tau methodology

Gate →	λ_{AND}	τ_{AND}	λ_{OR}	τ_{OR}
Expressions	$\prod_{j=1}^{n} \lambda_j \left[\sum_{i=1}^{n} \prod_{\substack{j=1 \\ i \neq j}}^{n} \tau_j \right]$	$\frac{\prod_{i=1}^{n} \tau_i}{\sum_{j=1}^{n} \left[\prod_{\substack{i=1 \\ i \neq j}}^{n} \tau_i \right]}$	$\sum_{i=1}^{n} \lambda_i$	$\frac{\sum_{i=1}^{n} \lambda_i \tau_i}{\sum_{i=1}^{n} \lambda_i}$

Table 2 Some reliability indices for repairable system with constant repair rate model

Reliability indices	Expressions
Mean time to failure	$MTTF_s = \frac{1}{\lambda_s}$
Mean time to repair	$MTTR_s = \frac{1}{\mu_s} = \tau_s$
Mean time between failures	$MTBF_s = MTTF_s + MTTR_s$
Expected number of failures	$ENOF_s(0,t) = \frac{\lambda_s \mu_s t}{\lambda_s + \mu_s} + \frac{\lambda_s^2}{(\lambda_s + \mu_s)^2}[1 - e^{-(\lambda_s + \mu_s)t}]$
Availability	$A_s(t) = \frac{\mu_s}{\lambda_s + \mu_s} + \frac{\lambda_s}{\lambda_s + \mu_s} e^{-(\lambda_s + \mu_s)t}$
Reliability	$R_s(t) = e^{-\lambda_s t}$

fuzzy arithmetic. In this technique expression of various reliability indices of the system in terms of system's components' failure rate and repair time are evaluated using Tables 1 and 2. Since system have complex structure, so evaluated reliability indices are nonlinear in nature. Also, system's components' failure and repair data i.e. input parameters are uncertain due to various practical reasons already discussed and hence represented by triangular fuzzy numbers [8, 9]. To finding system fuzzy reliability indices utilizing quantified failure and repair data in the form of fuzzy numbers, optimization problems (1) at each cut-level α is formulated and given as follows.

Minimize/Maximize:

$$\tilde{F}(\lambda_1, \lambda_2, , \lambda_n, \tau_1, \tau_2, , \tau_m) \text{ or } \tilde{F}(t/\lambda_1, \lambda_2, , \lambda_n, \tau_1, \tau_2, , \tau_m) \tag{1}$$

$$\begin{aligned} \textit{Subject to}: \; & \mu_{\lambda_i}(x) \geq \alpha, \\ & \mu_{\tau_j}(x) \geq \alpha, \\ & 0 \leq \alpha \leq 1, \\ & i = 1, 2, , n; \quad j = 1, 2, , m. \end{aligned}$$

where $\tilde{F}(\lambda_1, \lambda_2, , \lambda_n, \tau_1, \tau_2, , \tau_m)$ and $\tilde{F}(t/\lambda_1, \lambda_2, , \lambda_n, \tau_1, \tau_2, , \tau_m)$ are time independent and dependent fuzzy reliability indices respectively. The obtained minimum and maximum value of F are denoted by F_{min} and F_{max} respectively. The membership function values of $\tilde{F}$ at F_{min} and F_{max} are both α that is,

$$\mu_{\tilde{F}}(F_{min}) = \mu_{\tilde{F}}(F_{max}) = \alpha$$

The obtained optimization problem is non-linear in nature, needs some effective techniques and tools for its solution. Variety of methods and algorithms have been developed for solving nonlinear optimization problems and have been applied in various types of real life problems [10, 11]. Genetic algorithm (GA) is one of the most popular evolutionary algorithm. GA has been applied effectively to many different types of reliability optimization problems [1, 11]. GA is capable to solve nonlinear optimization problems without checking the convexity and differentiability of objective functions [11, 12]. Owing to these advantages, GA is used to solve the nonlinear optimization problem (1) for each cut-level α. After solving nonlinear optimization problems (1), we have fuzzy reliability indices with reduced range of uncertainty at each cut-level α.

2.3 NGABLT Technique

In NGABLT technique, first the failure rate (λ) and repair time (τ) of each component of the system is extracted from various sources like available historical records, logbooks, system analysts, etc. and integrated with expertise. Finally, extracted data are imprecise and vague in nature due to the previously stated reasons. To handle these uncertainties, fuzzy triangular numbers instead of crisp numbers are used to incorporate the uncertainties in the analysis [1–3, 5]. In the next step, reliability indices expressions are obtained with the help of fault tree model of the system and expressions given in Tables 1 and 2. For complex repairable industrial systems, these expressions are of non-linear in nature. So to simplify the calculation process, each reliability index expression is to be approximated by using ANN [13]. We are using a three layer ANN, different types of transfer functions attached with input,hidden and output layers will be used according to the nature of reliability indices.

For all the reliability indices, the transfer function used in the hidden layer is the tan-sigmoidal function. Since the reliability and availability values are in between 0 and 1, we choose the transfer function for the output layer to be the log-sigmoidal function. For other reliability indices (failure rate, repair time, MTBF and ENOF), the transfer function of the output layer is the linear function [4, 14]. Training algorithm is used to minimize the output error by adjusting network weights and biases. The standard backpropagation algorithm (gradient descent algorithm) is used. The input signals are sent forward and then the errors are propagated backward. The algorithm provides supervised learning with examples of input-output pairs to train the network and begins with random weights and biases, adjusted to minimize the errors. Once the training is complete, we have the approximated system reliability indices expressions in the form of matrix of weights attached with input, hidden and output layers of ANN [14].

After approximation of these reliability indices, fuzzified data ($\lambda^{'s}$ and $\tau^{'s}$) at cut level α are used as input variables in the approximated reliability indices for estimating these reliability indices at the same cut level α. The aim is that the computed reliability indices in the form of fuzzy membership functions should have optimized spread i.e. small range of prediction at each cut level α. In this search process, a nonlinear optimization problem is formulated. To find the lower and upper boundary values of any fuzzy reliability index at cut level α in the process of membership function construction, we need to solve the nonlinear optimization problems (1). Since the problem is nonlinear in nature and the function is approximated by ANN, so conventional methods are inadequate to solve these nonlinear optimization problems. To solve the nonlinear optimization problems (1), GA is used. After solving nonlinear optimization problem (1) for each cut-level α by using GA, we have fuzzy reliability indices.

In the present analysis binary coded GA is used. To solve the nonlinear optimization problems, system's components' failure rate ($\lambda^{'s}$) and repair time ($\tau^{'s}$) are encoded in strings of desired bit length l, that finally constitute a chromosome. The objective function for maximization problem (1) while the reciprocal of the objective function for minimization problem (1) are taken as the fitness function. Roulette wheel selection criterion is employed to choose better fitted chromosomes. One-point crossover and random point mutation are used in the present analysis. To stop the optimization process maximum number of generations and change in population fitness value are used. MATLAB 7.1 has been used for coding purpose.

3 Washing System Description

Here in a paper plat that comprises of subsystems namely chipping, feeding, pulping, washing, screening, bleaching, production of paper and collection, arranged in complex configuration is taken as a main system and the washing subsystem, an important functionary part of the paper plant, as a subject of discussion [5, 7]. The system consists of four main subsystems, defined as:

- **Filter (A)**. It consists of single unit which is used to drain black liquor from the cooked pulp.
- **Cleaners (B)**. In this subsystem three units of cleaners are arranged in parallel configuration. Each unit may be used to clean the pulp by centrifugal action. Failure of anyone will reduce the efficiency of the system as well as quality of paper.
- **Screeners (C)**. Herein two units of screeners are arranged in series. These are used to remove oversized, uncooked and odd shaped fibers from pulp through straining action. Failure of any one will cause the complete failure of the system.
- **Deckers (D)**. Two units of deckers are arranged in parallel configuration. The function f deckers is to reduce the blackness of pulp. Complete failure of decker occurs when both the components will fail.

The fault tree model of the washing system is shown in Fig. 1.

4 Result and Discussion

Under the information extraction phase, the data related to failure rates ($\lambda_i^{'s}$) and repair times ($\tau_i^{'s}$) of the main components of the washing system is collected from the present/historical records of the paper plant. The collected data is integrated with expertise of maintenance personnel and is given in the Table 3.

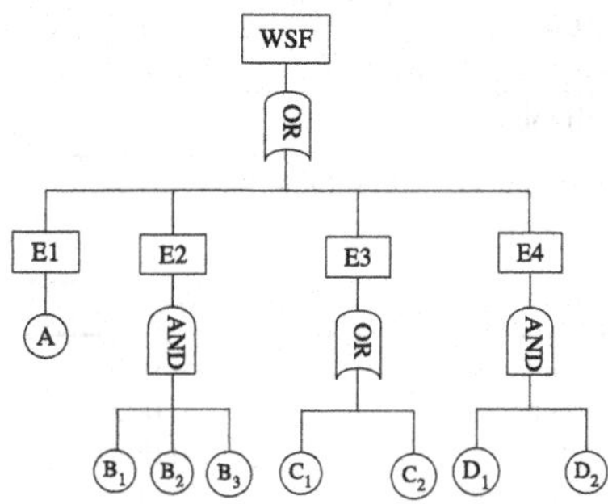

Fig. 1 Fault tree of washing system

Table 3 Failure rate and repair time data for washing system

Component	Failure rate(λ_i) (failures/h)	Repair time(τ_i) (h)
Filter ($i = 1$)	1×10^{-3}	3
Cleaners ($i = 2, 3, 4$)	3×10^{-3}	2
Screeners ($i = 5, 6$)	5×10^{-3}	3
Deckers ($i = 7, 8$)	5×10^{-3}	3

From the fault tree of the system,the minimal cut sets are $\{A\}$, $\{B_1, B_2, B_3\}$, $\{C_i\}_{i=1,2}$ and $\{D_1, D_2\}$, obtained by using matrix method [3]. Now, by making use of the Table1, the expressions for system failure rate (λ_s) and repair time (τ_s) take the following forms.

$$\lambda_s = \lambda_1 + \lambda_5 + \lambda_6 + \lambda_2\lambda_3\lambda_4(\tau_2\tau_3 + \tau_3\tau_4 + \tau_2\tau_4) + \lambda_7\lambda_8(\tau_7 + \tau_8)$$

$$\tau_s = \frac{(\lambda_1\tau_1 + \lambda_5\tau_5 + \lambda_6\tau_6 + \lambda_2\lambda_3\lambda_4\tau_2\tau_3\tau_4 + \lambda_7\lambda_8\tau_7\tau_8)}{\lambda_s}$$

Now using these expressions and Table 2, the system reliability indices expressions are obtained. After that, FLT, GABLT and NGABLT techniques have been applied to analyze the reliability of the system. The selected values of all the parameters for GA and ANN for applying GABLT and NGABLT techniques are given in Table 4. Following the basic steps of GABLT and NGABLT techniques, the fuzzy reliability indices for mission time $t = 10$(hrs) with left and right spreads are computed and depicted graphically in Fig. 2 for $\pm 15\,\%$ spreads along with FLT technique results. When NGABLT technique has been applied, the error curves obtained during approximation of all the reliability indices by ANN are plotted and shown in Fig. 3. The crisp and defuzzified values for all the three techniques with ± 15, ± 25 and $\pm 60\,\%$ spreads are computed and tabulated in Table 5. For defuzzification center of gravity method has been applied [15]. From Table 5, it is evident that defuzzified values change with change of spread. For example the failure rate of the system increases by 0.09, 0.18 and 6.36 % for FLT, GABLT and NGABLT respectively, when spread changes from ± 15 to $\pm 25\,\%$,

Table 4 GA and ANN parameters' values to apply GABLT and NGABLT techniques for washing system reliability analysis

Reliability indices	GABLT				NGABLT						
	Parameters for GA				Parameters for ANN			Parameters for GA			
	P_s	P_c	P_m	N_i	N_d	N_h	L_r	P_s	P_c	P_m	N_i
Failure rate	160	0.8	0.006	60	100	17	0.005	160	0.85	0.02	70
Repair time	160	0.8	0.005	80	110	25	0.008	160	0.85	0.02	80
MTBF	160	0.8	0.004	60	110	28	0.005	160	0.85	0.03	90
ENOF	160	0.8	0.006	60	120	28	0.002	160	0.85	0.02	90
Availability	160	0.8	0.005	70	100	27	0.007	160	0.85	0.02	80
Reliability	160	0.8	0.005	80	100	25	0.005	160	0.85	0.02	80

Notations:
P_s population size, P_c prob. of crossover, P_m prob. of mutation, N_i no. of iterations
N_d no. of training data, N_h no. of hidden layers, L_r learning rate

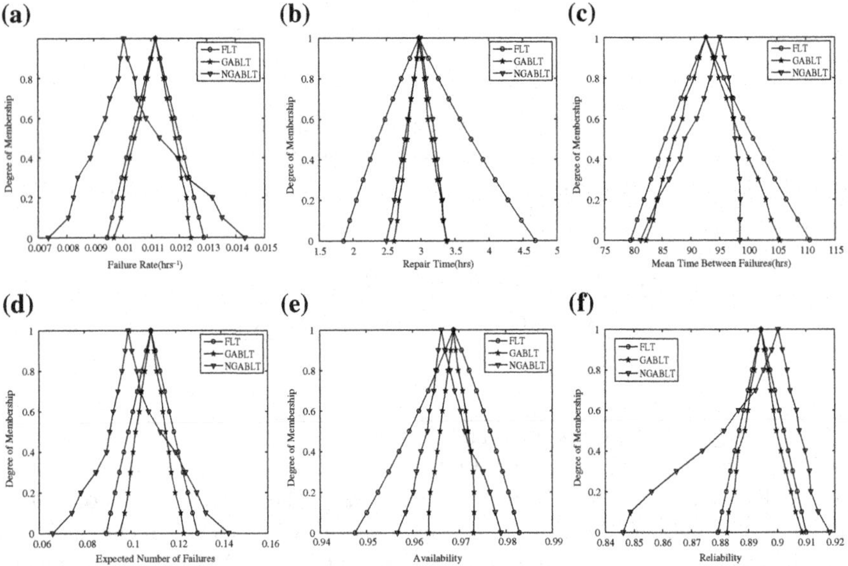

Fig. 2 Fuzzy reliability indices plots for washing system. **a** Failure rate, **b** repair time, **c** MTBF, **d** ENOF, **e** availability, **f** reliability

and it further increases by 0.54, 0.18 and 9.73 %, when spread changes from ±25 to ±60 %. From the results it is clear that the computed ranges of system reliability parameters considering GABLT results have compressed range of prediction in comparison of FLT technique results. On the other hand, computed ranges of system reliability parameters considering NGABLT results sometime have increased range of prediction (for failure rate, ENOF, reliability) and sometime

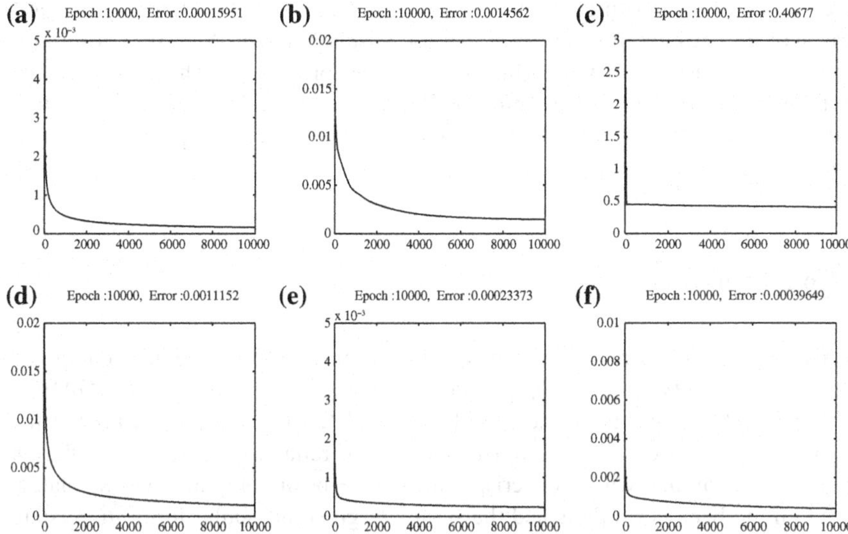

Fig. 3 Approximation error plots for washing system. **a** failure rate, **b** repair time, **c** MTBF, **d** ENOF, **e** availability, **f** reliability

Table 5 Crisp and defuzzified values of reliability indices for washing system

Reliability	Crisp	Defuzzified values at (spread)		
Indices		±15 %	±25 %	±60 %
Failure rate	0.01115	FLT: 0.01116	0.01117	0.01123
		GABLT: 0.01112	0.01114	0.01116
		NGABLT: 0.01053	0.01120	0.01229
Repair time	2.97975	3.12411	3.39742	6.48681
		2.98259	2.97592	2.95166
		2.96894	2.93378	2.80893
MTBF	92.6632	93.8539	96.0905	120.012
		93.3253	94.9451	112.433
		92.2845	92.5887	117.395
ENOF	0.10892	0.10918	0.10970	0.11761
		0.10917	0.10934	0.10956
		0.10342	0.09789	0.09553
Availability	0.96880	0.96701	0.96354	0.92599
		0.96867	0.96801	0.96757
		0.96769	0.96877	0.96901
Reliability	0.89448	0.89451	0.89454	0.89482
		0.89455	0.89472	0.89514
		0.88763	0.89035	0.89333

have reduced range of prediction (for repair time, MTBF, availability) in comparison of FLT results. It is observed that for washing system, NGABLT results do not perform consistently well while GABLT performs well. This is due to the errors involved in the approximation by ANN. Thus, it is inferred that if system analysts use GABLT results, then they may have less range of prediction which finally leads to more sound decisions.

5 Conclusion

In this chapter various reliability indices of a washing system have been computed in the form of fuzzy membership functions by using FLT, GABLT and NGABLT techniques. Depending upon the confidence level 'α', the analyst can predict the behavior of the system. The defuzzified values of reliability indices for different level of uncertainties with their crisp values have been computed and tabulated. Based on the behavioral plots and the summary given in tabular form, the system manager may analyze the critical behavior of the system and plan for suitable maintenance. It is also observed from the analysis that GABLT performs consistently well in comparison of NGABLT and FLT techniques. If system analysts use GABLT results then they may predict the system behavior with more confidence. However, NGABLT technique provides the flexibility of extension of its present form in future for the systems' whose components' functional dependencies are imprecisely known. These hybridized techniques can be applied to a wide range of industrial systems for helping the reliability engineers to gain valuable information and also to evaluate and implement various maintenance strategies.

References

1. Komal, Sharma, S.P., Kumar, D.: RAM analysis of repairable industrial systems utilizing uncertain data. Appl. Soft Comput. **10**(4), 1208–1221 (2010)
2. Sharma, S.P., Kumar, D., Komal, : Stochastic behavior analysis of the feeding system in a paper mill using NGABLT technique. Int. J. Qual. Reliab. Manage. **27**(8), 953–971 (2010)
3. Knezevic, J., Odoom, E.R.: Reliability modeling of repairable systems using Petri nets and Fuzzy Lambda-Tau Methodology. Reliab. Eng. Syst. Saf. **73**(1), 1–17 (2001)
4. Huang, H.Z., Zuo, M.J., Sun, Z.Q.: Bayesian reliability analysis for fuzzy lifetime data. Fuzzy Sets Syst. **157**(12), 1674–1686 (2006)
5. Sharma, R.K.: Analysis, design and optimization of QRM aspects in production systems. Indian Institute of Technology Roorkee, Roorkee, Uttrakhand, India (2006)
6. Rao, K.D., Kushwaha, H.S., Verma, A.K., Srividya, A.: Quantification of epistemic and aleatory uncertainties in level-1 probabilistic safety assessment studies. Reliab. Eng. Syst. Saf. **92**(7), 947–956 (2007)
7. Kumar, D.: Analysis and optimization of systems availability in sugar, paper and fertilizer Industries. University of Roorkee (Presently IIT Roorkee), Uttrakhand. India (1991)

8. Chen, S.M.: Fuzzy system reliability analysis using fuzzy number arithmetic operations. Fuzzy Sets Syst. **64**(1), 31–38 (1994)
9. Pedrycz, W.: Why triangular membership functions? Fuzzy Sets Syst. **64**(1), 21–30 (1994)
10. Tillman, F.A., Hwang, C.L., Kuo, W.: Optimization of Systems Reliability. Marcel Dekker, New York (1980)
11. Konak, A., Coit, D.W., Smith, A.: Multi-objective optimization using genetic algorithms: a tutorial. Reliab. Eng. Syst. Saf. **91**(9), 992–1007 (2006)
12. Goldberg, D.E.: Genetic Algorithm in Search, Optimization and Machine Learning. Addison-Wesley, Boston (1989)
13. Cybenko, G.: Approximation by superpositions of a sigmoidal function. Reliab. Eng. Syst. **2**(4), 303–314 (1989)
14. Kosko, B.: Neural Networks and Fuzzy System: A Dynamical Systems Approach to Machine Intelligence. Prentice-Hall, Englewood (1991)
15. Ross, T.J.: Fuzzy Logic with Engineering Applications, 2nd edn. Wiley, New York (2004)

Multi-objective Algorithms for the Single Machine Scheduling Problem with Sequence-dependent Family Setups

Marcelo Ferreira Rego, Marcone Jamilson Freitas Souza, Igor Machado Coelho and José Elias Claudio Arroyo

Abstract This work treats the single machine scheduling problem in which the setup time depends on the sequence and the job family. The objective is to minimize the makespan and the total weighted tardiness. In order to solve the problem two multi-objective algorithms are analyzed: one based on Multi-objective Variable Neighborhood Search (`MOVNS`) and another on Pareto Iterated Local Search (`PILS`). Two literature algorithms based on `MOVNS` are adapted to solve the problem, resulting in the `MOVNS_Ottoni` and `MOVNS_Arroyo` variants. Also, a new perturbation procedure for the `PILS` is proposed, yielding the `PILS1` variant. Computational experiments done over randomly generated instances show that `PILS1` is statistically better than all other algorithms in relation to the cardinality, average distance, maximum distance, difference of hypervolume and epsilon metrics.

M. F. Rego (✉) · M. J. F. Souza
Departament of Computing, Federal University of Ouro Preto, Ouro Preto, MG 35400-000, Brazil
e-mail: marcelofr@gmail.com

M. J. F. Souza
e-mail: marcone@iceb.ufop.br

I. M. Coelho
Institute of Computing, Fluminense Federal University, Niterói, RJ 24210-240, Brazil
e-mail: imcoelho@ic.uff.br

J. E. C. Arroyo
Departament of Computing, Federal University of Viçosa, Viçosa, MG 36571-000, Brazil
e-mail: jarroyo@dpi.ufv.br

V. Snášel et al. (eds.), *Soft Computing in Industrial Applications*, Advances in Intelligent Systems and Computing 223, DOI: 10.1007/978-3-319-00930-8_11,

1 Introduction

Scheduling problems have been extensively studied in the literature. This fact is due to at least two aspects. The first one is the practical interest, since there are various applications on this class of problems in industry field, as for example, textile [11], electronics [15] and iron [21]. The other aspect that interests the study of this kind of problem is the theoretical interest, once most of the scheduling problems belong to the class of $\mathcal{NP}$-hard problems [3].

Although the problem of scheduling jobs involves various objectives, in most of the researches in this field only one objective is considered. When more than one is considered, usually it is defined only one objective represented by the linear combination of involved objectives, thus, the problem is treated normally in a single-objective approach.

This work discusses the scheduling problem in single machines, in which the setup time of the machine depends on the scheduling and family of the jobs. The grouping of the jobs in the family occurs, for example, in the iron field. In [4], the author shows a process of manufacturing iron products (corner, rebar, bar, etc.) in the lamination sector, in which jobs are grouped in families in accordance with the similarity of the products. In this case, the products from same family are those which differ between themselves by the thickness. On those circumstances the setup time is so short and unimportant when compared to the processing time of jobs that is usual to consider it equivalent to zero. The advantage of making this grouping, thus, is that the jobs which belong to the same family, when processed sequentially, do not need setup time.

The problem in hand takes two objectives into account: makespan and total weighted tardiness minimization. It means that instead of looking for a solution which satisfies one or other objective separately, the main goal is to obtain a set of non-dominated solutions, this way each solution that belongs to this set is not worse than any other, considering both objectives simultaneously.

Noticing the computational complexity of the scheduling problems, the most used methods to solve them are metaheuristics. Reviews in literature show that methods inspired by the process of natural evolution, such as Non-dominated Sorting Genetic Algorithm II— `NSGA-II` [7] and Strength Pareto Approach— `SPEA2` [25] are among the most used when multi-objective optimization is concerned. On the other hand, recently it were discovered reports of successful applications of multi-objective methods based on local search, such as Multi-objective Variable Neighborhood Search—`MOVNS` [9] and Pareto Iterated Local Search—`PILS` [10], as shown in the works of [1, 18] and [20]. In this last, for example, it is observed a superiority of `PILS` over `SPEA2`.

Due to the good performance of the algorithms in similar problems, in this work the multi-objective algorithms `MOVNS` and `PILS` are tested to solve the scheduling problem at hand. The `MOVNS` algorithms of [1] and [20] were adapted to solve the problem, giving birth to the `MOVNS_Ottoni` and `MOVNS_Arroyo` variants, respectively. Furthermore, a new perturbation procedure is proposed for `PILS`, giving birth to the `PILS1` variant. Computational experiments showed that the last algorithm outperforms the other tested algorithms.

The rest of this work is organized as follows. In Sect. 2 the problem characteristics are described. Section 3 presents the proposed multi-objective algorithms, and in Sect. 4 the used test instances are described as well as the metrics used to assess and compare the developed algorithms. Also in Sect. 4, the results of the accomplished experiments are presented and analyzed. Section 5 concludes the work.

2 Problem Description

The problem in focus can be defined as follows: there is a set $J = \{1, 2, 3, \ldots, n\}$ with n jobs that have to be scheduled in a single machine at the starting point zero. Each job $j \in J$ has a processing time p_j, a due date d_j and a weight for tardiness β_j. The jobs are grouped in families f according to their characteristics and each family i has n_i jobs. A setup time s_{ik} is required between the execution of two consecutive jobs of different families i and k and, if they are from the same family, no setup time is necessary. Given a sequence π, for each job j a tardiness T_j is associated. Since C_j is the completion time of the job j, its tardiness is calculated by Eq. (1):

$$T_j = \max\ \{C_j - d_j, 0\} \tag{1}$$

The objectives of the problem in focus are to minimize the makespan $f_1(\pi)$ and the total weighted tardiness $f_2(\pi)$, simultaneously. The values of $f_1(\pi)$ and $f_2(\pi)$ are calculated by Eqs. (2) and (3):

$$f_1(\pi) = \max_{1 \leqslant j \leqslant n} \{C_j\} \tag{2}$$

$$f_2(\pi) = \sum_{1 \leqslant j \leqslant n} \beta_j T_j \tag{3}$$

3 Methodology

A solution is represented by a sequence $\pi = \{\pi_1, \pi_2, \ldots, \pi_k, \ldots, \pi_n\}$ in which π_k indicates the kth job to be done.

The proposed algorithms begin with a set of non-dominated solutions generated by four different heuristics based on the following dispatching rules [22]: Earliest Due Date (EDD) rule, generating a scheduling of jobs in a non-decreasing order of their due dates; Shortest Processing Time (SPT) rule, generating a scheduling of jobs in a non-decreasing order of their processing time; Longest Processing Time (LPT) rule, generating a scheduling of jobs in a non-increasing order of their processing time; Minimum Slack Time (MST) rule, generating a scheduling of jobs in a non-decreasing order of the difference between the due date and processing time.

In order to explore the solution space of the problem, insertion and exchange movements are applied changing the scheduling of the jobs, as follows.

Given a sequence $\pi = \{\pi_1, \pi_2, \ldots, \pi_n\}$, the insertion move of a job π_x consists in moving this job to a position y ($y \neq x$ e $y \neq x - 1$). The group of insertion movements in a sequence π defines the neighborhood $N^I(\pi)$ which is composed by $(n - 1)^2$ solutions.

Given a sequence $\pi = \{\pi_1, \pi_2, , \pi_n\}$, the exchanging move between two jobs π_x and π_y consists in moving the job π_x to the position y and the job π_y to the position x. The group of exchanging movements in a sequence π defines the neighborhood $N^T(\pi)$, formed by $\frac{n \times (n-1)}{2}$ solutions.

3.1 Algorithms Based on MOVNS

The Multi-objective Variable Neighborhood Search (MOVNS) is an optimization multi-objective algorithm proposed in [9] and the metaheuristic Variable Neighborhood Search—VNS [19].

Variants of MOVNS Algorithm In literature there are two variants of the MOVNS algorithm. The first one, named as MOVNS_Ottoni, was proposed by [20] and consists in adding an intensification procedure to the original MOVNS. The second variant of MOVNS, named MOVNS_Arroyo, was proposed by [1] and consists in adding another different intensification.

3.2 Algorithms Based on PILS

Pareto Iterated Local Search—PILSis an optimization multi-objective algorithm proposed by [10], with a structure based on the meta-heuristic Iterated Local Search—ILS [17]. PILS basic pseudo-code is presented in Algorithm 1.

Algorithm 1: PILS

```
   Input : Non-dominated set ND, k neighborhoods, stopping criterion
1  ND ← InitialSolution();
2  Select a solution s ∈ ND ;
3  while stopping criterion is not satisfied do
4      i ← 1 ;
5      while i < k ∧ stopping criterion is not satisfied do
6          foreach s′ ∈ N^i(s) do
7              ND ← non-dominated set of ND ∪ {s′} ;
8          end
9          if ∃s′ ∈ N^i(s) | s′ dominates s then
10             s ← s′ ;
11             Reorder the neighborhoods N^1, ..., N^k ;
12             i ← 1 ;
13         end
14         else
15             i++;
16         end
17     end
18     Mark s as visited ;
19     if ∃s′ ∈ ND | s′not yet been visited then
20         s ← s′;
21     end
22     else
23         Select a solution s′ ∈ ND ;
24         s″ ← Perturbation(s′) ;
25         s ← s″ ;
26     end
27 end
28 return ND ;
```

In Algorithm 1, initially it is obtained a set of non-dominated solutions (*ND*) (line 1), using four different heuristics described previously. After this, one of the solutions of the set *ND* is selected randomly (line 2). In each iteration of the external loop (lines 3–27) all neighbors of the current solution are explored (lines 5–17). If a neighbor solution dominates the current solution, then this neighbor solution becomes the new current solution, the neighborhoods are randomly reordered and the procedure returns to the first neighborhood of the new generated order. This procedure is repeated while there are non-visited solutions in set *ND*. After all solutions of set *ND* are visited—when the algorithm is in an local optimum concerning the explored neighborhood—is a solution is randomly selected from set *ND* (line 23) and a perturbation is applied (line 24), as described as follows. After this, all neighborhood of the current solution is explored (lines 5–17). In the case that all neighbors of the solution generated through the perturbation are dominated by any solution of set *ND*, then the perturbation procedure is repeated. The most external loop (lines 3–27) is repeated while the stopping criterion is not met.

A solution is perturbed in order to explore other local optima. The original perturbation strategy of PILS, from [10], works in the following way: initially a solution π from the set *ND* is randomly selected. Then, a position $j \leq n - 4$ and its four consecutively jobs of π are randomly chosen, i.e., positions $j, j+1, j+2$ and $j+3$. A perturbed solution s'' is then generated by applying an exchanging move on the jobs in the positions j and $j+3$, as well as the jobs in the positions $j+1$ and $j+2$. This way, the jobs before j and the jobs ahead of $j+3$ are kept in their respective positions after the perturbation application.

In this work, the perturbation procedure of a solution (line 24 from Algorithm 1) has been modified when concerning the proposal of [10]. The perturbation is applied in levels, varying from 1 to $(n/2 - 1)$. In each level p, $p+1$ modifications are made on the solution. This way, on the lowest pertubation level two exchanges are applied while on the highest level $n/2$ exchanges are made.

The level p of perturbation increases as the perturbation is not able to generate a non-dominated solution related to the set *ND*. The increasing is made by adding a unit of value to the current level of perturbation. When a non-dominated solution related to the set *ND* is found, the perturbation level returns to its lowest value, 1 in this case. If the perturbation level reaches its maximum value—$(n/2 - 1)$—and it is still not possible to generate a non-dominated solution in relation to the set *ND*, the perturbation level returns to its lowest value. The proposed procedure works as follows. A solution π from the set *ND* is randomly selected. Then it is chosen, also randomly, a subset of consecutive jobs of π on the positions $j, j+1, \ldots,$ $j+2p+1$. Then, exchanges are applied between the pairs of jobs $(j, j+2p+1), (j+1, j+2p), \ldots, (j+p, j+p+1)$. This way, the procedure makes $p+1$ exchanging moves on each call from the perturbation procedure. The PILS algorithm modified as such was named PILS1.

4 Computational Experiments

All algorithms were coded on C++ language and the tests were done in a Intel® Core™ 2 Quad 2.4 GHz with 6GB RAM.

The stopping criterion of each algorithm is a maximum CPU time proportional to the size of the instance. This criterion is common in literature and it has been established as $1000 \times n$ ms, in which n is the number of jobs of the instance. For each instance 30 tests were performed, each one with a different random seed.

4.1 Instances

To assess the algorithms, instances were generated in a random way and with uniform distribution. As in [14], the number of jobs is a integer number

$n \in \{60, 80, 100\}$, the number of families $f \in \{2, 3, 4, 5\}$ and the processing time is an integer number in the interval $[1, 99]$. The due date of the jobs was generated as in [2], being defined in the interval $(0, h \sum p_j)$, with $h \in \{0.5; 1.5; 2.5; 3.5\}$. Finally, the setup time between jobs families are integer numbers whose values belong to three classes of intervals: class S $[10, 20]$; class M $[51, 100]$ and class L $[101, 200]$.

The formation of such setup time intervals is a suggestion proposed in [13]. The class *S* setup time is relatively smaller than the average processing time. The class *M* setup time is close to the average processing time while the class *L* setup time is relatively bigger than the average processing time.

By combining the parameters of the number of jobs, number of families, number of intervals to the due dates and the number of classes of intervals to the setup time, an amount of 144 different instances were generated.Note that, for each *n*, 48 instances were generated.

4.2 Performance Assessment Metrics

The comparison of two sets of non-dominated points, *A* and *B*, obtained respectively through two optimization multi-objective algorithms is not a trivial task. Many performance assessment metrics have been proposed on literature [6, 8, 12, 23]. However, these metrics must be chosen in a proper way to make a fair comparison of algorithms.

In this work, five performance assessment metrics are used and they are called: cardinality [12], average distance [5], maximum distance [16], hypervolume [24] and epsilon [8]. In [8] it is shown that the hypervolume and epsilon metrics provide trustworthy measures, especially when two algorithms have similar performances.

The quality of a set of non-dominated points obtained by an algorithm, in a given instance, is assessed in relation to the set composed by all non-dominated points found during all experiments. This is called the set of reference points *R*.

4.3 Results

The results presented on the following tables were obtained by the developed algorithms with different evaluation metrics. In the first column of these tables is indicated the group *n* of 48 instances (with number of jobs *n*). In the other columns the results of each algorithm are presented, with 30 algorithm executions. For each metric the average and best results are presented so as the average of the results in all instances.

Table 1 Cardinality metric results

n	Algorithm									
	MOVNS		MOVNS_Ottoni		MOVNS_Arroyo		PILS		PILS1	
	Avg.	Best	Avg.	Best	Avg.	Best	Avg.	Best	Avg.	Best
60	2.18	12.89	2.79	15.40	3.21	17.24	2.78	10.08	32.82	61.07
80	0.25	3.41	0.56	8.67	0.54	9.86	0.36	4.61	11.64	50.10
100	0.07	2.13	0.15	3.97	0.37	8.36	0.15	2.55	4.40	53.87
Average	0.83	6.15	1.17	9.35	1.37	11.82	1.10	5.74	16.29	55.02

Table 2 Average distance metric

n	Algorithm									
	MOVNS		MOVNS_Ottoni		MOVNS_Arroyo		PILS		PILS1	
	Avg.	Best	Avg.	Best	Avg.	Best	Avg.	Best	Avg.	Best
60	4.18	2.44	3.90	2.22	3.78	2.18	4.72	3.27	1.20	0.25
80	5.55	2.62	4.92	2.32	4.83	2.27	6.22	3.82	1.74	0.35
100	7.12	3.34	6.08	2.68	5.92	2.42	7.19	4.14	2.85	0.39
Average	5.62	2.80	4.97	2.41	4.85	2.29	6.04	3.74	1.93	0.33

Table 3 Maximum distance metric

n	Algorithm									
	MOVNS		MOVNS_Ottoni		MOVNS_Arroyo		PILS		PILS1	
	Avg.	Best	Avg.	Best	Avg.	Best	Avg.	Best	Avg.	Best
60	10.22	7.57	10.05	7.73	9.86	7.53	11.07	9.38	5.43	1.03
80	9.16	6.20	8.68	6.01	8.58	5.80	10.00	7.80	5.23	1.05
100	8.71	4.77	7.85	4.03	7.67	3.87	9.05	5.89	4.68	1.04
Average	9.36	6.18	8.86	5.92	8.70	5.73	10.04	7.69	5.11	1.04

Table 1 presents the average and best results obtained in 30 runs of the algorithms considering the cardinality metric. As can be seen from Table 1, PILS1 algorithm is able to generate a superior number of non-dominated solutions compared to the other algorithms. Besides this, the number of reference solutions generated by the PILS1 algorithm is, in average, at least seven times higher than the one generated by any other algorithm.

Table 2 presents the average and best results in 30 runs of the algorithms considering the average distance metric.

From Table 2 we can conclude that the set of non-dominated solutions produced by the algorithm PILS1 is closer to the reference set than the other algorithms.

Table 3 presents the results obtained by the implemented algorithms considering the maximum distance metric.

Table 4 Hypervolume difference metric

n	Algorithm									
	MOVNS		MOVNS_Ottoni		MOVNS_Arroyo		PILS		PILS1	
	Avg.	Best	Avg.	Best	Avg.	Best	Avg.	Best	Avg.	Best
60	1060.92	556.08	996.74	498.32	969.70	482.68	1205.04	806.87	267.09	14.74
80	1375.61	660.69	1239.66	568.42	1212.16	534.70	1548.32	968.64	433.27	33.90
100	1618.76	767.46	1406.73	601.10	1367.94	527.21	1677.25	1015.85	705.22	45.01
Average	1351.76	661.41	1214.37	555.95	1183.26	514.86	1476.87	930.45	468.53	31.22

Table 5 Epsilon metric

n	Algorithm									
	MOVNS		MOVNS_Ottoni		MOVNS_Arroyo		PILS		PILS1	
	Avg.	Best	Avg.	Best	Avg.	Best	Avg.	Best	Avg.	Best
60	1.40	1.21	1.37	1.19	1.36	1.18	1.45	1.30	1.11	1.03
80	1.20	1.10	1.18	1.08	1.18	1.08	1.21	1.14	1.07	1.02
100	1.16	1.07	1.13	1.06	1.13	1.05	1.14	1.09	1.06	1.01
Average	1.25	1.13	1.23	1.11	1.22	1.11	1.27	1.18	1.08	1.02

As noticed in Table 3, the set of non-dominated solutions produced by the algorithm PILS1 is in a shorter distance from the reference set.

Table 4 presents the average and best results for the algorithms considering the hypervolume difference metric.

On Table 4 it is verified that the formed area between the points from the PILS1 algorithm solution set and the points from the non-dominated set are the smallest ones in comparison to the other algorithms. It means that the PILS1 algorithm produces a better cover of the reference set *R*.

Table 5 presents the average and best results obtained by the algorithms related to the epsilon metric.

From Table 5 it is noticed that the algorithm PILS1 is the one which produces the smallest values to the epsilon metric, indicating that the non-dominated solutions generated by this algorithm are closer to the reference set *R*.

We also apply the non-parametrical Kruskal–Wallis test in order to verify the statistical superiority of the PILS1 algorithm. According to this test, there is statistical difference between the pairs of algorithms: MOVNS × PILS1, MOVNS_Ottoni × PILS1, MOVNS_Arroyo × PILS1 and PILS × PILS1.

5 Conclusions

This work dealt with the scheduling problem in single machine where the setup time of the jobs depends on the sequence and on the family, and there are two optimization criteria to be satisfied: makespan and total weight tardiness minimization.

To solve this problem, five multi-objective algorithms based on Pareto Iterated Local Search (PILS) and Multi-objective Variable Neighborhood Search (MOVNS) were implemented. Of these algorithms, three are based on the MOVNS; one of them is the original algorithm and the other two variants of this one found in the literature. And on the other two remaining algorithms, one is the original PILS and the second is a variant proposed in this work, named PILS1. This variant consists in changing the perturbation strategy of PILS.

The algorithms were compared considering Cardinality, Average Distance, Maximum Distance, Hypervolume Difference and Epsilon metrics. The computational results performed in generated instances for the problem were validated by statistical analysis, thus showing that the PILS1 variant is superior to every other algorithm considering the assessed metrics. This way, it is clear the contribution of the perturbation procedure proposed in this work.

Acknowledgments The authors would like to thank CNPq and FAPEMIG for the financial support on the development of this work.

References

1. Arroyo, J.E.C., Ottoni, R.S., Oliveira, A.P.: Multi-objective variable neighborhood search algorithms for a single machine scheduling problem with distinct due windows. Electron. Notes Theor. Comput. Sci. **281**, 5–19 (2011)
2. Baker, K.R., Magazine, M.J.: Minimizing maximum lateness with job families. Eur. J. Oper. Res. **127**(1), 126–139 (2000)
3. Brucker, P.: Scheduling Algorithms. Springer, Berlin (2007)
4. Bustamante, L.M.: Minimização do custo de antecipação e atraso para o problema de sequenciamento de uma máquina com tempo de preparação dependente da sequência: aplicação em uma usina siderúrgica. Dissertação de mestrado, Programa de Pós-Graduação em Engenharia de Produção, Universidade Federal de Minas Gerais, Belo Horizonte (2007)
5. Czyżak, P., Jaszkiewicz, A.: Pareto simulated annealing—a metaheuristic technique for multiple-objective combinatorial optimization. J. Multi-Criteria Decis. Anal. **7**(1), 34–47 (1998)
6. Deb, K., Jain, S.: Running performance metrics for evolutionary multi-objective optimization. Technical report (2002). doi:10.1.1.9.159
7. Deb, K., Pratap, A., Agarwal, S., Meyarivan, T.: A fast and elitist multiobjective genetic algorithm: Nsga-ii. IEEE Trans. Evol. Comput. **6**(2), 182–197 (2002)
8. Fonseca, C.M., Knowles, J.D., Thiele, L., Zitzler, E.: A tutorial on the performance assessment of stochastic multiobjective optimizers. In: 3rd International Conference on Evolutionary Multi-Criterion Optimization (EMO), vol. 216 (2005)
9. Geiger, M.J.: Randomised variable neighbourhood search for multi objective optimisation. In: 4th EU/ME: Design and Evaluation of Advanced Hybrid Meta-Heuristics, pp. 34–42 (2004)
10. Geiger, M.J.: Improvements for multi-objective flow shop scheduling by pareto iterated local search. In: 8th Metaheuristics International Conference (MIC), pp. 195.1–195.10 (2009)
11. Gendreau, M., Laporte, G., Guimaraes, E.M.: A divide and merge heuristic for the multiprocessor scheduling problem with sequence dependent setup times. Eur. J. Oper. Res. **133**(1), 183–189 (2001)

12. Hansen, M.P., Jaszkiewicz, A.: Evaluating the quality of approximations to the non-dominated set. IMM, Department of Mathematical Modelling, Technical Universityof Denmark (1998)
13. Hariri, A.M.A., Potts, C.N.: Single machine scheduling with batch set-up times to minimize maximum lateness. Ann. Oper. Res. **70**, 75–92 (1997)
14. Jin, F., Gupta, J.N.D., Song, S., Wu, C.: Single machine scheduling with sequence-dependent family setups to minimize maximum lateness. J. Oper. Res. Soc. **61**(7), 1181–1189 (2010)
15. Kim, D.W., Kim, K.H., Jang, W., Chen, F.F.: Unrelated parallel machine scheduling with setup times using simulated annealing. Robot Comput. Integr. Manuf. **18**(3–4), 223–231 (2002)
16. Knowles, J., Corne, D.: On metrics for comparing nondominated sets. In: Proceedings of the Congress on Evolutionary Computation (CEC), vol. 1, pp. 711–716. IEEE (2002)
17. Lourenço, H.R., Martin, O., Stützle, T.: Iterated local search. Handbook of Metaheuristics, pp. 320–353, Springer, New York (2003)
18. Minella, G., Ruiz, R., Ciavotta, M.: A review and evaluation of multiobjective algorithms for the flowshop scheduling problem. INFORMS J. Comput. **20**(3), 451 (2010)
19. Mladenovic, N., Hansen, P.: Variable neighborhood search. Comput. Oper. Res. **24**(11), 1097–1100 (1997)
20. Ottoni, R.S., Arroyo, J.E.C., Santos, A.G.: Algoritmo vns multi-objetivo para um problema de programação de tarefas em uma máquina com janelas de entrega. In: Simpósio Brasileiro de Pesquisa Operacional (SBPO), pp. 1801–1812 (2011)
21. Tang, L., Wang, X.: Simultaneously scheduling multiple turns for steel color-coating production. Eur. J. Oper. Res. **198**(3), 715–725 (2009)
22. Valente, J.: An analysis of the importance of appropriate tie breaking rules in dispatch heuristics. Pesquisa Operacional **26**(1), 169–180 (2006)
23. Zitzler, E., Deb, K., Thiele, L.: Comparison of multiobjective evolutionary algorithms: empirical results. Evol. Comput. **8**(2), 173–195 (2000)
24. Zitzler, E., Thiele, L.: Multiobjective optimization using evolutionary algorithms—a comparative case study. In: Parallel Problem Solving from Nature-PPSN V, pp. 292–301. Springer, Berlin (1998)
25. Zitzler, E., Thiele, L.: Multiobjective evolutionary algorithms: a comparative case study and the strength pareto approach. IEEE Trans. Evol. Comput. **3**(4), 257–271 (1999)

Multi-Sensor Soft-Computing System for Driver Drowsiness Detection

Li Li, Klaudius Werber, Carlos F. Calvillo, Khac Dong Dinh, Ander Guarde and Andreas König

Abstract Advanced sensing systems, sophisticated algorithms and increasing computational resources continuously enhance active safety technology for vehicles. Driver status monitoring belongs to the key components of advanced driver assistance system which is capable of improving car and road safety without compromising driving experience. This paper presents a novel approach to driver status monitoring aimed at drowsiness detection based on depth camera, pulse rate sensor and steering angle sensor. Due to NIR active illumination depth camera can provide reliable head movement information in 3D alongside eye gaze estimation and blink detection in a non-intrusive manner. Multi-sensor data fusion on feature level and multilayer neural network facilitate the classification of driver drowsiness level based on which a warning can be issued to prevent traffic accidents. The presented approach is implemented on an integrated soft-computing system for

L. Li (✉) · K. Werber · C. F. Calvillo · K. D. Dinh · A. König
TU Kaiserslautern, Department of Electrical and Computer Engineering,
Institute of Integrated Sensor Systems, Erwin-Schrödinger-Str. Gebäude 12,
67663 Kaiserslautern, Germany
e-mail: lili@eit.uni-kl.de

K. Werber
e-mail: werber@rhrk.uni-kl.de

C. F. Calvillo
e-mail: calvillo@rhrk.uni-kl.de

K. D. Dinh
e-mail: khac@rhrk.uni-kl.de

A. König
e-mail: koenig@rhrk.uni-kl.de

A. Guarde
Faculty of Engineering of Bilbao, University of the Basque Country, Bilbao, Spain
e-mail: aguarde002@ikasle.ehu.es

V. Snášel et al. (eds.), *Soft Computing in Industrial Applications*,
Advances in Intelligent Systems and Computing 223, DOI: 10.1007/978-3-319-00930-8_12,

driving simulation (DeCaDrive) with multi-sensing interfaces. The classification accuracy of 98.9 % for up to three drowsiness levels has been achieved based on data sets of five test subjects with 588-min driving sequence.

1 Introduction

Drowsy driving is a serious problem that can affect anyone on the road. It was a major factor in 20 % of all accidents in the United States in 2006 according to the report of National Highway Traffic Safety Administration (NHTSA). A study by the Federal Highway Research Institute (BASt) in Germany showed that drowsy driving was the second most frequent cause of serious truck accidents on German highways. Due to severe damage caused by drowsy truck or bus drivers it is urgent to extend active safety to cope with driver drowsiness in commercial vehicles.

Driver Assistance Systems, in general, have been on the agenda of automotive and related industry for nearly two decades now. For instance, in the Electronic Eye [1] research program of the German Federal Ministry of Education and Research (BMBF), the topic had been pursued in the mid nineties, focusing both on CMOS sensing with high-dynamic range and high-speed as well as dedicated massively parallel digital computation platforms for applications, such as Sleep-Eye-Detectors or Overtake-Monitors (OTM) [2]. Advanced Driver Assistance System (ADAS) with the feature of driver drowsiness detection has been introduced by major automakers nowadays. Systems such as Driver Alert (Ford [3]), Attention Assist (Daimler [4]), Fatigue Detection System (Volkswagen [5]), Driver Monitoring System (Toyota [6]) and Driver Alert Control (Volvo [7]) are able to monitor driving behavior and issue alarm in a visual or audible manner if necessary. The details of such ADAS systems are summarized in Table 1. Depth information for automotive and robotic tasks was mostly obtained by stereo

Table 1 Summary of ADAS system with drowsiness detection

Vendor	System name	Integrated sensor	Technology and algorithm
Ford	Driver alert	Front and side mounted cameras	Camera based lane detection and tracking
Daimler	Attention assist	Steering sensor	High resolution steering sensor based driving behavior monitoring
Volkswagen	Fatigue detection system	Video camera	Driver monitoring based on head movement and facial features
Toyota	Driver monitoring system	CCD camera with infrared LED	Driver monitoring based on ocular measures
Volvo	Driver alert control and lane departure warning	CMOS Cameras	Car movement monitoring and lane tracking

camera setups. The advent of CMOS depth sensors, based on time-of-flight-principles, opened new possibilities and application fields from automotive, robotics, to HMI tasks. The Institute of Integrated Sensor System's research bases on such activities back to the end nineties, related to CMOS sensor system design and intelligent system design for, e.g., OTM or eye-tracking for 3D display applications [8]. Based on 3D depth sensors and sensor fusion approaches a continuation of the research, here a multi-sensor soft-computing system for driver drowsiness detection, is carried out.

After introduction a multi-sensor driving simulator—DeCaDrive is introduced which is a prototype for the presented multi-sensor soft-computing system aimed at driver drowsiness detection. The hardware setup of the presented system is addressed in Sect. 2. Software components such as multi-sensor feature computation and data fusion as well as neural network based pattern classification are discussed in Sect. 3. The system is validated and evaluated by presenting the experimental results in Sect. 4. Finally, with future perspectives the current work is concluded in Sect. 5.

2 Driving Scene Modeling and Hardware Setup

In order to investigate human driving behavior and to monitor vital signs of driver simultaneously a prototype of driving simulator was built. Initially the system was based on a single depth camera (**De**pth **Ca**mera based **Drive**) and afterwards has evolved to a multi-sensor soft-computing system incorporating PC based driving simulation and diversified sensing interfaces including depth camera, pulse rate sensor, blood oxygen saturation meter, steering angle sensor, tactile sensor and pressure sensor. In the presented system concept driving simulation, sensing and soft-computing are the key components.

Based on PC software different driving scenes for highway, city streets, country roads, etc. can be simulated. The current sensing subsystem consists of steering angle sensor, pulse oximeter so as to monitor steering behavior and pulse rate of driver respectively. In addition, a depth camera with active illumination, here Kinect sensor, is integrated in order to reliably provide visual cues of driver including eye gaze estimation and blink detection. The multi-sensing interfaces enable A/D conversion, sensor data streaming, time-based synchronization for multiple sensors and can be adapted to different simulation scenarios such as for passenger cars, buses or trucks. DeCaDrive hardware setup and the prototype system are illustrated in Fig. 1.

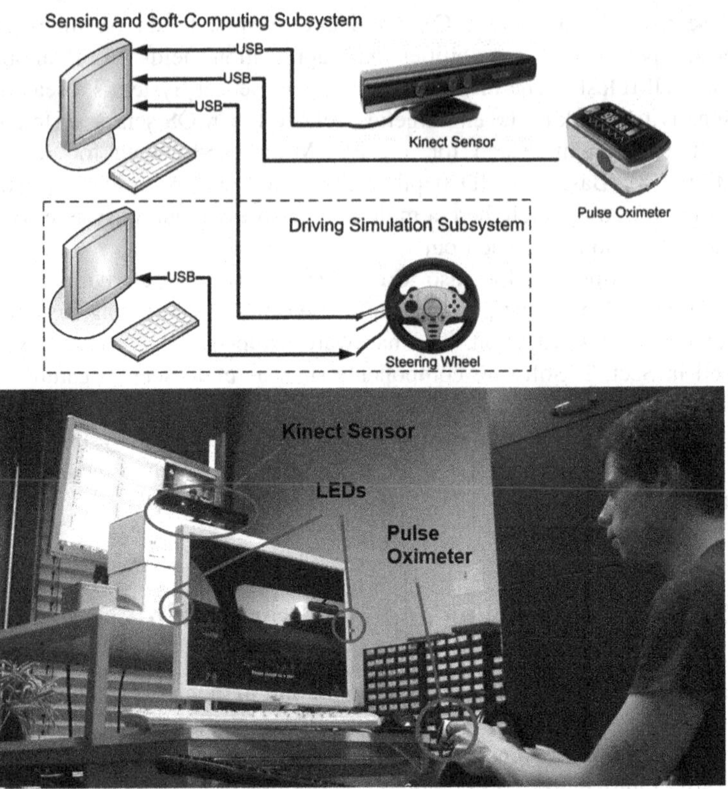

Fig. 1 DeCaDrive: a multi-sensor driving simulator, top to bottom, the details of hardware system setup, a test subject interacting with the prototype of driving simulator [9]

3 Software Components and Algorithms

3.1 Depth Camera Based Driver Status Monitoring

Depth camera can extend driver status monitoring in the 3rd dimension. By using state-of-the-art depth cameras vision systems are able to perceive distance and build up 3D profile of objects without compromising field-of-view, robustness to lighting conditions and computation performance, which are intrinsically limited in stereo vision systems [10]. As a low-cost commodity depth camera Kinect sensor is incorporated in the presented system due to its satisfactory depth sensing resolution, ease of use and the variety of software resource.

Depth and color image sequences are the raw data generated by Kinect sensor which are subsequently used in the following procedures: (1) head localization and face tracking; (2) head pose and face features computation; (3) eye tracking; (4) gaze estimation; (5) blink detection. The CANDIDE model [11] for facial

Fig. 2 Face and eye tracking based on Kinect sensor, left to right, CANDIDE-3 face model in Kinect Face Tracking Demo [12], face and eye tracking under poor illumination [13]

image coding has been widely used in computer vision. As an updated parameterized face model CANDIDE-3 consists of 113 vertices and 168 surfaces and has improved mouth and eyes modeling significantly compared to its earlier versions. With inbuilt CANDIDE-3 face model *Face Tracking* from Microsoft Kinect SDK [12] is adopted in this work with modifications so as to facilitate driver status monitoring. Figure 2 gives examples of face/eye tracking results with overlapped wire-frame model CANDIDE-3 and in poor lighting condition. Meanwhile the head movement can be captured in terms of pitch, yaw, roll around three axes in 3D coordinate system. Facial features such as locations of eyebrows, mouth, nose, and other facial components are computed as well.

Based on the outcome of eye tracking process gaze estimation is made by using a modified algorithm from [14]. Eye pupil and corners are detected to estimate the gaze direction. In addition, an algorithm based on adaptive eye templates has been developed to perform blink detection. The details of gaze estimation and blink detection algorithms being used in the presented system are discussed in [9]. Facial feature tracking results with eye pupil and corner detection are depicted in Fig. 3.

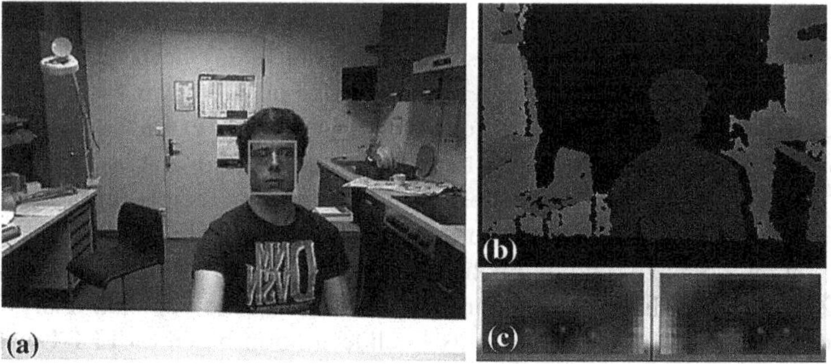

Fig. 3 Facial feature tracking based on Kinect sensor, **a** face tracking with eye locations, **b** distance information in grayscale depth image, **c** eye pupil and corner locations for left and right eyes respectively [9]

3.2 Steering Behavior and Physiological Measurements

Despite the dependency on driving experience, type of vehicle, road and weather conditions, etc. steering behavior indicated by steering wheel movement or by lane departure/line crossing is regarded as the most trustworthy measure of driving performance and has been widely employed in mainstream ADAS systems (see Table 1). The study in [15] shows correlation between steering wheel and lane position processes which can be jointly used for drowsiness detection.

Pulse rate is one of the measurable vital signals that can be used to check heart health and fitness level. Based on embedded pulse rate sensor on steering wheel the LF/HF ratio of heart rate variation, or say pulse rate, in frequency domain is suggested as an indicator for drowsiness detection in [16].

3.3 Feature Computation, Sensor Fusion, Pattern Classification

The features computed from steering angle sensor are described as follows. Steering reversals being related to micro-corrections indicate the frequency of lateral motion changes (left-right or right-left) within gap size θ. Depending on θ two features are taken into evaluation (Feature 1 and 3 in Table 2 with $\theta = 1°$ and $\theta = 3°$ respectively). Steering-same-side represents the frequency of steering motion in the same direction above threshold ϑ which indicates lane changing or curve turning movements. Feature 2 and 4 in Table 2 are computed based on $\vartheta = 12°$ and $\vartheta = 32°$ respectively. Feature 5 and 7 reflect mean and standard deviation of steering wheel positions within a measurement time frame. Feature 6 gives the percentage of micro-corrections being taken to the overall steering motion. Feature 8 represents the steering velocity. Feature 9 to 11 are frequency domain analysis of steering statistics based on FFT. Parameters for feature computation are depending on the steering wheel specification (e.g., wheel size, sensor resolution, etc.) and system setup. Please refer to [9, 15] for more details.

As mentioned in Sect. 3.1 various visual clues including head movement, eye gaze direction and ocular measures, can be provided by depth camera, here Kinect sensor. In the current system the following features are extracted: mean head position in 3D coordinate system of the depth camera within a measurement time frame (see Feature 12 to 14 in Table 2); mean head orientation measures, i.e., pitch, yaw and roll (Feature 15 to 17); frequency domain analysis for head translation on three axes and its Euclidean norm based on FFT LowBand (Feature 18 to 21) and FFT HighBand (Feature 22 to 25); frequency domain analysis for head rotation around three axes and its Euclidean norm based on FFT LowBand (Feature 26 to 29) and FFT HighBand (Feature 30 to 33); translation speed (Feature 34) and rotation speed (Feature 35); mean of eyebrow positions relative to

Table 2 List of features being computed from multiple sensor measurements

Feature	Sensor	Description	Feature	Sensor	Description
1, 3	Steer wheel	Steering reversals 1°, 3°	18–21	Kinect	x-, y-, z-, Norm- FFT-LowBand
2, 4	Steer wheel	Steering SameSide 12°, 32°	22–25	Kinect	x-, y-, z-, Norm- FFT-HighBand
5	Steer wheel	Std of position	26–29	Kinect	Pitch-, Yaw-, Roll-, Norm-FFT-LowBand
6	Steer wheel	Low steering percentage	30–33	Kinect	Pitch-, Yaw-, Roll-, Norm-FFT-HighBand
7	Steer wheel	Mean of absolute position	34, 35	Kinect	Translation, rotation speed
8	Steer wheel	Steering velocity	36	Kinect	Mean eyebrow position
9–11	Steer wheel	FFT-LowBand, -MidBand, -HighBand	37, 38	Kinect	Mean blink frequency, duration
12–14	Kinect	Mean x-, y-, z- Head Position	39	Pulse oximeter	LF/HF
15–17	Kinect	Pitch-, Yaw-, Roll- Head orientation	40	Pulse oximeter	Mean pulse rate

left and right eyes (Feature 36); mean eye blink frequency (Feature 37) and blink duration (Feature 38).

With predefined low frequency band 0.04–0.15 Hz (LF) and high frequency band 0.15–0.4 Hz (HF) the LF/HF ratio (see Sect. 3.2) of pulse rate course within a measurement time frame is computed in Feature 39. The mean pulse rate is provided by Feature 40.

Data sets of different sensors are synchronized on the same time base and fused on the feature level for pattern classification process. Different features are treated equally in this work even though a proper weighting scheme can be applied.

In the current system driver drowsiness detection is modeled as a three-class pattern classification problem. Due to advantage of learning complex, nonlinear, high-dimensional patterns a multilayer feedforward neural network is trained for classification purpose. Two learning algorithms have been evaluated in this work, i.e., scaled conjugate gradient (SCG) algorithm [17] and Levenberg–Marquardt (LM) algorithm [18]. The classification results are described in Sect. 4.

4 Experimental Results

Five test subjects in the experiments are all male between 22 and 25 years old (mean: 23.6, std: 1.1). They all have a valid driver's license for at least 4 years and up to 7 years. Two of them are regular drivers and the other three drive occasionally. They did not drink any alcohol before test. One-hour driving simulation was conducted on five test subjects respectively. Afterwards a 588-min driving sequence of all test subjects is captured. Despite that sensor data evaluation can be performed during system runtime all the sensor measurements of driving simulation are time-based synchronized and recorded for more thorough offline analysis.

To detect and classify driver drowsiness the ground truth (GT), or say, the target class of drowsiness level is defined as: not drowsy, a little drowsy and deep drowsy. Two criteria, i.e., self-rated score (subjective) and measured response time (objective) are combined to assess the drowsiness level and establish the ground truth (see Fig. 4). In this work sensor data of 353 time frames/windows with approximately 100-s measurement time per window are evaluated. Each measurement consists of multiple sensor inputs including steering angle measures, head movement, ocular measures and pulse rate values. 40 features (see Table 2) are extracted from different sensor inputs and afterwards are fused to construct an input vector for classification process.

The classifier based on multilayer feedforward neural network is trained in a supervised manner by using scaled conjugate gradient algorithm (SCG) and Levenberg-Marquardt algorithm (LM), respectively. The classification results are carried out by performing 10-fold cross-validation process. Tables 3 and 4 give a comparison between two training algorithms in terms of confusion matrix. The classification accuracy (ACC) with dependency on the number of hidden neurons are illustrated in Fig. 5. With 80 hidden neurons the classifier based on LM

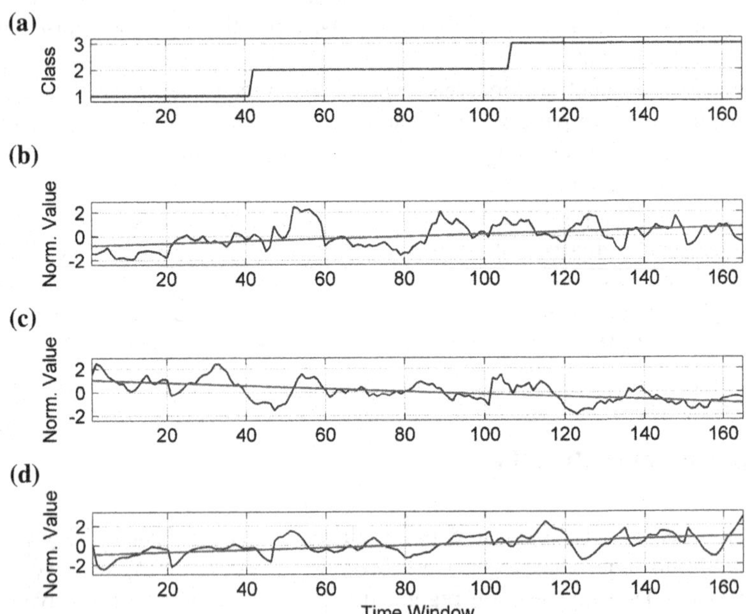

Fig. 4 Ground truth definition and feature extraction of 165 measurement time windows, **a** ground truth, **b** blink frequency (Feature 37), **c** low steering percentage (Feature 6) and **d** mean pulse rate (Feature 40)

Table 3 Confusion matrix of classification results based on Scaled Conjugate Gradient (SCG) algorithm

SCG		Target class (GT)			
40		I	II	III	∑
Output class (PR)	I	89	3	2	94.7 %
		25.2 %	0.8 %	0.6 %	5.3 %
	II	2	160	4	96.4 %
		0.6 %	45.3 %	1.1 %	3.6 %
	III	1	8	84	90.3 %
		0.3 %	2.3 %	23.8 %	9.7 %
	∑	96.7 %	93.6 %	93.3 %	94.3 %
		3.3 %	6.4 %	6.7 %	5.7 %

Table 4 Confusion matrix of classification results based on Levenberg–Marquardt (LM) algorithm

LM		Target class (GT)			
80		I	II	III	∑
Output class (PR)	I	91	1	0	98.9 %
		25.8 %	0.3 %	0.0 %	1.1 %
	II	1	170	2	98.3 %
		0.3 %	48.2 %	0.6 %	1.7 %
	III	0	0	88	100 %
		0.0 %	0.0 %	24.9 %	0.0 %
	∑	98.9 %	99.4 %	97.8 %	98.9 %
		1.1 %	0.6 %	2.2 %	1.1 %

Note GT and PR represent Ground Truth and Predicted Result respectively. Class I, II, III indicate driver drowsiness levels: not drowsy, a little drowsy, deep drowsy. ∑ gives the aggregated results for specific rows, columns or for the overall statistics in the matrix.

algorithm achieves the ACC result of 98.9 % with high performance and modest memory consumption in our experiments. Compared to the state-of-the-art, e.g., 89 % accuracy for Class II—a little drowsy and 99 % accuracy for Class III—deep drowsy in [19] our approach yields superior results.

Due to the experimental environment and system setup the steering angle sensor is able to provide continuous measurements of steering wheel, while the measurements of pulse oximeter and depth camera may be invalid or even missing in a specific time frame. In our experiments the following combinations of sensor inputs and the corresponding feature sets are investigated: (1) steering wheel (Sw) with 11 features; (2) steering wheel and pulse oximeter (SwPo) with 13 features; (3) steering wheel and depth camera (SwKn) with 38 features; (4) all available sensors (SwKnPo) with 40 features; (5) a subset with 8 features (Feature 6, 12, 13, 14, 36, 37, 39, 40 in Table 2) being selected from the full feature set by applying sequential feature selection algorithm (SFS). In this case backward heuristic search

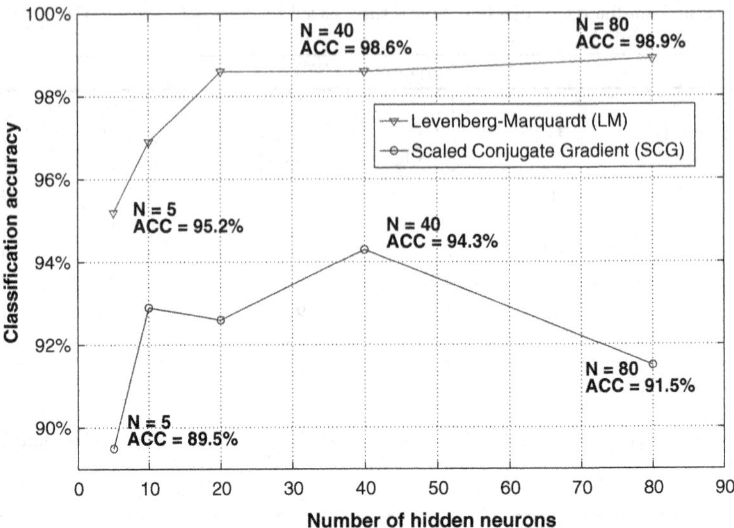

Fig. 5 Comparison of classification accuracy between SCG and LM algorithms

is used to minimize the feature set while preserving the feature quality in terms of overlap and separability. The classification results of SFS feature set in comparison with the outcome based on other possible feature sets being mentioned above are summarized in Fig. 6. By using SFS feature set the classifier is able to achieve the same accuracy with only 20 neurons specified in the hidden layer.

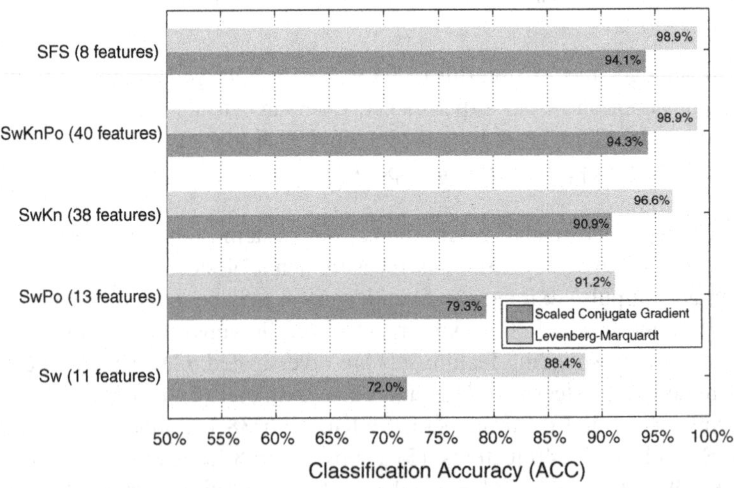

Fig. 6 Drowsiness level classification accuracy (ACC) with dependency on selected features

5 Conclusion and Future Work

Objectively detecting the tiredness or drowsiness still remains a challenge nowadays. Non-intrusive, accurate and robust driver drowsiness detection is one of the final goals of advanced driver assistance. In this paper, we present a novel system approach to driver drowsiness detection based on multi-sensor data fusion and soft-computing algorithms. By observing head movement and facial features with depth camera, by monitoring steering behavior and pulse rate of driver the presented system is able to classify three different drowsiness levels with up to 98.9 % accuracy based on data sets of five test subjects. The robustness of the presented approach needs to be validated with more statistics and with data from real vehicles. The variance of drivers, vehicles, road and weather conditions can be compensated by adaptive learning. Feature selection can be optimized with sophisticated heuristics, e.g., genetic algorithm (GA) and particle swarm optimization (PSO). Advanced classification techniques such as support vector machine (SVM) can be incorporated in the system as well for further optimization. In addition, we plan to integrate more miniaturized embedded sensors with wireless technology, e.g., EEG and ECG sensors, to improve the effectiveness and robustness of the system.

Acknowledgments The authors would like to thank Abhaya C. Kammara for giving support to construct the DeCaDrive system. The help from students in ISE are gratefully appreciated.

References

1. von Seelen, W.: Elektronisches Auge OPEL, Multimodaler Sensor zur Fahrzeugführung: Teilprojekt: Architektur, Rundumsicht und Objekterkennung; Abschlubericht zum 30. Juni 1997. Univ., Inst. für Neuroinformatik (1997)
2. Skribanowitz, J., Knobloch, T., Schreiter, J., König, A.: VLSI Implementation of an application-specific vision chip for overtake monitoring, real time eye tracking, and visual inspection. In: MicroNeuro, vol. 99, pp. 4552 (1999)
3. Ford Motor Company: http://www.ford.com. last visited: 02.11.2012
4. Daimler, A.G.: http://www.daimler.com. last visited: 02.11.2012
5. Volkswagen, A.G.: http://www.volkswagen.com. last visited: 02.11.2012
6. Toyota Motor Corporation: http://www.toyota.com. last visited: 02.11.2012
7. Volvo Car Corporation: http://http://www.volvocars.com. last visited: 02.11.2012
8. Li, L., Xu, Y., König, A.: Robust depth camera based multi-user eye tracking for autostereoscopic displays. In: 9th International Multi-Conference on SSD, pp. 1–6 (2012)
9. Werber, K.: Untersuchung von Fahrerassistenzsystemen zur Fahrer- Zustands- und Absichtserkennung mit Multisensorik. In: Diplomarbeit, ISE, TU Kaiserslautern (2012)
10. Li, L., Xu, Y., König, A.: Robust depth camera based eye localization for human-machine interactions. In: KES'2011, vol. 6881, p. 424 (2011)
11. Li, H., Roivainen, P., Forchheimer, R.: 3-D motion estimation in model-based facial image coding. IEEE Trans. PAMI **15**(6), 545–555 (1993)
12. Face Tracking: http://msdn.microsoft.com. last visited: 28.10.2012

13. Guarde, A.: Kinect based eye tracking for driver drowsiness detection. In: Studienarbeit, Institute of Integrated Sensor Systems (ISE), TU Kaiserslautern (2012)
14. Matsumoto, Y., Zelinsky, A.: An algorithm for real-time stereo vision implementation of head pose and gaze direction measurement. In: 4th IEEE International Conference on Automatic Face and Gesture Recognition, pp. 499–504 (2000)
15. Sherman, P.: The potential of steering wheel information to detect driver drowsiness and associated lane departure. Technical Report, Iowa State University (1996)
16. Yu, X.: Real-time nonintrusive detection of driver drowsiness: Final Report. Technical Report, University of Minnesota (2009)
17. Møller, M. F.: A scaled conjugate gradient algorithm for fast supervised learning. Neural Netw. **6**(4), 525–533 (1993)
18. Moré, J.J.: The Levenberg-Marquardt algorithm: implementation and theory. Numerical Anal. Dundee (1977)
19. Feng R., Zhang, G., Cheng, B.: An on-board system for detecting driver drowsiness based on multi-sensor data fusion using Dempster-Shafer theory, In: ICNSC, pp. 897–902 (2009)

A New Evolving Tree for Text Document Clustering and Visualization

Wui Lee Chang, Kai Meng Tay and Chee Peng Lim

Abstract The Self-Organizing Map (SOM) is a popular neural network model for clustering and visualization problems. However, it suffers from two major limitations, *viz.*, (1) it does not support online learning; and (2) the map size has to be predetermined and this can potentially lead to many "trial-and-error" runs before arriving at an optimal map size. Thus, an evolving model, i.e., the Evolving Tree (ETree), is used as an alternative to the SOM for undertaking a text document clustering problem in this study. ETree forms a hierarchical (tree) structure in which nodes are allowed to grow, and each leaf node represents a cluster of documents. An experimental study using articles from a flagship conference of Universiti Malaysia Sarawak (UNIMAS), i.e., the *Engineering Conference* (ENCON), is conducted. The experimental results are analyzed and discussed, and the outcome shows a new application of ETree in text document clustering and visualization.

1 Introduction

Clustering is a task of assigning data objects into a number of groups (or clusters) so that the objects in the same cluster share the same similarities than to those in other clusters [1]. It converts sets of non-linear data into a human and/or machine understandable format, which can be very useful for unsupervised learning systems. Examples of some famous clustering tools are the Self-Organizing Map (SOM) [2, 3], k-mean clustering [4], and fuzzy c-mean clustering [5, 6]. With respect to SOM, it is an artificial neural network that maps high-dimensional data

W. L. Chang · K. M. Tay (✉)
Faculty of Engineering, Universiti Malaysia Sarawak, Sarawak, Malaysia
e-mail: kmtay@feng.unimas.my

C. P. Lim
Centre for Intelligent Systems Research, Deakin University, Geelong, Australia

V. Snášel et al. (eds.), *Soft Computing in Industrial Applications*,
Advances in Intelligent Systems and Computing 223, DOI: 10.1007/978-3-319-00930-8_13,

onto a low-dimensional grid of nodes [2, 7], and retains the relationship of the data as faithfully as possible. From the literature, various applications of SOM, e.g., speech recognition [8, 9], feature extraction [10], robotic arm [11], noise reduction in telecommunication [12], and textual documents clustering [13], have been reported. Indeed, various extensions for SOM, e.g., hierarchical search [14, 15], growing SOM [16, 17], growing hierarchical SOM (GHSOM) [18], and evolving tree (ETree) [19], have been proposed over the years. In general, these approaches increase the flexibility of SOM and improve the learning time for processing large data samples.

With respect to text document clustering (also known as text categorization), it is a process to group similar text documents into group(s), based on their similarity [20]. The use of clustering tools in text document clustering is not new. Examples include the naive Bayes-based document clustering model [21], WEBSOM [22], and support vector machines-based for imbalanced text document classification [23]. These approaches allow a collection of documents to be clustered and visualized. Regardless of the popularity of these approaches, it is not sure how these approaches can be extended to online learning. Thus, it is important to develop an evolving text document clustering model for the following reasons: (1) new documents are generated or created everyday; and (2) it is not practical to perform re-training of a model while a new document appears. Thus, an evolving or online learning system is useful for this problem.

The focus of this paper is on the use of an evolving system, i.e., ETree, as an alternative to SOM, for document clustering. Instead of SOM (as in WEBSOM [22]), ETree allows a text document clustering tool to have evolving feature. Besides, it serves as a solution to a few shortcomings of WEBSOM, i.e., its learning time [19], and the difficulty in determining the map size before learning [19]. Even through ETree has been proposed as an alternative to SOM, its application is still limited. To the best of our knowledge, it is a new attempt to use ETree in document clustering. In addition, we adopt a general application framework for an evolving system [24] (details are explained in Sect. 2.3) to allow an ETree-based document clustering model to be implemented in practice. In our proposed approach, some features of WEBSOM, e.g., text pre-processing, are retained. A case study with information/data from a flagship conference of Universiti Malaysia Sarawak (UNIMAS), i.e., the *Engineering Conference* (ENCON), is experimented to evaluate the effectiveness of the proposed approach.

The organization of this paper is as follows. The background of ETree and its implementation is provided in Sect. 2. In Sect. 3, the use of the ETree as a text document clustering and visualization tool is described. An experimental study to evaluate the usefulness of ETree in text document clustering is presented in Sect. 4, with the results analyzed and discussed. A summary of concluding remarks is included in Sect. 5.

2 Background

In this section, the structure and learning algorithm of ETree are first explained. Then, an application framework for implementing ETree is described.

2.1 The Structure of ETree

Figure 1 depicts an example of the ETree structure. The tree structure consists of n_{node} nodes, and each node is denoted as $N_{l,j}$, where $l = 1, 2, 3, \ldots, n_{node}$ is the identity of a node and $j = 0, 1, 2, \ldots$ is its parent node. As an example, the parent node for $N_{2,1}$ is $N_{1,0}$, or $N_{2,1}$ is a child node for $N_{1,0}$. There are three types of nodes, i.e., root node, trunk nodes, and leaf nodes. The root node is the first created node (i.e., $N_{1,0}$) located at the top layer of the tree structure and with no parent node (i.e., $j = 0$). A trunk node is a parent node (labeled in white in Fig. 1). A leaf node, located at the bottom layer of the tree, is the node with no child node (labeled in black in Fig. 1).

The tree structure can be described by its tree depth. As an example, the tree depth is 4 in Fig. 1. The distance between two leaf nodes can be measured by the number of trunks that connect them together with the shortest path. As an example, the tree distance between $N_{4,2}$ and $N_{7,3}$ is 3, i.e. $d(N_{4,2}, N_{7,3}) = 3$, and the connected trunks are $N_{2,1}, N_{1,0}$, and $N_{3,1}$. Every leaf node is associated with two attributes, i.e., a weight vector, i.e., w_l , and a hit counter, i.e., $b_l(t)$. $w_l(t)$ is a n-dimension real data vector, i.e., $[\mu_{l,1}, \mu_{l,2}, \ldots, \mu_{l,n}] \in \mathbb{R}^n$, while $b_l(t) = 0, 1, 2, \ldots$ is a counter that records the number of times $N_{l,j}$ being selected as the Best Matching Unit (BMU).

The degree of matching for a data vector, $x(t) = [\xi_1, \xi_2, \ldots, \xi_n]$ corresponding to which is associated with a weight vector (i.e., $w_l = [\mu_{l,1}, \mu_{l,2}, \ldots, \mu_{l,n}]$) can be obtained by the Euclidian distance between $x(t)$ and w_l , as in Eq. 1. The lower the distance, the higher the degree of matching.

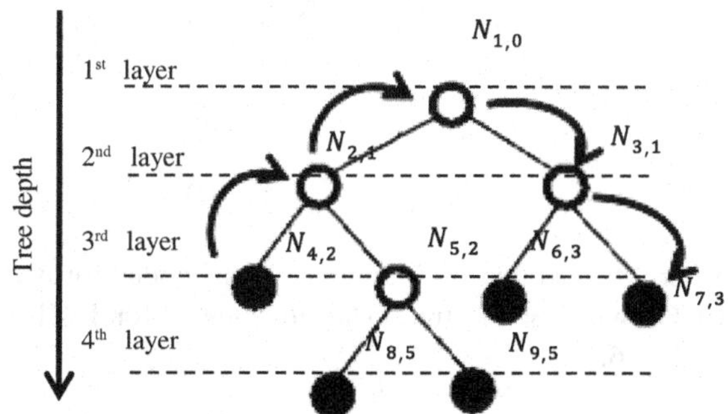

Fig. 1 The structure of ETree

$$dist(x(t), w_l(t))) = ||x(t) - w_l(t)|| \tag{1}$$

2.2 The Learning Algorithm of ETree

Two pre-determined parameters, i.e., the splitting threshold (i.e., $b_{splitting}$) and the number of split node (i.e., n_{child_node}) are considered. Assuming a new data (i.e. $x(t)$) is provided to ETree. The learning algorithm to update the tree comprises three steps, as follows:

(a) Determination of the BMU from root to leaf. An BMU determination process is a top-down process, starting at the 1st layer node (root node) towards the bottom layer nodes (leaf nodes). From the root node, its children nodes at the 2nd layer are matched with $x(t)$ based on Eq. 1. The child with the minimum distance is the best matched child. Then, the children nodes of the best matched child at the next layer are matched in the same manner. This process is repeated until a leaf node is obtained. The best matched leaf node is the BMU, and is denoted as N_{BMU} and $BMU \in [1, 2, 3, \ldots, n_{node}]$.
(b) Updating of the leaf node. The weight vectors for each of the leaf nodes are updated using the Kohonen learning rule, as follows.

$$w_l(t+1) = w_l(t) + h_{BMU\_l}(t)[x(t) - w_l(t)] \tag{2}$$

where $h_{BMU\_l}(t)$ is a neighborhood function between N_{BMU} and $N_{l,j}$, which is calculated using Eq. 3.

$$h_{BMU\_l}(t) = \propto (t) exp\left(\frac{-d(N_{BMU}, N_{l,j})^2}{2\sigma^2(t)}\right) \tag{3}$$

where $\propto (t)$ is the learning rate, $\sigma(t)$ is the Gaussian kernel width, and $d(N_{BMU}, N_{l,j})$ is the tree distance between N_{BMU} and $N_{l,j}$. The learning rate and the Gaussian kernel width are reduced monotonically with time, as in Eqs. 4 and 5, respectively.

$$\propto (t) = 1 - \frac{t}{n_{leaf}(t)} \tag{4}$$

$$\sigma(t) = \frac{1}{t} \tag{5}$$

where n_{leaf} is the total number of leaf nodes formed in the tree structure.
(c) Growing of the tree. The hit counter for BMU, i.e, b_{BMU}, is updated, as in Eq. 6.

$$b_{BMU}(t+1) = b_{BMU}(t) + 1 \tag{6}$$

If $b_{BMU}(t+1) = b_{splitting}, N_{BMU}$ is split into n_{child_node} child nodes. N_{BMU} becomes a trunk node. The weight vectors for these children nodes are initialized as $w_{BMU}(t+1)$. One of its children nodes is randomly chosen as the BMU. The leaf node is further updated with the updating procedure in Step (b).

2.3 A Framework for ETree Implementation Implementation

Figure 2 shows a general application framework for implementing an evolving system. The procedure is as follows. First, some initial (from previous training) models are available. The models are refined or expanded based on incremental feedback to produce a pool of new models. New data are provided to the pool of new models. The pool of new provides prediction or classification results for the new data. The result is examined with the operator and the internal algorithm, and feedback with respect to the quality of the result is generated. The feedback is sent as a reference to refine the models. This procedure allows the model inside the framework to expand without the need for re-learning.

3 Text Document Clustering and Visualization Using the ETree

The ETree is used as a text document clustering and visualization tool in this study. The ETree and its learning algorithm (as described in Sects. 2.1 and 2.2) as well as the application framework for implementing an evolving system (as described in Sect. 2.3) are synthesized. Some notations from WEBSOM [22] are detained in our proposed system. Figure 3 depicts the proposed procedure to construct the ETree-based text document clustering and visualization tool.

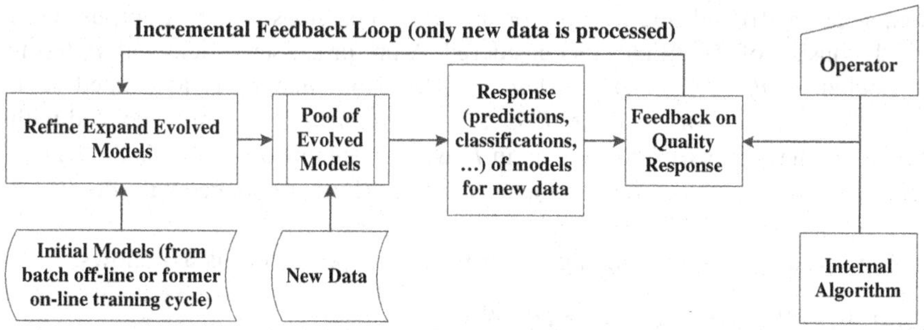

Fig. 2 An application framework for ETree with incremental learning (*source* [24])

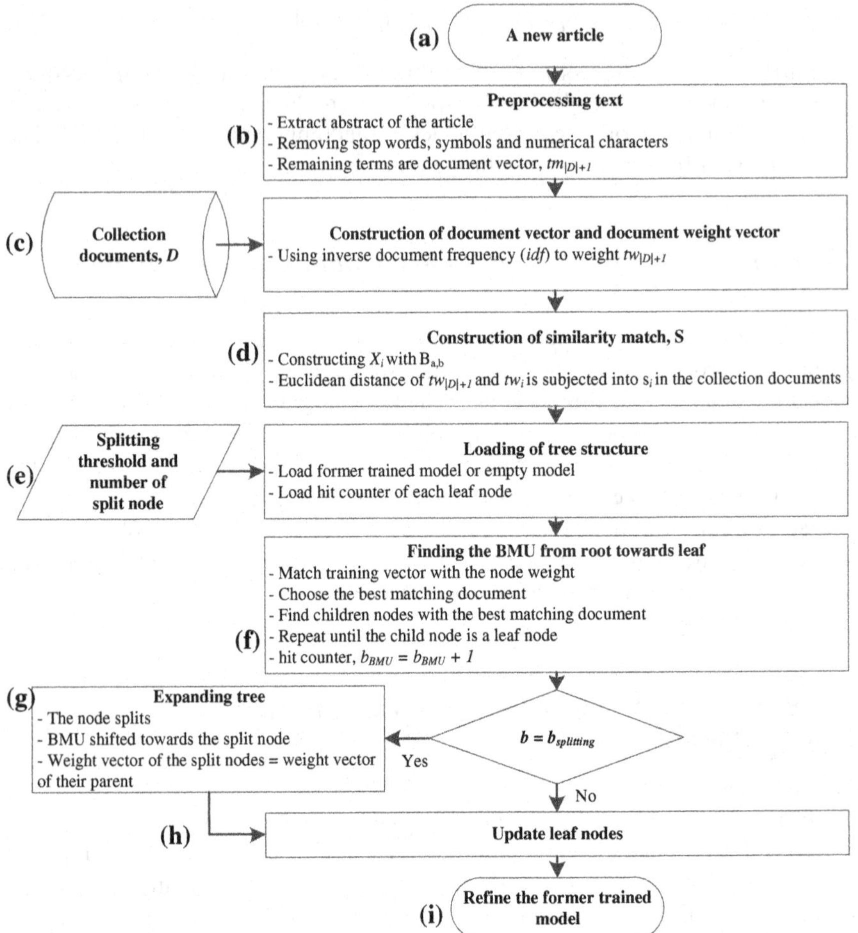

Fig. 3 The evolving tree-based text document clustering and visualization tool

The proposed procedure can be divided into two stages: updating article terms (i.e., steps (a–d)) and tree learning or updating (i.e., steps (e–i)). A corpus (i.e.,) which consists of $|D|$ articles, is considered. A pre-processed article with n_i terms is labeled as $d_i, \{d_i\} \in D, i = 0, 1, 2, \ldots, |D|$. Note that d_i is associated with $tm_{i,k_i}, df_{i,k_i}, tf_{i,k_i}$, and tw_{i,k_i} , where $k_i = 1, 2, \ldots, n_i$. tm_{i,k_i} is a term associated with d_i. df_{i,k_i} is the number of documents in D which contains tm_{i,k_i}. tf_{i,k_i} is a count of occurrence for tm_{i,k_i} in d_i. tw_{i,k_i} is the term weight (further obtained with Eq. 7) for tm_{i,k_i} in d_i.

The updating and learning algorithm for the proposed tool is as follows.

(a) A new article (i.e., $d_{|D|+1}$) is provided.

(b) The article is pre-processed, and the abstract is extracted. Less informative words (as defined in a list of stop words) are removed. Symbols and numerical characters are removed too. The remaining terms are considered as a document vector, where $\{tm_{|D|+1,1}, tm_{|D|+1,2,\ldots}, tm_{|D|+1,n_{|D|+1}}\} \in tm_{|D|+1}$.

(c) Each term (i.e., $tm_{|D|+1,k_{|D|+1}}$) is further associated with several attributes, i.e., $df_{|D|+1,k_{|D|+1}}, tf_{|D|+1,k_{|D|+1}}$, and $tw_{|D|+1,k_{|D|+1}}$. $tw_{|D|+1,k_{|D+1|}}$ is obtained with Eq. 7. The inverse document frequency approach is used to determine the importance of $tm_{|D|+1,k_{|D|+1}}$ in $d_{|D|+1}$, among D.

$$tw_{|D|+1,k_{|D|+1}} = tf_{|D|+1,k_{|D|+1}} \times log\frac{|D|}{df_{|D|+1,k_{|D|+1}}} \tag{7}$$

(d) A similarity of matching between $d_{|D|+1}$ and d_i , is obtained. A binary description i.e., B , is an n_i-by- $n_{|D|+1}$ matrix. Every element of B is denoted by $B_{a,b}$, where $a = 1, 2, 3, \ldots, n_i$ and $b = 1, 2, 3, \cdots, n_{|D|+1}$. $B_{a,b}$ is determined with Eq. 8. A n_i-by-1 matrix, i.e., X_i, is further obtained with Eq. (9). X_i indicates the significance for each of the terms in $tm_{|D|+1}$, in comparison with the terms in d_i.

$$B_{a,b} = \begin{cases} 1, if\ tm_{|D|+1,b}\ is\ tm_{i,a} \\ 0, if\ else. \end{cases} \tag{8}$$

$$X_i = tw_{|D|+1,1}B_{a,1} + tw_{|D|+1,2}B_{a,2} + \cdots + tw_{|D|+1,n_{|D|+1}}B_{a,n_{|D|+1}} \tag{9}$$

In this study, Euclidean distance between X_i and tw_i, $s(d_{|D|+1}, d_i)$ is obtained with Eq. 10. It is used as a measure of similarity between $d_{|D|+1}$ and d_i.

$$s(d_{|D|+1}, d_i) = ||X_i - tw_i|| \tag{10}$$

(e) A (empty) tree structure with its predefined parameters (i.e., $b_{splitting}, n_{child\_node}$) is loaded. Each of the node, i.e., $N_{l,j}$, is attributed with a set of document id (i.e., d_l, where $l = 1, 2, 3, \ldots, n_{node}$, and $d_l = [d_{l,1}, d_{l,2}, \ldots, d_{l,doc_l}] \in D$), a weight vector (i.e., $w_l(|D|)$, and a hit counter (i.e., $b_l(|D|)$).

(f) The BMU at a leaf node is determined (as explained in Sect. 2.2). From the root node, the associated documents in its children nodes at the 2nd layer are matched with $d_{|D|+1}$ based on Eq. 10. A cumulative similarity measure between the associated documents in $N_{l,j}$ with respect to $d_{|D|+1}$ is obtained using Eq. 11.

$$S_{l,d_{|D|+1}} = \sum\nolimits_{p=1}^{p=doc_l} S(d_{|D|+1}, d_{l,p}) \tag{11}$$

The child with the minimum $S_{l,d_{|D|+1}}$ is the best matched child. Then, the children nodes of the best matched child at the next layer are matched in the same manner as described. This process is repeated until a leaf node is obtained. The best matched leaf node is the BMU, and is denoted as $N_{BMU}, l = BMU$. The hit counter is further updated as in Eq. 12.

$$b_{BMU}(|D|+1) = b_{BMU}(|D|) + 1 \quad (12)$$

(g) If $b_{BMU} = b_{splitting}$, then N_{BMU} is split into n_{child_node} children nodes. In this study, $n_{child\_node} = 2$. The weight vectors for these children nodes are initialized to $\mathrm{w}_{BMU}(|D|)$. The best matched document in N_{BMU} corresponding to $d_{|D|+1}$ is considered as the only document for the new BMU (randomly selected from these two new leaf nodes). The remaining documents are clustered to the other node.
(h) The leaf nodes are updated using the Kohonen learning rule (Eqs. 2–5).
(i) Finally, the previously trained tree is refined or updated.

4 Case Study

A case study was conducted to evaluate the effectiveness of the ETree as a text document clustering and visualization tool. ENCON is a flagship conference organized by Faculty of Engineering, UNIMAS. In this study, the abstracts of 50 randomly selected articles from ENCON 2008 were used for experimentation to observe changes of the tree. A predefined list of stop words, which consisted of 119 words, was used. The articles were fed into the ETree in sequence. The parameters used were: $b_{splitting} = 10, 15$ and 20 and $n_{child\_node} = 2$. In this study, a laptop with i7-3612QM quad core processer, and Windows7 (64-bit) was used. Matlab 2012A was used for software development. The experimental results in terms of tree structure and computational complexity are analyzed and discussed, as follows.

4.1 (a) Growing (evolving) of tree

Figure 4a–c show the growing behavior of ETree for $b_{splitting} = 10$. As shown in Fig. 4a, two children nodes (i.e., $N_{2,1}$ and $N_{3,1}$) for the trunk node were first created. $N_{2,1}$ was further split to $N_{4,2}$ and $N_{5,2}$, as shown in Fig. 4b. The tree structure for these 50 abstracts is shown in Fig. 4c. The tree depth (tree layers) and tree size (number of nodes) is expanding from (a) to (c). Each leaf node represents a cluster of articles which share similarities in their abstract. Besides, the Kohonen learning rule was used to update the weight vector of the clusters. Leaf nodes that are located in the same branch share a higher similarity than those of other branches. For example, $N_{4,2}$ and $N_{5,2}$ are located at the same branch. Thus, $N_{4,2}$ and $N_{5,2}$ are more similar to each other, as compared to $N_{4,2}$ and $N_{3,1}$.

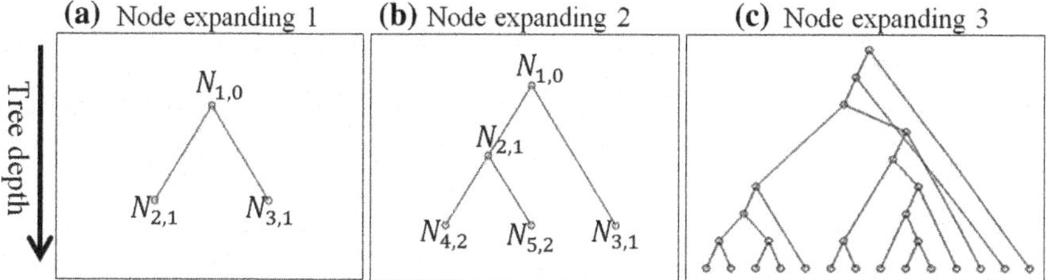

Fig. 4 Growing/Evolving behavior of the tree nodes

Table 1 Tree structures with different

$b_{splitting}$	Number of clusters	Tree size	Tree depth
10	14	27	8
15	5	9	4
20	3	5	2

Table 1 summarizes the number of clusters created, tree size (i.e., total number of nodes), and tree depth for and $b_{splitting} = 10, 15\ and\ 20$. The results show that the number of clusters, tree size, and tree depth decreased as $b_{splitting}$ increased. This is because $b_{splitting}$ decides the maximum number of documents at each node. The higher the value of $b_{splitting}$, the lower the tendency a node would split.

4.2 (b) Computation complexity

Figure 5a, b summarize the time required to update the tree structure (steps (a–i) in Sect. 3), as a new abstract was provided for $b_{splitting} = 10$ and $b_{splitting} = 10, 15$ and 20, respectively. The time axis shows the time (in seconds) required to update the tree structure when a new abstract was fed. As an example, referring to Fig. 5a, when an abstract labeled as #10 was fed to the system and $b_{splitting} = 10$, it took 1.4 s to update the tree structure. From Fig. 5b, the time required generally increased as the number of abstracts in the tree structure increased.

4.3 (c) Remarks

With the proposed approach, articles from ENCON 2008 could be clustered and visualized as a tree structure. Such clustering and visualizations is useful in aiding the conference organization committee to analyze the manuscripts received, i.e., to detect plagiarism and to ease the work of designing the conference tracks. In short,

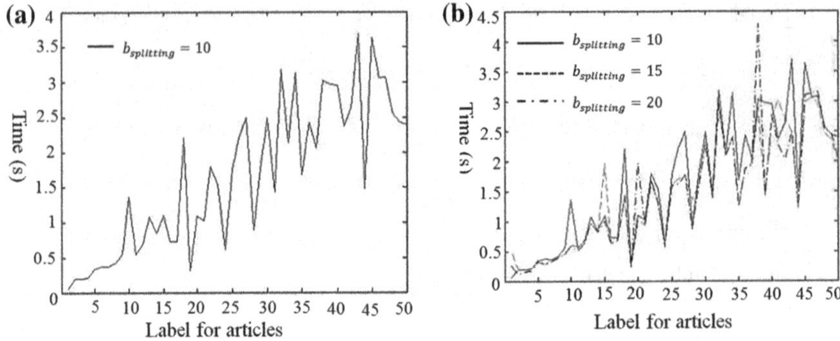

Fig. 5 The learning time with (**a**) $b_{splitting} = 10$ (**b**) $b_{splitting} = 10, 15, 10$

the proposed approach constitutes to a new decision support supporting tool for conference organizer. Besides, the proposed procedure could be useful with a larger number of articles with an expected increase in the computation complexity.

5 Summary

A new ETree-based text document clustering and visualization tool with evolving feature has been described. An experiment with data/information from ENCON 2008 was conducted. Promising results with an acceptable results and computation complexity (time) have been obtained. For future work, an ETree with dynamic n_{child_node} setting will be developed. In addition, other potential applications (e.g., image and signal processing) of ETree will be further investigated.

References

1. Rui, X., Donald, C.W.: Clustering. In: IEEE Series on Computational Intelligence. Wiley, Hoboken (2009)
2. Kohonen, T.: Self-organizing maps, 3rd edn. Springer, Berlin (2001)
3. Vesanto, J., Alhoniemi, E.: Clustering of the self-organizing map. IEEE Trans. Neural Netw. **11**(3), 586–600 (2000)
4. Chang, W.C., Luo, J., Kelvin, J.P.: Image segmentation via adaptive K-mean clustering and knowledge-based morphological operations with biomedical applications. IEEE Trans. Image Process. **7**(12), 336–344 (1998)
5. Rezaee, R., Lelieveldt, B.P.F., Reiber, J., H., C.: A new cluster validity index for the Fuzzy c-Mean. Pattern Recogn. Lett. **19**(3–4), 237–246 (1998)
6. James, C.B., Robert, E., William, F.: FCM: The fuzzy C-means clustering algorithm. Comput. Geosci. **10**(2–3), 191–203 (1984)

7. Kohonen, T., Kaski, S., Lagus, K., Salojärvi, J., Honkela, J., Paatero, V., Saarela, A.: Self organization of a massive document collection. IEEE Trans. Neural Netw. **11**(3), 574–585 (2000)
8. Kohonen, T., Simula, O., Visa, A.: Engineering applications of the self-organizing map. Proc. IEEE **84**(10), 1358–1384 (1996)
9. Kohonen, T., Somervuo, P.: Self-organizing maps of symbol strings. Neurocomputing **21**, 19–30 (1998)
10. Mao, J., Jain, A.K.: Artificial neural networks for feature extraction and multivariate data projection. IEEE Trans. Neural Netw. **6**(2), 296–317 (1995)
11. Buessler, J.L., Kara, R., Wira, P., Kihl, H., Urban, J.P.: Multiple self-organizing maps to facilitate the learning of visuo-motor correlations. In: IEEE International Conference on Systems, Man, and. Cybernetics **3**, pp. 470–475 (1999)
12. Kohonen, T., Raivio, K., Simula, O., Henriksson, J.: Start-up behaviour of a neural network assisted decision feedback equalizer in a two-path channel. In: IEEE International Conference on, Communications, ICC92, vol. 3, pp. 1523–1527 (1992)
13. Lagus, K., Kaski, S., Kohonen, T.: Mining massive document collections by the WEBSOM method. Inf. Sci. **163**(1–3), 135–156 (2003)
14. Chung, C.H., Shu, H.L., Wei, S.T.: Apply extended self-organizing map to cluster and classify mixed-type data. Neurocomputing **74**, 3832–3842 (2011)
15. Tai, W.S., Hsu, C.C., Chen, J.C.: A mixed-type self-organizing map with a dynamic structure. In: The 2010 International Joint Conference On Neural Networks (IJCNN), pp. 1–8 (2010)
16. Matharage, S., Alahakoon, D., Rajapakse, J., Pin, H.: Fast growing self organizing map for text clustering. In: Lecture Notes in Computer Science, Neural Information Processing, Vol. 7063/2011, pp. 406–415 (2011)
17. Kuo, R.J., Wang, C.F., Chen, Z.Y.: Integration of growing self-organizing map and continuous genetic algorithm for grading lithium-ion battery cells. Appl. Soft Comput. **12**, 2012–2022 (2012)
18. Huang, S.Y., Tsaih, R.H.: The prediction approach with growing hierarchical self-organizing map. In: The 2012 International Joint Conference On Neural Networks (IJCNN), pp. 1–7 (2012)
19. Pakkanen, J., Iivarinen, J., Oja, E.: The evolving tree—analysis and applications. IEEE Trans. Neural Netw. **17**(3), 591–603 (2006)
20. Fabrizio, S.: Text categorization. Alessandro, Z. (ed.) Text Mining and its Applications, pp. 109–129. WIT Press, Southampton (2005)
21. Lewis, D.: Naïve Bayes at forty: The independence assumption in information retrieval. Lect. Notes Compu. Sci. **1398**, 4–15 (1998)
22. Azcarraga, A.P., Yap, T.J., Tan, J., Chua, T.S.: Evaluating keyword selection methods for WEBSOM text archives. IEEE Trans. Knowl. Data Eng. **16**(3), 380–383 (2004)
23. Liu, T., Loh, H.T., Sun, A.: Imbalanced text classification: A term weighting approach. Expert Syst. Appl. **36**, 690–701 (2009)
24. Lughofer, E.: Evolving fuzzy systems—methodologies, advanced concepts and applications. 1st edn. Springer (2011)

Brain–Computer Interface Based on Motor Imagery: The Most Relevant Sources of Electrical Brain Activity

Alexander A. Frolov, Dušan Húsek, Václav Snášel, Pavel Bobrov, Olesya Mokienko, Jaroslav Tintěra and Jan Rydlo

Abstract Examined are sources of brain activity, contributing to EEG patterns which correspond to motor imagery during training to control brain–computer interface (BCI). To identify individual source contribution into EEG recorded during the training, Independent Component Analysis (ICA) was employed. Those independent components, for which the BCI system classification accuracy was at maximum, were treated as relevant to performing the motor imagery tasks. Activities of the three most relevant components demonstrate well exposed event related desynchronization (ERD) and event related synchronization (ERS) of the mu-rhythm during imagining of contra- and ipsilateral hand and feet movements. To reveal neurophysiological nature of these components we solved the inverse EEG problem in order to localize the sources of brain activity causing these

A. A. Frolov (✉) · P. Bobrov · O. Mokienko
Institute for Higher Nervous Activity and Neurophysiology of Russian Academy of Sciences, Butlerova 5a, Moscow, Russian Federation
e-mail: aafrolov@mail.ru

A. A. Frolov · D. Húsek · V. Snášel · P. Bobrov
VSB Technical University of Ostrava, 17. listopadu 15/2172 708 33 Ostrava, Czech Republic
e-mail: dusan@cs.cas.cz

V. Snášel
e-mail: vaclav.snasel@vsb.cz

P. Bobrov
e-mail: p-bobrov@yandex.ru

D. Húsek
Institute of Computer Science, Academy of Sciences of the Czech Republic, Pod Vodárenskou Věží 2, Prague 8, Czech Republic

J. Tintěra · J. Rydlo
Institute for Clinical and Experimental Medicine, Vídeňská 1958/9, Praha, Czech Republic
e-mail: jaroslav.tintera@ikem.cz

J. Rydlo
e-mail: Jan.Rydlo@ikem.cz

V. Snášel et al. (eds.), *Soft Computing in Industrial Applications*,
Advances in Intelligent Systems and Computing 223, DOI: 10.1007/978-3-319-00930-8_14,

components to appear in EEG. Individual geometry of brain and its covers provided by anatomical MR images, was taken into account when localizing the sources. The sources were located in hand and feet representation areas of the primary somatosensory cortex (Brodmann areas 3a). Their positions were close to foci of BOLD activity obtained in fMRI study.

1 Introduction

A brain–computer interface (BCI) provides a direct functional interaction between the human brain and the external device. The most prevalent BCI systems are based on the discrimination of EEG patterns related to execution of different mental tasks. By agreement with the BCI operator each mental task is associated with one of the commands to the external device. Then to produce the commands, the operator switches voluntary between corresponding mental tasks. If BCI is dedicated to control device movements then psychologically convenient mental tasks are motor imaginations. Another advantage of these mental tasks is that their performance is accompanied by the easily recognizable EEG patterns. Moreover, motor imagination is considered now as an efficient rehabilitation procedure to restore movement after paralysis [3]. Thus, namely the analysis of BCI performance based on motor imaginary is the object of the present study.
The most stable electrophysiological phenomenon accompanying motor performance is the decrease of EEG mu-rhythm recording from the central electrodes located over the brain areas representing the involved extremity [13]. This decrease (Event Related Desynchronization, ERD) occurs also when the subject observes the movement of another person [17] and during motor preparation and imagination [14]. In the state of motor relaxation the increase of EEG mu-rhythm is observed [13] which is called Event Related Synchronization (ERS). The exposure of ERD and ERS in specific brain areas during motor imagination of different extremities is the reason of the high efficiency of EEG patterns classifying. To classify the patterns we used the simplest Bayesian classifier based on the analysis of EEG covariance matrixes [6, 7].

Until now the most widespread technique to localize brain functions is fMRI study which provides high spatial but low temporal resolution. By contrast EEG study provides high temporal but low spatial resolution. The most prospective seems to be the combination of these techniques [5] especially if to take into account the fast progress in methods of solving inverse EEG problem [8, 10] focused on localization of sources of brain activity by calculation of distribution of electric potential over head surface. One of the approaches towards the integration of these techniques was suggested in our previous work [7]. Here we develop the approach and apply it to the analysis of more mental states used for BCI control.

2 Methods

Eight subjects (4 male, 4 female) aged from 25 to 65 participated in the study. The experiment with each subject was conducted for 10 experimental days, the one series per day. Each series consisted of training and testing sessions. The first, training, session was designed to train BCI classifier. The following, testing session was designed to provide subjects with the output of the BCI classifier in real time to enhance their efforts to imagine a movement. The subjects had to perform one of the four instructions presented on a screen of a monitor: to relax and to imagine the movement of the right or left hand or feet. The movement which they were asked to imagine was a handgrip or feet pressure.

Subject was sitting in a comfortable chair, 1 m from a 17" LCD monitor, and was instructed to fix a gaze on a motionless circle (1 cm in diameter) in the middle of the screen. Four gray markers were placed around the circle to indicate the mental task to be performed. The change of the marker color into green signaled the subject to perform the corresponding mental task. Left and right markers corresponded to left and right hand movement imagining respectively. The lower marker corresponded to feet movement imagining and the upper one corresponded to relaxation. Each command was displayed for 10 s. Each clue was preceded by a 4 s warning when the marker color changed into blue.

Four such instructions presented in random order constituted a block, one block constituted a training session and nine blocks a testing session. Thus each subject received 10 blocks of instructions at each experimental day.

During the testing sessions the result of classification was presented to a subject by color of the central circle. The circle became green if the result coincided with the instruction and its brightness increased with the increase of classifying confidence. During the instruction "relax" the presentation of quality of classification was switched off, not to attract the subject's attention.

EEG was recorded by 48 active electrodes using g.USBamp and g.USBamp API for MATLAB (g-tec, Graz, Austria) with sampling frequency 256 Hz and filtered by notch filter to suppress power supply noise.

The data of the last BCI session were used for solving inverse EEG problem, that is for localizing the sources of EEG signals inside the brain. We solved it taking into account individual geometry of brain and its covers provided by MRI data.

The fMRI examinations were conducted with 3T MR scanner (Siemens Trio Tim, Erlangen, Germany). During fMRI recording instructions to relax or to imagine hand or feet movements were presented to the subject without EEG recording. We believe that the subjects who are trained to control BCI are able to produce similar patterns of the brain activity during both EEG and fMRI experiments.

The quality of BCI performance was estimated by the results of on-line classifying during the testing session of each experimental day and offline by the data obtained during both training and testing sessions. For offline analysis the data were additionally filtered within 5–30 Hz bandpass. Then seven blocks of 10 were randomly chosen for classifier learning, i.e. for calculation of covariance matrices for all mental states. Recordings of the remaining three blocks were split into epochs of 1 s length. These epochs were used for classifier testing. Fifty such classification trials were made. Averaging over all classification trials resulted in $L \times L$ confusion matrix $\mathbf{P} = (p_{ij})$, where L=4 is the number of used mental tasks. Each coefficient of confusion matrix p_{ij} is an estimate of probability to recognize the i-th mental task in case the instruction is to perform the j-th mental task.

We chose Cohen's κ as an index of classification efficacy. In case of perfect classifying $\kappa = 1$. If classification is random then $\kappa = 0$.

To identify the sources of brain activity the most relevant to BCI performance we used Independent Component Analysis (ICA) which provides representation of a multidimensional EEG signal $\mathbf{X}(t)$ (where components of $\mathbf{X}(t)$ represent electric potentials recording from N individual electrodes at the head surface) as a superposition of activities of independent components ξ:

$$\mathbf{X}(t) = \mathbf{W}\xi(t) = \mathbf{W}_1\xi_1 + \mathbf{W}_2\xi_2 + \cdots + \mathbf{W}_N\xi_N \tag{1}$$

Columns $\mathbf{W}_i$ of matrix $\mathbf{W}$ specify the contribution of the corresponding independent component (or source) into each of the electrodes and the components ξ_i of the vector $\xi(t)$ specify sources intensity in each time point. The combination of active sources is supposed to be specific and individual for each mental task. Thus their activities in many tasks can be treated as independent. To represent the signal $\mathbf{X}$ in the form (1) we used algorithm RUNICA (MATLAB toolbox EEGLab, [4]).

To reveal the sources of brain activity the most significant for BCI performance κ was calculated in dependence on the number N_{cmp} of ICA components used for mental state classifying. For each N_{cmp} we found the optimal combination of components providing the highest κ. Since the total number 2^N (where $N = 48$) of possible component combinations is extremely large we used exhaustive search to find the optimal combination of components only for $N_{cmp} = 3$. To find the optimal combination of components for $N_{cmp} > 3$ we used a "greedy" algorithm which added components one by one starting from the optimal combination of three components [7]. The case when all ICA components are used, i.e. $N_{cmp} = N$, directly corresponds to classifying the original signal X [7].

With respect to the EEG analysis, it is reasonable to assume that independent sources of electrical brain activity recorded at the head surface are current dipoles distributed over the neocortex. As shown below our experiments confirm this assumption and at least for the sources the most relevant for BCI performance the distribution of electrical potential over the head produced by each of these sources could be actually interpreted in terms of electrical field produced by single current dipole. Thus, for each of such sources its location was searched in a single dipole approximation. In other words, position and orientation of a single current dipole were searched which provided maximal matching between patterns of EEG distribution on the head surface given by ICA and by a dipole.

The pattern of EEG distribution for dipole with given position and orientation was calculated by solving the direct EEG problem. It was solved by the finite element method (FEM) which allows to take into account individual geometry of the brain and its covers. To generate the FEM meshes from the MRI data, MR images were segmented into five sub-regions: white matter, gray matter, cerebrospinal fluid (CSF), skull and scalp. The segmentation of the different tissues within the head was made by means of SPM8 New Segmentation Tool. To construct the FE models of the whole head the FEM mesh generation was performed using tetrahedral elements with inner-node spacing of 2 mm. Thus, the total number of nodes amounted to about four millions. Electrical conductivities were assigned to the tissues segmented in accordance with each tissue type: 0.14 S/m for white matter, 0.33 S/m for gray matter, 1.79 S/m for CSF, 0.0132 S/m for skull, and 0.35 S/m for scalp [9, 19]. To solve the EEG forward problems, the FEM mesh along with electrical conductivities were imported into the commercial software ANSYS (ANSYS, Inc., PA, USA).

3 Results

For all subjects values of index κ averaged over all experimental days are shown in Fig. 1. The subjects are ranged according to mean value of κ computed for on-line classification. Mean on-line and offline estimates of BCI performance are shown to be very close. However on-line estimations are more variable. This is reasonable

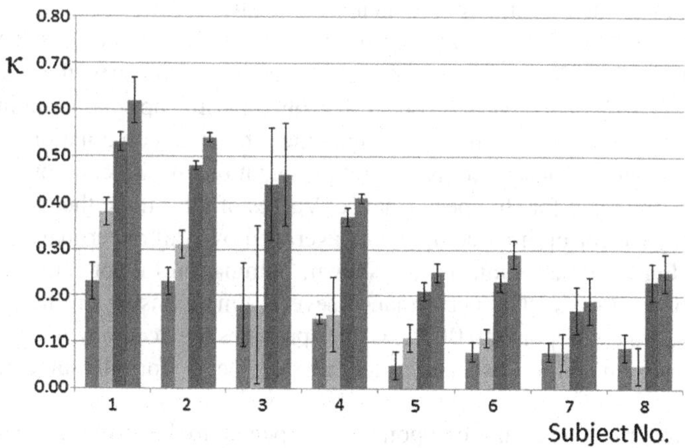

Fig. 1 Index κ of BCI control accuracy for all subjects averaged over all experimental days (means and standard errors). *Blue* bars—on-line, *green* bars—offline, *red*—three the most relevant components, *grey*—optimal components. The subjects are ranged according to κ computed offline. The most relevant components and optimal components were calculated to maximize κ

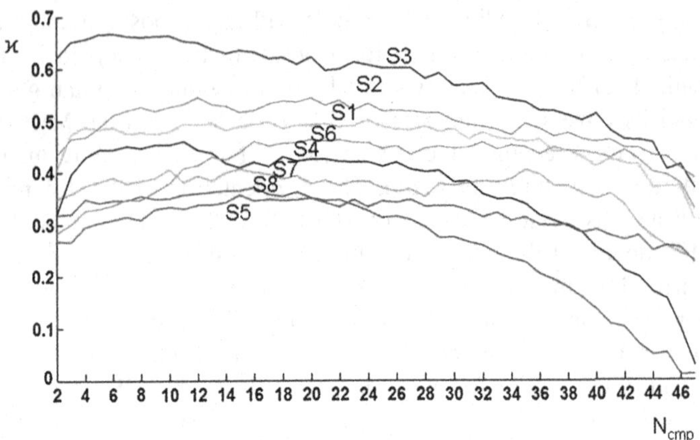

Fig. 2 Index κ of BCI control accuracy in dependence on the number N_{cmp} of ICA components used for mental states classifying. The optimal combination of three components ($N_{cmp} = 3$) was obtained by the exhausted search. The other optimal combinations were obtained by the "greedy" algorithm. Each curve represents the data for each individual subject obtained for the last day of BCI training

because on-line estimation is based on one classification trial obtained directly during experiment performance, while offline estimation is based on 50 classification trials as described above. Therefore, confusion matrix computed offline is more confident than that computed on-line.

Figure 2 demonstrates the dependency of κ on the number N_{cmp} of ICA components for all subjects on the last experimental day. For each N_{cmp} index κ is shown for individual optimal component combination.

As shown in Fig. 2, κ dependence on N_{cmp} is not monotonic. κ reaches maximum when some ICA components are discarded. However the classifying accuracy for $N_{cmp} = 3$ is very close to that obtained for optimal combination of components providing maximal κ. Thus the obtained combinations of three components can be considered as good representations of the components which are the most relevant for BCI performance. Values of κ for three the most relevant components and for optimal components averaged over all experimental days are shown in Fig. 1 for all subjects. As shown, elimination of not relevant (noisy) components improves BCI performance several times. Basing on the most relevant and optimal components BCI classifier provides the accuracy of mental state recognition significantly exceeding the random level for all subjects (t-test, $P<0.001$).

The three most relevant components happened to be not identical for all experimental days and all subjects but some of them appeared very repeatedly. Four such components which appeared most often over all subjects and all experimental days are shown in Fig. 3. The features of these components are shown in terms of their contribution into EEG electrodes (topoplots) and their

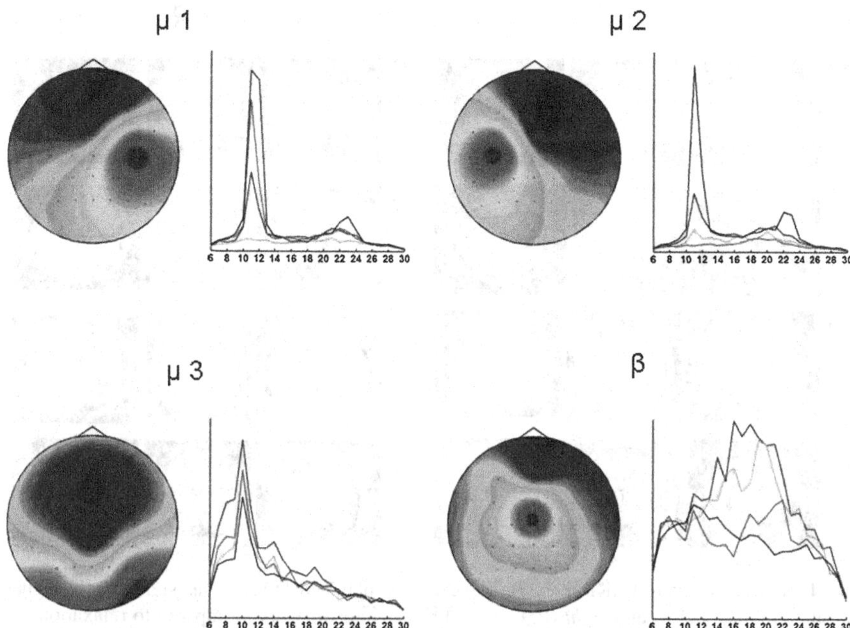

Fig. 3 Topoplots and spectrograms for four sources of electrical brain activity the most relevant to BCI performance. Sources $\mu1$, $\mu2$ and $\mu3$ demonstrate ERD of mu-rhythm during imagination of *left* hand, *right* hand and *feet* movement. β demonstrates ERD in the band 6–30 Hz. Spectral frequency is given in Hz, *blue* lines-relaxation, *red* lines-*right* hand motor imagery, *green* lines-*left* hand motor imagery, *black* lines-*feet* motor imagery

spectrograms for four considered mental states: relaxation, right or left hand movement imagination and feet movement imagination. The data relates to the last experimental day of Subject one who showed the best BCI control on average.

Three of these components ($\mu1$, $\mu2$ and $\mu3$) demonstrate well exposed Event Related Desynchronization (ERD) of mu-rhythm. For the component denoted $\mu1$ mu-rhythm is suppressed during the left hand motor imagery. Its focus is in the right hemisphere presumably above the primary sensorimotor areas presenting the left hand. Respectively, for the components $\mu2$ and $\mu3$ mu-rhythm is suppressed during the right hand and feet motor imagery and their foci are presumably above the areas presenting right hand and feet.

It is worth to note that the topoplots of four described components were very stable for all experimental days but the manifestation of ERD and ERS was rather variable. Namely the variability of their manifestation determined the variability of BCI control quality. However in any case ERD and ERS are much better exposed in terms of ICA components than in direct EEG recording. The level of ERD can be estimated as $r = S_{im}/S_{rel}$ where S_{im} and S_{rel} are the maximal spectral densities in the alpha band during the motor imagination and relaxation, respectively. For Subject one on average over all experimental days and imagination of both hands r

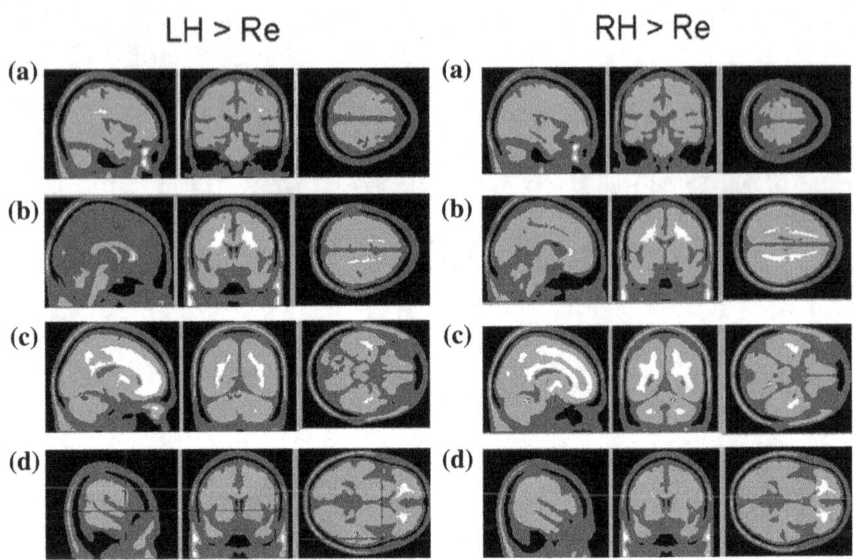

Fig. 4 Results of group fMRI analysis. Voxels for which BOLD level was significantly higher during *left* (LH $>$ Re) and *right* (RH $>$ Re) hand motor imagery compared to relaxation are shown in *red*. For *left* hand motor imagery: **a** Brodmann areas 3 and 4, **b** Supplementary motor area, **c** Cerebellum, **d** Brodmann area 22, *left* insula and *right* insula. For *right* hand motor imagery: **a** Brodmann area 4, **b** Brodmann area 24, **c** Cerebellum, **d** - Brodmann area 22 and *right* insula

amounted to 0.2 in terms of ICA components and to 0.69 for two central electrodes C3 and C4 where ERD was maximally exposed. Thus ICA allowed to rectify ERD and ERS due to excluding the component not exposing these changes of the brain activity.

Although the shown ICA components happened to be rather stable and repeatable over all subjects and all experimental days one could expect that they are only the formal results of some mathematical transformations of the actual experimental data and have no physiological sense. To clarify their sense we show, first, that their contribution to EEG recordings can be explained by the current dipole sources of brain activity located in the sensorimotor cortical areas, and, second, that locations of these sources coincide with locations of brain activity identified in fMRI study.

Results of fMRI group analysis are shown in Fig. 4. Figure demonstrates voxels for which BOLD level was significantly higher during left (LH $>$ Re) and right (RH $>$ Re) hand motor imagery compared to relaxation.

Dipole positions obtained for components $\mu 1$ and $\mu 2$ happened to be very close to the sensorimotor "hand areas" marked in Fig. 4a. Over all subjects and over two sources $\mu 1$ and $\mu 2$ the discrepancy between dipole positions and fMRI foci amounted only 15 ± 2 mm. For all subjects the dipoles corresponding to components $\mu 1$ and $\mu 2$ were located at the bottom of the central sulcus, i.e. at the area

3a responsible for proprioceptive sensation. According to subjects' reports this corresponds to their internal feeling of the imagined movement.

On average over all subject and over the three most relevant components $\mu 1$ and $\mu 2$ the residual variance of the single dipole approximation amounted only 1 %.

4 Discussion

Generally, the difficulties in interpreting the original EEG signals are due to the overlapping of activities coming from different brain sources, due to the distortion of the current flows caused by the inhomogeneity in the conductivity of the brain and its covers and due to uncertainty not only in dipole source locations but also in the dipole orientation which determines the relation between its position and the EEG amplitude maxima which it produces at the head surface. These difficulties result in common notion that EEG data provide high temporal but very low spatial resolution comparing to fMRI data. Last years there were many efforts to match these techniques to enhance both resolutions [11]. One of the approaches is presented here. It allowed us to find the location of the sources of the brain activity which are the most relevant for motor imagination. Two of them denoted as $\mu 1$ and $\mu 2$ happened to be localized at the bottom of central sulcus close to the Brodmann area 3a responsible for proprioceptive sensation. Thus, the experience of imagery (at least for motor imagery) involves perceptual structures despite the absence of perceptual stimulation.

There were many efforts to reveal whether SM1 is active during motor imagery basing on EEG data [2, 12, 14]. Particularly, in [12, 14] the conclusion that it activates was based on the observation that ERD is maximally exposed at the electrodes related to SM1 activity. However as mentioned above it is difficult to prescribe electrical activity to some particular brain area on the base of original EEG data. We believe that our approach allows to do this more substantiated. The reasonable and well interpreted results were obtained here due to solving the inverse EEG problem with the data refined by ICA simultaneously using the most realistic head model. Besides the foci of fMRI activity which were associated with three sources of EEG activity the most relevant for BCI performance (foci in primary sensorimotor areas 3 and 4 shown in Fig. 4) we observed many other foci. Among them are foci in cerebellum, superior temporal area 22, ventral anterior cingulate area 24 and insula also shown in Fig. 4. Thus motor imagery involves rather wide brain networks. According to the literature it can involve also superior and inferior parietal lobule, pre-frontal areas, inferior frontal gyrus, secondary somatosensory area and basal ganglia (see, for example [5, 18]). We also obtained many other ICA components which were relevant to motor imagination except four main components $\mu 1$, $\mu 2$, $\mu 3$ and β. Since we obtained good relation between these components and fMRI data the natural goal of our future research is to reveal the relations between other fMRI foci and other ICA components.

Acknowledgments This work was supported by the European Regional Development Fund in the IT4Innovations Centre of Excellence project (CZ.1.05/1.1.00/02.0070) and by the Bio-Inspired Methods: research, development and knowledge transfer project, reg. no. CZ.1.07/2.3.00/20.0073 funded by Operational Programme Education for Competitiveness, co-financed by ESF and state budget of the Czech Republic, by Institute of Computer Science from its long-term strategic development financing budget RVO:67985807, and by RFBR grant 11-04-12025.

References

1. Bashashati, A., Fatourechi, M., Ward, R., Birch, G.: A survey of signal processing algorithms in brain–computer interfaces based on electrical brain signals. J. Neural Eng. **4**, R32 (2007)
2. Beisteiner, R., Hollinger, P., Lindinger, G., Lang, W., Berthoz, A.: Mental representations of movements. Brain potentials associated with imagination of hand movements. Electroecephalogr. Clin. Neurophysiol. **96**, 183–193 (1995)
3. Birbaumer, N., Cohen, L.G.: Brain–computer interfaces: communication and restoration of movement in paralysis. J Physiol. **579**, 621–636 (2007)
4. Delorme, A., Makeig, S.: EEGLAB: an open source toolbox for analysis of single-trial EEG dynamics. J. Neurosci. Methods **134**, 9–21 (2004)
5. Formaggio, E., Storti, S.F., Cerini, R., Fiaschi, A., Manganotti, P.: Brain oscillatory activity during motor imagery in EEG-fMRI coregistration. Magn. Reson. Imaging **28**(10), 1403–1412 (2010)
6. Frolov, A., Husek, D., Bobrov, P.: Comparison of four classification methods for brain computer interface. Neural Netw. World **21**(2), 101–115 (2011)
7. Frolov, A., Husek, D., Bobrov, P., Korshakov, A., Chernikova, L., Konovalov, R., Mokienko, O.: Sources of EEG activity most relevant to performance of brain–computer interface based on motor imagery. NNW **1**(12), 21–37 (2012)
8. Grech, R., Cassar, T., Muscat, J., Camilleri, K.P., Fabri, S.G., Zervakis, M., Xanthopoulos, P., Sakkalis, V., Vanrumste, B.: Review on solving the inverse problem in EEG source analysis. J. NeuroEngineering and Rehabil. **5**(25), 1–33 (2008)
9. Kim, T.S., Zhou, Y., Kim, S., Singh, M.: EEG distributed source imaging with a realistic finite-element head model. IEEE Trans Nucl Sci. **49**, 745–752 (2002)
10. Lee, W.H., Liu, Z., Mueller, B.A., Limb, K., He, B.: Influence of white matter anisotropic conductivity on EEG source localization: comparison to fMRI in human primary visual cortex. Clin. Neurophysiol. **120**(12), 2071–2081 (2009)
11. Mulert, C., Lemieux, L. (eds.): EEG-fMRI. Physiological Basis, Techniques and Application. Springer, New York (2010)
12. Neuper, C., Scherer, R., Reiner, M., Pfurtscheller, G.: Imagery of motor actions: differential effects of kinesthetic and visual-motor mode of imagery in single-trial EEG. Cogn. Brain Res. **25**, 668–677 (2005)
13. Pfurtscheller, G., Neuper, C.: Event-related synchronization of mu rhythm in the EEG over the cortical hand area in man. Neurosci. Lett. **174**, 93–96 (1994)
14. Pfurtscheller, G., Neuper, C., Flotzinger, D., Pregenzer, M.: EEG-based discrimination between imagination of right and left hand movement. Electroencephalogr. and Clin. Neurophysiol. **103**, 642–651 (1997)
15. Pfurtscheller, G., Neuper, C.: Motor imagery activates primary sensorimotor area in humans. Neurosci. Lett. **239**, 65–68 (1997)
16. Pfurtscheller, G., Neuper, C.: Motor imagery and direct brain–computer communication. Proc. IEEE **82**(7), 1123–1134 (2001)
17. Pineda, J.A.: The functional significance of mu rhythm: translating "seeing" and "hearing" into "doing". Brain Res. Rev. **50**, 57–68 (2005)

18. Solodkin, A., Hlustik, P., Chen, E.E., Small, S.L.: Fine modulation in network activation during motor execution and motor imagery. Cereb. Cortex **14**(11), 1246–1255 (2004)
19. Wolters, C.H., Anwander, A., Tricoche, X., Weinstein, D., Koch, M.A., MacLeod, R.S.: Influence of tissue conductivity anisotropy on EEG/MEG field and return current computation in a realistic head model: a simulation and visualization study using high-resolution finite element modeling. Neuroimage **30**, 813–826 (2006)

A Single Input Rule Modules Connected Fuzzy FMEA Methodology for Edible Bird Nest Processing

Chian Haur Jong, Kai Meng Tay and Chee Peng Lim

Abstract Despite of the popularity of the fuzzy Failure Mode and Effects Analysis (FMEA) methodology, there are several limitations in combining the Fuzzy Inference System (FIS) and the Risk Priority Number (RPN) model. Two main limitations are: (1) it is difficult and impractical to form a complete fuzzy rule base when the number of required rules is large; and (2) fulfillment of the monotonicity property is a difficult problem. In this paper, a new fuzzy FMEA methodology with a zero-order Single Input Rule Modules (SIRMs) connected FIS-based RPN model is proposed. An SIRMs connected FIS is adopted as an alternative to the traditional FIS to reduce the number of fuzzy rules required in the modeling process. To preserve the monotonicity property of the SIRMs-connected FIS-based RPN model, a number of theorems in the literature are simplified and adopted as the governing equations for the proposed fuzzy FMEA methodology. A case study relating to edible bird nest (EBN) processing in Sarawak (together with Sabah, known as the world's number two source area of bird nest after Indonesia) is reported. In short, the findings in this paper contribute towards building a new fuzzy FMEA methodology using the SIRMs connected FIS-based RPN model. Besides that, the usefulness of the simplified theorems in a practical FMEA application is demonstrated.

1 Introduction

Failure Mode and Effects Analysis (FMEA) is a popular tool for quality assurance and reliability improvement [1]. In FMEA, a failure mode occurs when a component, system, subsystem, or process fails to meet the designated intent [2, 3].

C. H. Jong · K. M. Tay (✉)
Universiti Malaysia Sarawak, Kota Samarahan, Malaysia
e-mail: kmtay@feng.unimas.my

C. P. Lim
Centre for Intelligent Systems Research, Deakin University, Geelong, Australia

V. Snášel et al. (eds.), *Soft Computing in Industrial Applications*,
Advances in Intelligent Systems and Computing 223, DOI: 10.1007/978-3-319-00930-8_15,

Traditionally, the Risk Priority Number (RPN) model (i.e. Eq. (1)) is used to rank failure modes [2]. Three risk factors, i.e. Severity (S), Occurrence (O) and Detect (D) [3–5], are multiplied to produce an RPN score. S and O are the frequency and seriousness (effects) of a failure mode, and D is the effectiveness to detect a failure mode before it reaches the customer [2].

$$RPN = S \times O \times D \tag{1}$$

Fuzzy concept was incorporated to FMEA methodology to allow uncertainty and imprecise information to be included [1]. Bowles and Paláez suggested using a Fuzzy Inference System (FIS) to aggregate S, O, and D ratings (namely an FIS-based RPN model), instead of a simple product function [5]. An FIS-based RPN was introduced, for the following reasons. (i) It allows expert knowledge and experience to be incorporated [4, 6]; (ii) It is robust against uncertainty and vagueness [7]; (iii) It allows a nonlinear relationship between the RPN score and the three risk factors to be formed [5]; and (iv) The three risk factors can be captured qualitatively, instead of quantitatively [5]. Indeed, the FIS-based RPN model has been successfully applied to a variety of domains, e.g. engine system [1], semiconductor manufacturing [4], maritime [8], and nuclear engineering systems [6].

Nevertheless, an FIS-based RPN model suffers from two major shortcomings *viz*., (i) combinatorial rule explosion [9]; and (ii) monotonicity property fulfillment [10–12]. The first shortcoming suggests that an FIS-based RPN model requires a large number of fuzzy rules, and it is a tedious process to gather a complete fuzzy rule base in practice [4]. The second limitation suggests that it is essential to fulfill the monotonicity property [10–12]. For an FIS-based RPN that fulfills the monotonicity property, $dRPN/dx \geq 0$, where $x \in [S, O, D]$. Fulfillment of the monotonicity property is difficult, and yet important to ensure the validity of the RPN scores [11].

In this paper, the use of a relatively new fuzzy inference framework (i.e. a zero-order Single Input Rule Modules (SIRMs) connected FIS), as an alternative to the traditional FIS model (e.g., Mamdani or Sugeno FIS) in FMEA, is studied. An SIRMs connected FIS-based RPN is proposed, as a solution to the first shortcoming. The SIRMs connected FIS was proposed by Yubazaki et al. [13] for a plural input fuzzy control to reduce the number of fuzzy rules required in FIS modeling. Functional-type SIRMs connected FIS was further proposed by Seki et al. [14], in which the consequences are generalized as functions. As a solution to the second shortcoming, the theorems in [15] are simplified and adopted as a part of the proposed fuzzy FMEA methodology. The theorems [15] suggest that an SIRMs connected FIS is able to satisfy the monotonicity property, if the fuzzy membership functions are *compare-able* (refer to Definition 1), and a set of monotonically-ordered fuzzy rules is available. Hence, in this study, the fuzzy membership functions are designed in such that they are *compare-able*. A constraint is further imposed to ensure that the fuzzy rules follow a monotonic order. An application of the proposed approach to Edible Bird Nest (EBN) processing is further demonstrated. Data and information gathered from several swiftlets farms and EBN production plants in Sarawak, Malaysia are used to assess the efficacy of the proposed approach.

2 Background

2.1 The Zero-Order SIRMs Connected FIS Model

A zero-order SIRMs connected FIS model with n inputs (i.e. $y = f(\bar{x})$), where $\bar{x} = (x_1, x_2, \ldots x_n)$ is considered. It consists of n fuzzy rule modules, as in Fig. 1. Note that SIRM-i represents the i th rule module, where x_i is the sole variable in the antecedent, where $i = 1, 2, \ldots, n$. $R_i^{j_i}$ is the j th rule in SIRM-i, where $j_i = 1, 2, \ldots, m_i$. A fuzzy rule can be viewed as a mapping from $A_i^{j_i}$ to $c_i^{j_i}$, i.e. $R_i^{j_i} : A_i^{j_i} \rightarrow c_i^{j_i} \cdot c_i^{j_i}$ is a variable or a fuzzy singleton.

The output of SIRM-i, i.e. $y_i(x_i)$, is obtained using Eq. (2). The membership function for $A_i^{j_i}$ is denoted as $\mu_i^{j_i}(x_i)$. The zero-order SIRMs connected FIS model further aggregates the output for each rule module with a weighted addition, as in Eq. (3), where ω_i is a numerical value that reflects the relative importance of the i th rule module.

$$y_i(x_i) = \frac{\sum_{j_i=1}^{m_i} \left[\mu_i^{j_i}(x_i) \times c_i^{j_i}\right]}{\sum_{j_i=1}^{m_i} \left[\mu_i^{j_i}(x_i)\right]} \tag{2}$$

$$y = \sum_{i=1}^{n} \left[\omega_i y_i(x_i)\right] \tag{3}$$

2.2 The α-Level Set and Comparable Fuzzy Sets

Definition 1 [15, 16]. Consider two convex fuzzy sets, i.e. A and B, in the **R** domain, with universe of discourse ranged from $\underline{X}$ to $\overline{X}$, as shown in Fig. 2. The α-level sets of A and B are defined as follows.

SIRM − 1:	$\{R_1^{j_1}: if\ x_1\ is\ A_1^{j_1}\ then\ y_1^{j_1} = c_1^{j_1}\}_j^{m_1}$
……..	……………………
SIRM − i:	$\{R_i^{j_i}: if\ x_i\ is\ A_i^{j_i}\ then\ y_i^{j_i} = c_i^{j_i}\}_j^{m_i}$
……..	……………………
SIRM − n:	$\{R_n^{j_n}: if\ x_n\ is\ A_n^{j_n}\ then\ y_n^{j_n} = c_n^{j_n}\}_j^{m_n}$

Fig. 1 Fuzzy rules for a zero-order SIRMs connected FIS model

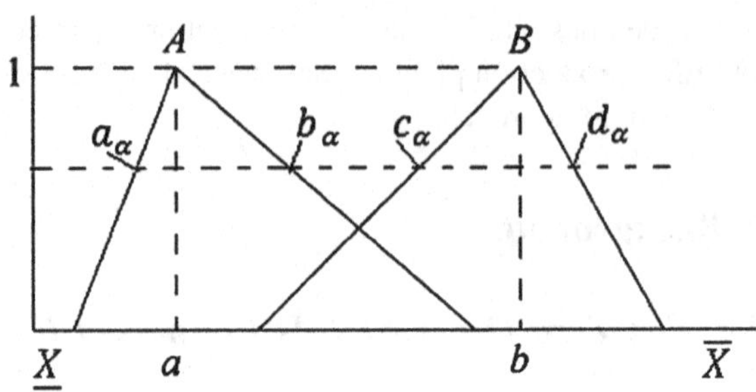

Fig. 2 Comparable fuzzy sets with a fuzzy ordering

$$[A]^{\alpha}= [a_{\alpha}, b_{\alpha}], [B]^{\alpha}= [c_{\alpha}, d_{\alpha}] \tag{4}$$

Definition 2 [15, 16]. A fuzzy ordering A≼B fuzzy ordering exists and there are comparable, if the following condition is satisfied.

$$\alpha_{\alpha} \leq c_{\alpha}, b_{\alpha} \leq d_{\alpha}, \alpha \in [0, 1]$$

2.3 The Monotonicity Property

Let $f(\bar{x})$ denote an n-input function, where $\bar{x} = (x_1, x_2, \ldots x_n) \in X_1 \times X_2 \times \ldots X_n$. The i th input in $\bar{x}$ is represented by x_i, where $x_i \in X_i, i = 1, 2, \ldots n$. A sequence, $\bar{s}$, denotes a subset of $\bar{x}$, whereby x_i is excluded from $\bar{s}$, i.e. $\bar{s} \subset \bar{x}; x_i \notin \bar{s}$. The monotonicity property of $f(\bar{x})$ can be formally written as follows.

Definition 3 An SIRMs-connected FIS model is said to fulfill the monotonicity property between its output, y, and its input, x_i , when y monotonically increases as x_i increases, i.e. $f(\bar{s}, x_i) \geq f(\bar{s}, x_i')$, where $x_i > x_i' \in X_i$.

Theorem 1 By simplifying the theorems in [15, 16], a zero-order SIRMs connected FIS model (as in Sect. 2.1) fulfills the monotonicity property if the following conditions are satisfied.

Condition 1: The fuzzy membership functions for the domain are compare-able, i.e. $A_i^{j_i} \preccurlyeq A_i^{j_i+1}$, $j_i = 1, 2, \ldots, m_i - 1$.

Condition 2: $c_i^{j_i} \leq c_i^{j_i+1}$, $j_i = 1, 2, \ldots, m_i - 1$.

3 The Proposed Fuzzy FMEA Methodology

3.1 The Proposed Fuzzy FMEA Procedure

The proposed SIRMs-based FMEA methodology is summarized in Fig. 3. The details are as follows.

1. Define the scale tables for S, O, and D.
2. Construct the membership functions for each input factor (i.e. S, O and D). *Condition 1* is used as the governing equation.
3. Gather expert knowledge to construct the fuzzy rule base. *Condition 2* is imposed in the construction phase.
4. Construct the SIRMs connected FIS-based RPN model.
5. Study the intent, purpose, goal, objective for the product/process. Generally, it is identified by studying the interaction among the component/process flow diagram and is followed by a task analysis.
6. Identify the potential failures of a product/process, which include problems, concerns, and opportunities for improvement.
7. Identify the consequence of each failure to other components or the other processes, operation customers, and government regulations.
8. Identify the potential root causes of the potential failures.
9. Identify the method/procedure to detect/ prevent the potential failures.
10. Evaluate S, O, and D based on the predefined scale tables.
11. Calculate the fuzzy RPN (FRPN) scores using the SIRMs connected FIS-based RPN model.
12. Make any necessary corrections. Go back to step (5) if needed.
13. End.

3.2 Modelling of the SIRMs Connected FIS-Based RPN Model

The SIRMs connected FIS-based RPN model has three inputs (i.e. S,O and D) and one output, i.e. the FRPN score. It consists of three fuzzy rule modules, as shown in Fig. 4. Note that *SIRM*-1, *SIRM*-2 and *SIRM*-3 are the rule modules for S, O, and D, respectively. The FRPN score is obtained using Eq. (5). The inference outputs of *SIRM*-1, *SIRM*-2 and *SIRM*-3 i.e. $FRPN_1$, $FRPN_2$, and $FRPN_3$, respectively, are combined. Note that ω_S, ω_O, and ω_D are the degree of importance of each rule module towards the combined FRPN score. In this study, we set $\omega_S = \omega_0 = \omega_D = 1/3$, i.e. all three inputs are equally important.

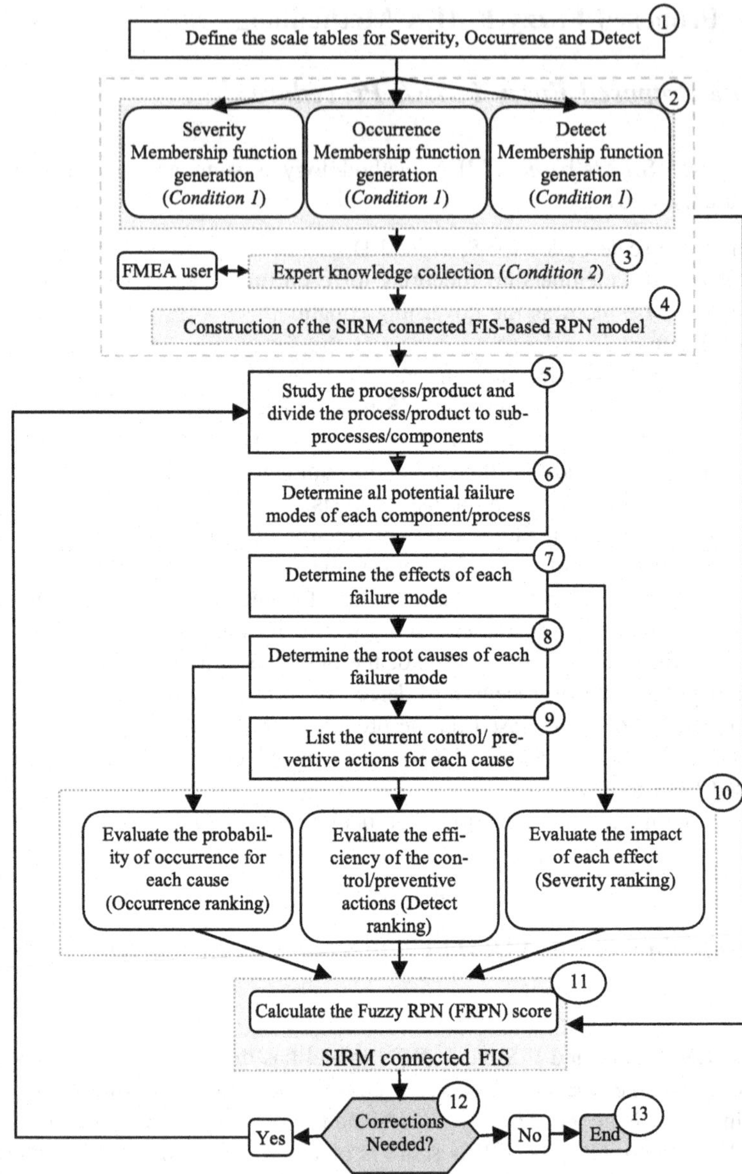

Fig. 3 The proposed fuzzy FMEA methodology

$SIRM-1$:	$\left\{R_1^{j_1}: if\ \mathbf{S}\ is\ A_1^{j_1}\ then\ FRPN_1^{j_1} = c_1^{j_1}\right\}_{j_1=1}^{m_1}$
$SIRM-2$:	$\left\{R_2^{j_2}: if\ \mathbf{O}\ is\ A_2^{j_2}\ then\ FRPN_2^{j_2} = c_2^{j_2}\right\}_{j_2=1}^{m_2}$
$SIRM-3$:	$\left\{R_3^{j_3}: if\ \mathbf{D}\ is\ A_3^{j_3}\ then\ FRPN_3^{j_3} = c_3^{j_3}\right\}_{j_3=1}^{m_3}$

Fig. 4 Fuzzy rules for an SIRMs connected FIS-based RPN model

$$\begin{aligned} FRPN &= f(S,O,D) = \omega_S \times FRPN_1 + \omega_O \times FRPN_2 + \omega_D \times FRPN_3 \\ &= \omega_S \frac{\sum_{j_1=1}^{m_1}\left[\mu_1^{j_1}(S) \times c_1^{j_1}\right]}{\sum_{j_1=1}^{m_1}\left[\mu_1^{j_1}(S)\right]} + \omega_O \frac{\sum_{j_2=1}^{m_2}\left[\mu_2^{j_2}(O) \times c_2^{j_2}\right]}{\sum_{j_2=1}^{m_2}\left[\mu_2^{j_2}(O)\right]} \\ &+ \omega_D \frac{\sum_{j_3=1}^{m_3}\left[\mu_3^{j_3}(D) \times c_3^{j_3}\right]}{\sum_{j_3=1}^{m_3}\left[\mu_3^{j_3}(D)\right]} \end{aligned} \tag{5}$$

To demonstrate the applicability of the proposed model, information from EBN processing is used. The scale tables for S and O (for EBN processing) are summarized in Tables 1 and 2, respectively. The fuzzy membership functions for each S, O, and D are designed according to the scale tables, subject to *Condition* 1 (as in Sect. 2.3).

The membership functions for S and O are depicted in Figs. 5 and 6 respectively. Form these figures, we can see that *Condition* 1 is satisfied.

Table 1 The scale table for severity

Linguistic value	Ranking	Description
Very low	1–2	• Effect of the potential failure mode is not obvious and can be ignored • Excellent yield and excellent product quality
Low	3–4	• Very minor impact on the production yield • Failures cause a minor impact on the EBN food production process control • Products' cosmetic and packaging.
Medium	5–7	• Failure leads to minor security breaches issue of the farm, habitat of the swiftlets is affected by some of the pest and enemy of swiftlets • A drop in the population of swiftlets • A minor impact on the production yield
High	8–9	• Failure leads to serious security breaches issue of the farm • Safety of the swiftlets is threatened • Major impact on the production yield
Very High	10	• Failure will cause an impact on food safety and quality • Compliance to law • Major impact on the reputation of the company and the products • Failure in the yield management

Table 2 The scale table for occurrance

Linguistic value	Ranking	Description
Extremely Low	1	• The failure happens at least once ever
Very Low	2–3	• The failure happens at least once within 6–12 months
Low	4–5	• The failure happens at least once within 1–6 months
Medium	6–7	• The failure happens at least once within 1–30 days
High	8–9	• The failure happens at least once within 1–8 hours
Very High	10	• The failure could happen many times within an hour

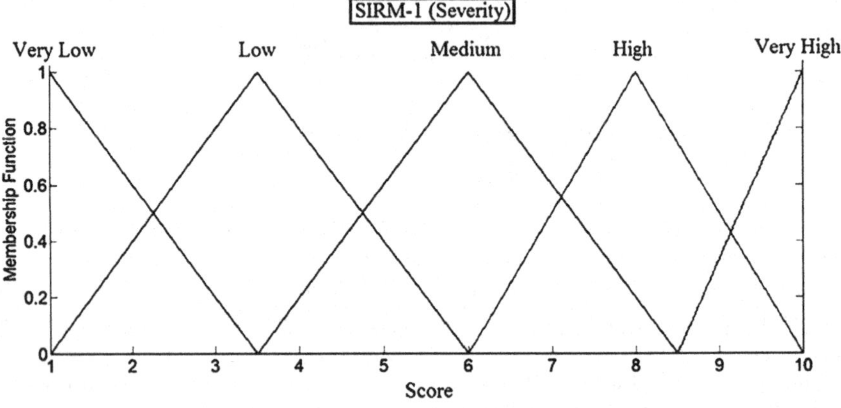

Fig. 5 The membership function for severity

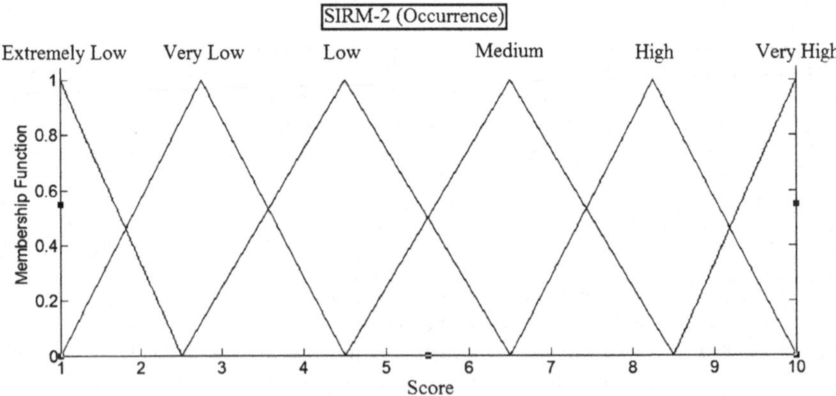

Fig. 6 The membership function for occurrence

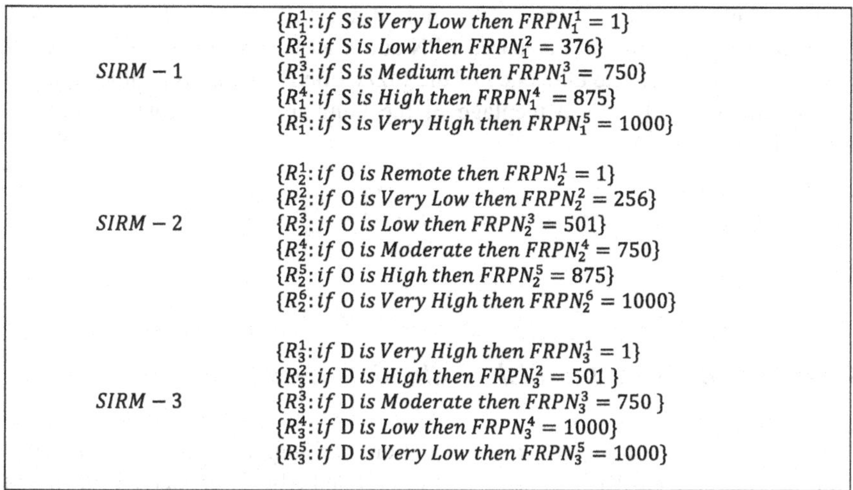

$SIRM-1$
$\{R_1^1: if\ S\ is\ Very\ Low\ then\ FRPN_1^1 = 1\}$
$\{R_1^2: if\ S\ is\ Low\ then\ FRPN_1^2 = 376\}$
$\{R_1^3: if\ S\ is\ Medium\ then\ FRPN_1^3 = 750\}$
$\{R_1^4: if\ S\ is\ High\ then\ FRPN_1^4 = 875\}$
$\{R_1^5: if\ S\ is\ Very\ High\ then\ FRPN_1^5 = 1000\}$
$SIRM-2$
$\{R_2^1: if\ O\ is\ Remote\ then\ FRPN_2^1 = 1\}$
$\{R_2^2: if\ O\ is\ Very\ Low\ then\ FRPN_2^2 = 256\}$
$\{R_2^3: if\ O\ is\ Low\ then\ FRPN_2^3 = 501\}$
$\{R_2^4: if\ O\ is\ Moderate\ then\ FRPN_2^4 = 750\}$
$\{R_2^5: if\ O\ is\ High\ then\ FRPN_2^5 = 875\}$
$\{R_2^6: if\ O\ is\ Very\ High\ then\ FRPN_2^6 = 1000\}$
$SIRM-3$
$\{R_3^1: if\ D\ is\ Very\ High\ then\ FRPN_3^1 = 1\}$
$\{R_3^2: if\ D\ is\ High\ then\ FRPN_3^2 = 501\}$
$\{R_3^3: if\ D\ is\ Moderate\ then\ FRPN_3^3 = 750\}$
$\{R_3^4: if\ D\ is\ Low\ then\ FRPN_3^4 = 1000\}$
$\{R_3^5: if\ D\ is\ Very\ Low\ then\ FRPN_3^5 = 1000\}$

Fig. 7 Fuzzy rules for the SIRMs connected FIS-based RPN model from an EBN processing expert

The fuzzy rules are designed in such a way that *Condition 2* (as in Sect. 2.3) is satisfied. To evaluate the effectiveness of the proposed approach, a set of fuzzy rules gathered from an EBN processing expert, as summarized in Fig. 7, is used. As an example, for SIRM-1, the fuzzy singletons are 1, 376, 750, 875, and 1000. Since $1 \leq 376 \leq 750 \leq 875 \leq 1000$, *Condition 2* is satisfied.

4 Results and Discussion

To evaluate the proposed procedure, real data and information were gathered from several EBN sites in Sarawak, Malaysia. Two swiftlets farms located in Sarikei (2°7′4.13″N and 111°31′16.36″E) and Asajaya (1°32′28″N and110°30′52″E), and two EBN production plants located in Batu Kawah (1°31′10″N and 110°19′37″E) and Baki (1°13′40″N and 110°30′24″E), were visited for data/information collection.

4.1 An Analysis on Fuzzy Rule Reduction

In this case study, there are 5, 6, and 5 membership functions, for S, O, and D, respectively. With the conventional FIS-based RPN model, the number of fuzzy rules required for an FIS-based RPN model is $\prod_{i=1}^{3} m_i$. With the proposed SIRMs connected FIS-based RPN model, the number of fuzzy rules required is $\sum_{i=1}^{3} m_i$.

In short, the numbers of fuzzy rules required are 150 ($5 \times 6 \times 5$) and 16 ($5 + 6 + 5$) for the conventional FIS-based RPN and SIRMs connected FIS-based RPN models, respectively. The efficiency of the fuzzy rule reduction is evaluated using Eq. (6). In this study, the percentage of the number of fuzzy rules reduced is $(150 - 16)/150 \times 100 = 89.33\,\%$.

$$\textit{Percentage of fuzzy rules reduced} = \frac{\prod_{i=1}^{3} m_i - \sum_{i=1}^{3} m_i}{\prod_{i=1}^{3} m_i} \times 100\,\% \qquad (6)$$

4.2 An Analysis on the Risk Evaluation Results

Table 3 summarizes the failure risk evaluation, ranking, and prioritization results with the conventional RPN model (i.e. Eq. (1)) and SIRMs connected FIS-based RPN models. Columns "Sev" (S), "Occ" (O), and "Det" (D) show the S, O, and D ratings that describe each failure mode. The failure risk evaluation and prioritization outcomes from the conventional RPN model are explained in columns "RPN" and "RPN Rank", respectively. For the SIRMs connected FIS-based RPN model, its risk evaluation and prioritization outcomes are summarized in columns "FRPN" and "FRPN Rank", respectively.

Table 3 Failure risk evaluation, ranking, and prioritization results using the RPN and SIRMs connected FIS-based RPN models for EBN processing

Failure mode	Rating scores			RPN models			
	Sev	Occ	Det	RPN	RPN rank	FRPN	FRPN rank
1	1	1	1	1	1	1	1
2	2	1	1	2	2	51	2
3	3	1	1	3	3	101	3
4	3	1	2	6	5	141	4
5	4	1	2	8	6	190	5
6	5	1	1	5	4	201	6
7	5	1	4	20	7	309	7
8	6	1	4	24	8	359	8
9	6	1	6	36	10	417	9
10	3	10	1	30	9	434	10
11	4	9	1	36	10	458	11
12	3	7	4	84	12	471	12
13	4	10	1	40	11	484	13
14	3	8	4	96	14	491	14
15	9	1	10	90	13	642	15

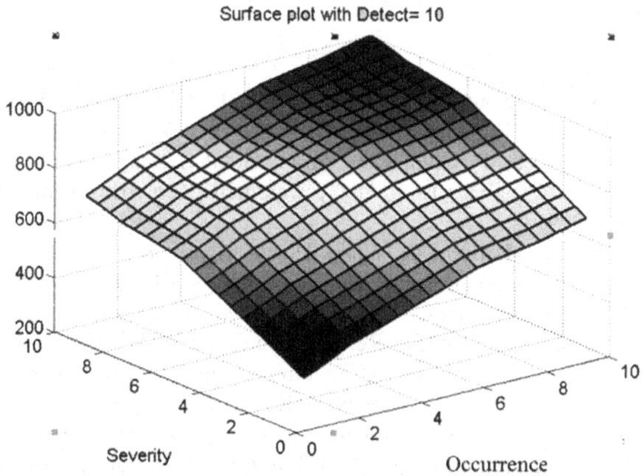

Fig. 8 A surface plot of FRPN versus severity and occurrence (with detect = 10)

From Table 3, the fulfillment of the monotonicity property can be observed. As an example, failure modes #1, #2, and #3 have the same O and D ratings, i.e. 1 and 1, respectively. The S scores are 1, 2, and 3, for each of the failure modes, respectively. With the SIRMs connected FIS-based RPN model, the FRPN scores are 1, 51, and 101, respectively; hence satisfying the monotonicity property. To easily observe the monotonity property of the overall model (as shown in Table 3), a surface plot can be used. Figure 8 depicts a surface plot for FRPN versus S and O, with D = 10. A monotonic surface can be observed easily.

5 Summary

In this paper, a new fuzzy FMEA methodology with a zero-order SIRMs connected FIS-based RPN model has been proposed. The theorems in [15, 16] have been simplified and adopted as a set of governing equations in the proposed fuzzy FMEA methodology. The proposed approach constitutes a new fuzzy FMEA methodology with a reduced fuzzy rule base, which satisfies the monotonicity property. A case study relating to EBN processing has been examined. The results have shown that the proposed approach is able to reduce the number of fuzzy rules effectively and yet, to satisfy the monotonicity property. This paper also contributes to a new application of fuzzy FMEA to food processing (i.e. EBN processing).

For future work, more experiments will be conducted. The applicability of the SIRMs connected FIS-based RPN model in other domains, e.g., transportation systems, will be conducted. Besides that, the fuzzy FMEA methodology can be extended for group risk assessment or decision making. Application of a monotonicity index/test [17, 18] in a zero-order SIRMs connected FIS-based RPN will be further evaluated.

References

1. Xu, K., Tang, L.C., Xie, M., Ho, S.L., Zhu, M.L.: Fuzzy assessment of FMEA for engine systems. Reliab. Eng. Systety Safe. **75**, 17–29 (2002)
2. Wang, Y.M., Chin, K.S., Poon, G.K.K., Yang, J.B.: Risk evaluation in failure mode and effects analysis using fuzzy weighted geometric mean. Expert Syst. Appl. **36**, 1195–1207 (2009)
3. Liu, J., Martínez, L., Wang, H., Rodríguez, R.M., Novozhilov, V.: Computing with words in risk assessment. Int. J. Comput. Int. Syst. **3**, 396–419 (2010)
4. Tay, K.M., Lim, C.P.: Fuzzy FMEA with a guided rules reduction system for prioritization of failures. Int. J. Qual. Reliab. Manag. **23**, 1047–1066 (2006)
5. Bowles, J.B., Pelaez, C.E.: Fuzzy logic prioritization of failures in a system failure mode, effect and criticality analysis. Reliab. Eng. Syst. Safety **50**, 203–213 (1995)
6. Guimarães, A.C.F., Lapa, C.M.F.: Fuzzy FMEA applied to PWR chemical and volume control system. Progr. Nucl. Energy **44**, 191–213 (2004)
7. Pillay, A., Wang, J.: Modified failure mode and effects analysis using approximate reasoning. Reliab. Eng. Syst. Safe. **79**, 69–85 (2003)
8. Yang, Z., Bonsall, S., Wang, J.: Fuzzy rule-based Bayesian reasoning approach for prioritization of failures in FMEA. IEEE Trans. Reliab. **57**, 517–528 (2008)
9. Jin, Y.C.: Fuzzy modeling of high-dimensional systems: complexity reduction and interpretability improvement. IEEE Trans. Fuzzy Syst. **2**, 212–21 (2000)
10. Tay, K.M., Lim, C.P.: On monotonic sufficient conditions of fuzzy inference systems and their applications. Int. J. Uncertain. Fuzz. **19**, 731–757 (2011)
11. Tay, K.M., Lim, C.P.: On the use of fuzzy inference techniques in assessment models: part I— theoretical properties. Fuzzy Optim. Decis. Making **7**, 269–281 (2008)
12. Tay, K.M., Lim, C.P.: On the use of fuzzy inference techniques in assessment models: part II: industrial applications. Fuzzy Optim. Decis. Making **7**, 283–302 (2008)
13. Yubazaki, N., Yi, J.Q., Hirota, K.: SIRMs (single input rule modules) connected fuzzy inference model. J. Adv. Comput. Intell. **1**, 22–29 (1997)
14. Seki, H., Ishii, H., Mizumoto, M.: On the generalization of single input rule modules connected type fuzzy reasoning method. IEEE Trans. Fuzzy Syst. **16**, 1180–1187 (2008)
15. Seki, H., Ishii, H., Mizumoto, M.: On the monotonicity of fuzzy-inference method related to T-S inference method. IEEE Trans. Fuzzy Syst. **18**, 629–634 (2010)
16. Seki, H., Tay, K.M.: On the monotonicity of fuzzy inference models. J. Adv. Comput. Intell. Intell. Inform. **16**, 592–602 (2012)
17. Tay, K.M., Lim, C.P.: Optimization of Gaussian fuzzy membership functions and evaluation of the monotonicity property of Fuzzy Inference Systems. In: IEEE International Conference on Fuzzy Systems, pp. 1219–1224 (2011)
18. Tay, K.M., Lim, C.P., Teh, C.Y., Lau, S.H.: A monotonicity index for the monotone fuzzy modeling problem. In: IEEE International Conference on Fuzzy Systems, pp. 1–8 (2012)

A Novel Energy-Efficient and Distance-Based Clustering Approach for Wireless Sensor Networks

M. Mehdi Afsar and Mohammad-H. Tayarani-N.

Abstract Hierarchical architecture is an effective mechanism to make the Wireless Sensor Networks (WSNs) scalable and energy-efficient. Clustering the sensor nodes is a famous two-layered architecture which is suitable for WSNs and has been extensively explored for different purposes and applications. In this paper, a novel clustering approach called the Energy-Efficient Distance-based Clustering (EEDC) protocol is proposed for WSNs. Selecting the cluster heads in the proposed EEDC is performed based on a hybrid of residual energy and the distances among the cluster-heads. At first, the nodes with the most residual energy are elected and form an initial set of cluster-head candidates. Then the candidates with a suitable distance to other neighbour candidates are elected as the cluster-heads. The proposed algorithm is fast with a low time complexity. The proposed EEDC offers a long lifetime for the network, and at the same time, a proper level of fault tolerance. Different simulation experiments are done on different states and the algorithm is compared to some well-known clustering approaches. The experiments suggest that, in terms of longevity, the EEDC presents better performance than the existing protocols.

1 Introduction

Wireless Sensor Networks (WSNs) have gained worldwide attention in recent years, particularly with the proliferation of Micro-Electro-Mechanical Systems (MEMS) technology which has facilitated the development of smart sensors [1]. Such networks are composed of a large number of tiny sensors that can be used for

---

M. M. Afsar (✉)
Sama Technical Vocational Training College, Islamic Azad University,
Mashhad Branch, Mashhad, Iran
e-mail: m.afsar@qiau.ac.ir

M.-H. Tayarani-N.
University of Southampton, Southampton, UK
e-mail: tayarani@ieee.org

V. Snášel et al. (eds.), *Soft Computing in Industrial Applications*,
Advances in Intelligent Systems and Computing 223, DOI: 10.1007/978-3-319-00930-8_16,

various applications [2]. Hierarchical and multi-layered architectures are suitable approaches for large networks, as they make such networks scalable and fully-connected [3]. Clustering is a well-known two-layered architecture in WSNs which has been extensively explored in the past few years by researchers for different purposes. In the clustering approaches, the nodes are divided into different groups, and then some nodes, called the cluster-heads, are elected to represent the head of each group. The Cluster-heads are then responsible for gathering data from regular nodes, and aggregating and transmitting them to the Base Station (BS).

One of the first major attempts in the area of clustering in homogeneous WSNs, is the LEACH [4] method proposed by Heinzelman et al. The LEACH protocol uses a random probabilistic approach for selecting the cluster-heads which assures a proper load balancing among all the nodes. The HEED [5] protocol is another method which uses a hybrid method for the cluster-head election. The residual energy is used in this method as the first parameter to select an initial set of cluster-head candidates with respect to high residual energy. The second parameter is the node degree or proximity to the neighbours, which is used to break the ties in the cluster head selection process. The ACE [6] protocol is another clustering approach, which is based on emergent algorithm and uses the node degree as the main factor to form clusters with reduced overlapping. Another clustering approach is the FLOC [7] protocol which uses the state transitions for election of cluster-heads. Hierarchical clustering with main focus on longevity of the network lifetime has been investigated in EEHC [8].

Distance based clustering is another method which has been investigated in several works [9–11]. In [9], a distance-vector method called the CODA has been proposed. The CODA considers the distance of the nodes to the BS for the cluster-head election. In [10] a clustering approach based on the residual energy and a local competition for the cluster-head election (EECS) is proposed. The EECS is an extension on the LEACH protocol which supports a distance-based method for the cluster formation. In [11], a clustering approach based on neighbours (EECABN) is proposed. In that work, a novel weight for electing the cluster-heads was introduced which considers several factors, like the distance between the nodes and the BS, the node neighbours, and the residual energy.

Although in recent works the distance between the nodes and the BS is utilized for the cluster formation, the distance among the cluster-heads has not been discussed in the previous research yet. This encouraged us to present a novel clustering approach to provide all the requirements; a method which utilizes both the residual energy and the distance. In this paper, we propose a novel Energy-Efficient Distance-based Clustering, or EEDC, to overcome the problems existing in the previous works. The proposed EEDC algorithm selects the cluster-heads by a hybrid of residual energy and the distance among the cluster-heads.

The rest of this paper is organized as follows. In Sect. 2 the network model and the objective of clustering are discussed. In Sect. 3, we present the proposed EEDC protocol and describe it in details. The performance evaluation and experiments are presented in Sect. 4, where we compare EEDC to some state of the art clustering approaches. And finally the paper is concluded in Sect. 5.

2 Problem Statement

We start with the description of the network model used in this paper, and then the objectives of the proposed clustering scheme are presented.

Network Model: In the network studied in this paper, we assume a homogeneous WSN, where the nodes are uniformly randomly dispersed throughout the area. The nodes in the network are scattered within a square area, where the length of the sides of which is represented by M in this paper. We assume that all the nodes can communicate with the BS with enough energy, and also can use different power levels for communications. Nodes and the BS are stationary and no mobility is supported. Although the BS can be located farther from the field, we investigate a network that the BS is approximately located at the center of the field. All the nodes can communicate with their neighbours located at the cluster range of the nodes, in one single-hop. This is assumed that all the nodes are synchronized at least once at the beginning of each phase. For the sake of simplicity, wireless transmission channel is assumed to be secure; thus, there is no fading, message loss or etc. Also this is assumed that none of the nodes are equipped by the GPS devices and are unaware of their locations in the field. Similar to the work done in [4], we assumed that the entire network operational time is divided into some *rounds*, at the beginning of each of which, the clusters are formed, and in the remainder of the round, the data are gathered, aggregated and transmitted to the BS. In our clustered network, we assumed the cluster-head to be awake in all the time of a round, and the ordinary nodes (or cluster members) can go to sleep except in their time slots.

The Clustering Objectives: The main aim of our proposed EEDC scheme is to achieve better network lifetime in contrast with the previous works, and also to have a *reliable* clustering approach. Accordingly, we have proposed EEDC approach to reach the following objectives: (1) The clustering must be completely distributed, (2) The clustering should be efficient in complexity of message and time, (3) The cluster-heads should be well-distributed across the network, (4) The load balancing should be done well, (5) The clustered WSN should be fully-connected.

3 The EEDC Approach

In this section we describe the proposed EEDC approach. The EEDC approach utilizes adaptive clustering scheme. A clustering scheme is called an adaptive scheme, if over time, the number of clusters varies and the nodes membership evolves [12]. The proposed EEDC approach consists of four main phases; they are the cluster-head election, the cluster formation, the route update and the data transmission. In other words, at first the clusters are formed; then the main operations of network begin. These phases are described in more details in the following.

1. Cluster-head Election Phase: The most important part of each clustering scheme is the cluster-head election. For the cluster-head election, the proposed EEDC uses a hybrid scheme of residual energy and distance among the cluster-heads.

The cluster-head election phase is done in two steps: the *local competition* and the *distance condition*.

Algorithm 1 Distributed pseudo code for cluster-head election in EEDC

```
Cluster-head election phase
  Local competition
    calculate the $P_{CCH}(i)$ probability
    broadcast the CCH-Inf message to the $R_{comp}$ range
    wait for $t_{wait}$ seconds to receive CCH-Inf message
    IF the CCH-Inf message is received THEN
      evaluate the received CCH-Inf messages
      IF $\forall j, P_{CCH}(i) > P_{CCH}(j)$ THEN
        broadcast the CCH-ADV messages to the higher power levels
      ELSEIF $\forall j, P_{CCH}(i) \geq P_{CCH}(j)$ and $\exists j, P_{CCH}(j) = P_{CCH}(i)$ THEN
          IF $\forall j, d_i > d_j$ THEN
            broadcast the CCH-ADV message to the higher power levels
          ELSEIF $\forall j, d_i \geq d_j$ and $\exists j, d_i = d_j$ THEN
              IF $\forall j, node\ ID_i > node\ ID_j$ THEN
                broadcast the CCH-ADV messages to the higher power levels
              ELSE
                  wait $t_{wait}$ seconds for the CH-ADV message
              ENDIF
          ELSE
              wait $t_{wait}$ seconds for the CH-ADV message
          ENDIF
      ELSE
          wait $t_{wait}$ seconds for the CH-ADV message
      ENDIF
    ELSE
        broadcast the CCH-ADV message to all the nodes in $R_c$ range
    ENDIF
  Distance Condition
    IF the current node is a cluster-head candidate THEN
      wait for $t_{wait}$ seconds to receive the CCH-ADV message
      IF the CCH-ADV message received THEN
        evaluate the received messages and calculate the distance
        IF $\forall j, \sqrt{(X_{CCH}(i) - X_{CCH}(j))^2 + (Y_{CCH}(i) - Y_{CCH}(j))^2} \geq D_{thr}$ THEN
          broadcast the CH-ADV messages to all the nodes in the $R_c$ range
        ELSE
            IF $\forall j, P_{CCH}(i) > P_{CCH}(j)$ THEN
              broadcast the CH-ADV message to all the nodes in the $R_c$ range
            ELSEIF $\forall j, P_{CCH}(i) \geq P_{CCH}(j)$ and $\exists j, P_{CCH}(j) = P_{CCH}(i)$ THEN
                IF $\forall j, node\ ID_i > node\ ID_j$ THEN
                  broadcast the CH-ADV message to all the nodes in the $R_c$ range
                ELSE
                    send a Join-Req message to the cluster-head candidate
                ENDIF
            ELSE
                send a Join-Req message to the cluster-head candidate
            ENDIF
        ENDIF
      ELSE
          broadcast the CH-ADV messages to the $R_c$ range
      ENDIF
    ELSE
        wait $t_{wait}$ seconds for the CH-ADV message
    ENDIF
```

1. Local competition: In our proposed method the nodes compete in a competition scheme to be elected as the cluster head candidate. At first the probability of each node being selected as the cluster-head candidate is found. To do so, this probability, $P_{CCH}(i)$, is determined proportional to the remainder energy of node i as,

$$P_{CCH}(i) = \frac{E_{residual}(i)}{E_{initial}(i)}. \tag{1}$$

Each node calculates this probability, and then broadcasts a message to the other nodes, called *CCH-Inf*. This message includes the node ID, the probability $P_{CCH}(i)$, and the node degree d_i (the number of neighbours in a certain range of the node i). In the proposed competition scheme, we define a competition range called R_{comp}; this range should be reasonable, that is it should not be too long to overload the network and should not be too short to increase the number of cluster-head candidate advertisements. Then each node should wait t_{wait} seconds to receive *CCH-Inf* message from all its neighbours. Note that the waiting time, t_{wait}, should not be too short as some nodes may not receive the message *CCH-Inf*, and and it should not be too long as it increases the time complexity. The node i waits for t_{wait} seconds and receives the message *CCH-Inf* from all its neighbours. Then it compares its probability of being elected, $P_{CCH}(i)$, with that of its neighbours. If it found its P_{CCH} greater than P_{CCH} probability of all its neighbours, then it elects itself as cluster-head candidate and then broadcasts a *CCH-ADV* message to higher power levels, described in distance condition step. Otherwise, it waits for *CCH-ADV* message. Although this is very unlikely to see two nodes with the same energy in range of R_{comp} in the network operations, we can use the node degree as the tie break in the beginning of operations when all nodes have the same energy. This reduces cluster-head candidate advertisements and thus the time complexity. That is if two nodes are located in range of R_{comp} from each other, and have the same residual energy, the node with the higher degree is selected as cluster-head candidate. For completeness, if two nodes are located in range of R_{comp} from each other and have the same residual energy and node degree, the node with higher node ID is elected as cluster-head candidate.

2. *Distance condition:* After electing the cluster-head candidates, each cluster-head candidate should pass the distance condition to get elected as a cluster-head as follow. This means that before being elected as the cluster-heads, they have to be at the farthest distance from the other cluster-head candidates possible. Therefore, before being elected, we check if they are far enough from the other cluster-head candidates. Each elected cluster-head candidate should broadcast the information about its node ID and the $P_{CCH}(i)$ probability in *CCH-ADV* message to the higher power levels and waits for t_{wait} seconds. Doing so, the other cluster-head candidates located in those levels can hear this message. Note that this message is ignored via ordinary nodes (i. e., non cluster-head nodes). When a cluster-head candidate hears this message, based on the signal power, it calculates the distance between the sender and itself. If this distance is greater than or equal

to a threshold distance, D_{thr}, it ignores the message, otherwise the receiver cluster-head candidate checks energy level of sender by checking $P_{CCH}(i)$ probability. If it found the energy level of sender greater than its own energy, then the receiver cluster-head candidate becomes an ordinary node and sends a *Join-Req* message to the sender candidate cluster-head. This message can be considered as an acknowledgement to the sender that the mentioned node satisfies the distance condition. After waiting for t_{wait} seconds, if the cluster head candidate receives no *CCH-ADV* message, it elects itself as the cluster-head, and then broadcasts *CH-ADV* message to the nodes in its range of R_c. Similar to the previous step, if two cluster-head candidates are in the same range, the distance between them is less than D_{thr} and they have the same energy, then the cluster-head candidate with the higher node ID is elected as the cluster-head. The pseudo code of the cluster-head election phase of the proposed EEDC is presented in Algorithm 1.

2. Cluster Formation Phase: After the election phase, is the cluster formation phase. In this phase, each ordinary node joins the nearest cluster-head by sending a *Join-Req* message to the cluster-head. To do so, each ordinary node finds its nearest cluster-head, based on the received signal power of the cluster-heads. When The cluster-head receives this message, adds the node specifications to its table of cluster members. Note that there is no limitation on the number of members for the cluster-heads. After the cluster formation phase, the cluster-heads establish a TDMA protocol, and send a time schedule to each of their members. In this protocol, a transmission time slot is assigned to each node, during which the nodes can send their messages [4]. This protocol also ensures communications with no collision.

3. Route Update Phase: In this phase, the paths between the cluster-heads to the BS are specified. In this paper we assume that the BS has no restriction (like energy constraint). At the beginning of this phase, the BS broadcasts a *Route* message to the whole network with enough energy so that all the nodes can hear the message. Based on the power of the received message, each node estimates its distance to the BS. Then the nodes produce a cost proportional to the delay in the time it takes for the message to get to the BS, and adds this cost to the message. Each node then forwards the message to all the nodes locating in the range of R_c around it. The node may receive several messages from different nodes. If multiple messages are received, the node checks the messages and selects the node with the lowest cost as its next-hop node, and forwards the message with the cost to all the nodes in the range of R_c around it. When the costs between every pair of nodes in the network are determined, some shortest path algorithms like *Dijkstra* or *Bellman-Ford* could be utilized for finding the shortest path among the cluster-heads.

4. Data Transmission Phase: When routes are established, the ordinary nodes sense the field, generate some data and send them to the cluster-heads, where the data are gathered and aggregated. Then these data, via the multi-hop path through a predetermined path between the cluster-heads, are transmitted to the BS. Once a round is finished, the clustering process is triggered in the next round. These operations are repeated until the power of all nodes in the network is depleted.

4 Performance Evaluation

In this section we evaluate the performance of the proposed EEDC via several simulation experiments. At first the simulation setup is explained and then the results are presented.

Simulation Setup: We use two scenarios for simulations. In the first scenario, 400 nodes are uniformly and randomly dispersed in a field of size $200 \times 200\,m$. To study the effect of scale on the performance of EEDC, in the second scenario, 800 nodes are uniformly and randomly dispersed in a field of size $400 \times 400\,m$. We assume that the BS is located at the center of the field. Each result is the average over 50 simulations. In our simulations, we have used the first radio model proposed in [4]. The other simulation parameters are summarized in Table 1.

In this paper two sets of simulations are performed. In the first set of simulations, we evaluate the performance of EEDC for different values of factors D_{thr} and R_{comp}. In the second one, we compare the performance of EEDC with the well-known clustering protocols LEACH and HEED. Performance of the EEDC is evaluated with the following metrics:

1. *Network lifetime:* We have used FND (First Node Dies) and HNA (Half of the Nodes Alive) metrics as the network lifetime.
2. *Average dissipated energy*: This is the average energy which is dissipated by all the nodes of the network in each round.
3. *Average energy of the cluster-heads to the members:* This is the average residual energy of all the elected cluster-heads divided by the average residual energy of all members (i. e. non-cluster-head) in each round.

1. First set-EEDC with variant factors: Figure 1a and b represent the average dissipated energy per round for the two scenarios discussed above. The data suggest that for the first scenario, the network has the lowest energy consumption, when the parameters D_{thr} and R_{comp} are set to 20 and 25 m, respectively. Also, for the second scenario, this is achieved when the parameters are set to 20 and 30 m respectively. Hence, we can conclude that for large scale networks, in order to achieve a better energy consumption, larger clusters are more suitable. The results of the dissipated energy indicate that for very small or very large D_{thr} and R_{comp}, the energy consumption in the whole network increases.

The average energy of the cluster-heads divided by the ordinary nodes or members is depicted in Fig. 1c and d. As expected, when the range of R_{comp}

Table 1 Simulation Parameters

Parameter	Value	Parameter	Value
ϵ_{fs}	10 pJ/bit/m^2	E_{elec}	50 nJ/bit
ϵ_{mp}	0.0013 pJ/bit/m^4	E_{DA}	5 nJ/bit/signal
Initial Energy	$2\,J$	d_0	87 m
Data Frame	500 Byte	R_c	25 m

increases, the average energy of the cluster-heads divided by that of the members also increases. This is because when the range increases, a bigger number of nodes participate in the competition. Since it is the nodes with the highest residual energy which are elected as the cluster-head candidates, a bigger number of participants in the competition, means the cluster heads with higher energy. In both scenarios, this ratio is never less than 1; therefore, the *reliability* in EEDC approach is improved. This is because the elected cluster-heads have higher residual energy than their members, thus having enough energy to transmit their data to the next-hop cluster-head.

Figure 1e and f present the experiments performed on the network lifetime. According to these data, FND metric always reaches the best result when R_{comp} is 20 m, and when D_{thr} is 25 m for the first scenario and 30 m for second one. Also, as is seen in Fig. 1g and h, this is the case for the HNA metric.

2. Second set- comparing EEDC with other approaches: In the second set of simulations, we compare our proposed algorithm to the basic clustering protocols namely LEACH and HEED. The energy consumption in each round of the first scenario in the whole network for 30 randomly selected rounds of the three protocols is depicted in Fig. 1i. According to the data presented in this figure, EEDC has less energy consumption than the other two protocols. The main reason for this is the suitable number and distribution of the clusters in the network. As expected, LEACH has a variant energy consumption, relevant to the pendulous number of its clusters in consecutive rounds. Although, HEED has distributed clusters across the network properly, as the number of clusters in HEED is large, energy consumption in the whole network increases. Therefore, EEDC has the lowest energy consumption among the two protocols and HEED has more energy consumption in contrast with the other protocols.

Figure 1j shows the average energy of cluster-heads to the members. It is clear that the EEDC approach has a suitable ratio which never becomes less than 1. However, in the LEACH protocol, this ratio is variant. This means that the *reliability* in LEACH is so unsuitable, as there is always the possibility of the data of a cluster be lost due to the death of a low energy cluster-head. This weakness is fixed in the HEED protocol, where HEED includes the remainder energy of the nodes when electing the *tentative* cluster-heads.

The network lifetime for three protocols is depicted in Fig. 1k and l. The data indicate that in terms of FND metric, the EEDC protocol outperforms both the LEACH and HEED protocols. In the first scenario, EEDC improves the FND metric in contrast with the LEACH and HEED protocols by about 95 and 125 %. This is about 100 and 150 % for the second scenario respectively. This improvement on FND metric is because the EEDC method elects the nodes with the highest residual energy as the cluster-heads. Also, in this approach, the load balancing in the network is performed properly, which provides a longer time between the beginning of the operations until the time the first node dies (FND). The network lifetime with HNA metric is depicted in Fig. 1(l). Again, the proposed EEDC protocol has a better network lifetime with HNA metric. The EEDC

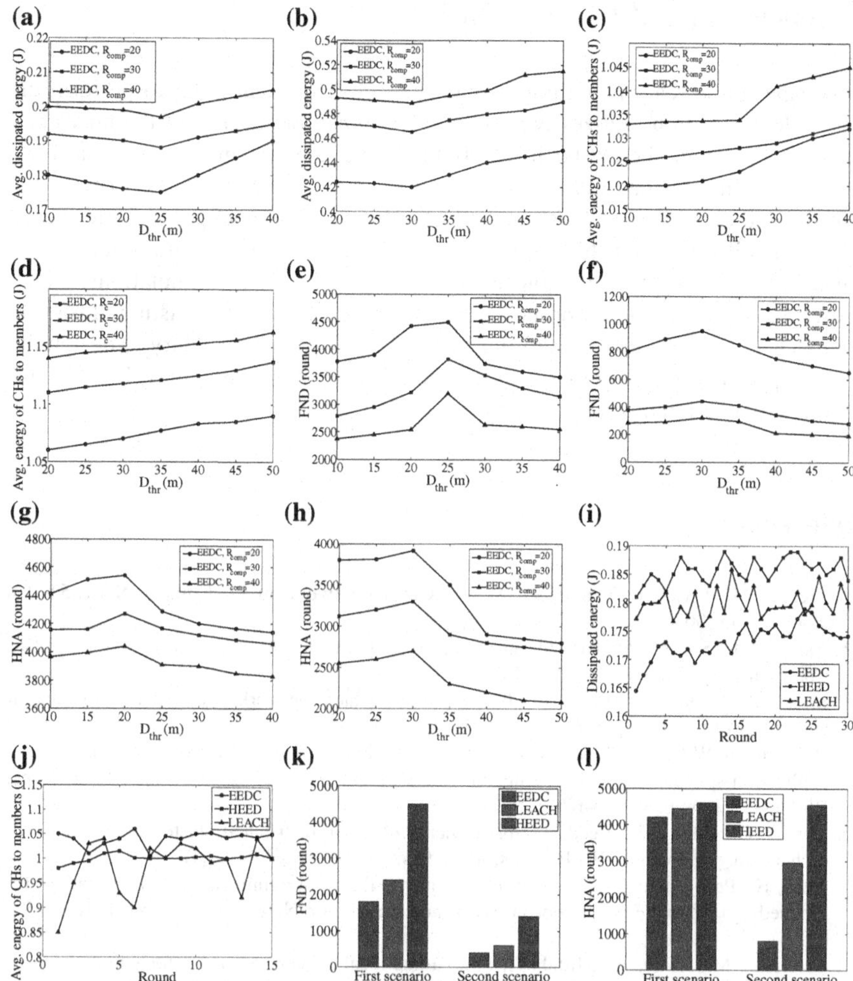

Fig. 1 Simulation results of proposed EEDC approach (**a**) First set. Dissipated energy in first scenario. (**b**) First set. Dissipated energy in second scenario. (**c**) First set. Avg. energy of the cluster-heads to the members in first scenario. (**d**) First set. Avg. energy of the cluster-heads to the members in second scenario. (**e**) First set. FND metric in first scenario. (**f**) First set. FND metric in second scenario. (**g**) First set. HNA metric in first scenario. (**h**) First set. HNA metric in second scenario. (**i**) Second set. Dissipated energy in first scenario. (**j**) Second set. Avg. energy of the cluster-heads to the members in first scenario. (**k**) Second set. FND metric. (**l**) Second set. HNA metric

improves the HNA metric in contrast with the LEACH and HEED by about 5 and 10 % in the first scenario, and about 45 and 100 % in the second scenario, respectively.

5 Conclusion and Future Work

This paper proposes a novel clustering approach, which, in order to perform the cluster-head election, utilizes two factors: the remaining energy of the nodes and the distance among the cluster-heads. In this paper, the remaining energy has been used as the first factor to elect the nodes with the highest residual energy as the cluster-head candidates. In this method also the distance among the cluster-heads has been utilized to reach a well-distributed clustered WSN with suitable size clusters. Performance evaluation has been done via different simulations, which shows the advantage of the proposed EEDC approach over the existing clustering approaches. In the future, in addition to the energy-efficiency, we try to design the EEDC approach in a way that it meets the other WSN requirements, like the full coverage of the monitored area.

References

1. Yick, J., Mukherjee, B., Ghosal, D.: Wireless sensor network survey. Comput. Netw. **52**(12), 2292–2330 (2008)
2. Akyildiz, I., Su, W., Sankarasubramaniam, Y., Cayirci, E.: A survey on sensor networks. Commun Mag IEEE **40**(8), 102–114 (2002)
3. Kleinrock, L., Kamoun, F.: Hierarchical routing for large networks, performance evaluation and optimization. Comput. Netw. **1**(3), 155–174 (1977)
4. Heinzelman, W.B., Ch, IEEE, Chandrakasan, A.P., Balakrishnan, M.H., Balakrishnan, H., An application-specific protocol architecture for wireless microsensor networks. IEEE Trans. Wireless Commun. **1**, 660–670 (2002)
5. Younis, O., Fahmy, S.: Heed: a hybrid, energy-efficient, distributed clustering approach for ad hoc sensor networks. IEEE Trans. Mob. Comput. **3**, 366–379 (2004)
6. Chan, H., Perrig, A.: Ace: an emergent algorithm for highly uiniform cluster formation. In: Proceedings of the First European Workshop on Sensor Networks, pp. 154–171, EWSN (2004)
7. Demirbas, M., Arora, A., Mittal, V.: Floc: a fast local clustering service for wireless sensor networks. In: Workshop on Dependability Issues in Wireless Ad Hoc Networks and Sensor Networks, DIWANS/DSN (2004)
8. Bandyopadhyay, S., Coyle, E.: An energy efficient hierarchical clustering algorithm for wSensor networks. In: INFOCOM: Twenty-Second Annual Joint Conference of the IEEE Computer and Communications. vol. 3, PP. 1713–1723, IEEE Societies (2003)
9. Lee, S., Yoo, J., Chung, T.: Distance-based energy efficient clustering for wireless sensor networks .In 29th Annual IEEE International Conference on Local Computer Networks,Vol. 2004, PP. 567–568 (2004)
10. Ye, M., Li, C., Chen, G., Wu. J., Al, M. Y. E.: Eecs: an energy efficient clustering scheme in wireless sensor networks. In: Proceedings of the IEEE International Performance Computing and Communications Conference, pp. 535–540 (2005)
11. Wei, Z.: Energy efficient clustering algorithm based on neighbors for wireless sensor networks. J. Shanghai Univ. (Engl Ed) **15**, 150–153 (2011)
12. Abbasi, A., Younis, M.: A survey on clustering algorithms for wireless sensor networks. Comput. Commun. **30**(14–15), 2826–2841 (2007)

Characterization of Coronary Plaque by Using 2D Frequency Histogram of RF Signal

Satoshi Nakao, Kazuhiro Tokunaga, Noriaki Suetake and Eiji Uchino

Abstract Tissue characterization of plaque in coronary arteries by using histogram-based frequency spectrum in window is proposed. Radio frequency (RF) signals, observed by the intravascular ultrasound catheter rotating in the coronary artery, are used for the tissue characterization. The conventional methods only use the frequency spectrum at the point of tissue of concern. However, in the proposed method the 2D histogram, concerning frequency and spectral band intensity, created from the window matrix of RF signals, is employed. The accuracy of the tissue charactrization has been improved compared with the conventional methods which only use the statistical information of the frequency spectrum.

1 Introduction

A heart disease, such as angina or myocardial infarction, is caused by the plaque accumulated on an inner surface of a coronary artery. In particular, fibrous and lipid tissues in the plaque are intimately-involved in the heart disease. If the lipid tissue is covered with a thin and brittle fibrous tissue, this plaque is easy to break.

S. Nakao (✉) · N. Suetake · E. Uchino
Yamaguchi University, 1677-1 Yoshida, Yamaguchi 753-8512, Japan
e-mail: s018vc@yamaguchi-u.ac.jp

N. Suetake
e-mail: suetake@yamaguchi-u.ac.jp

E. Uchino
e-mail: uchino@yamaguchi-u.ac.jp

K. Tokunaga · E. Uchino
Fuzzy Logic Systems Institute, 680-41 Kawazu, Iizuka 820-0067, Japan
e-mail: tokunaga@flsi.or.jp

V. Snášel et al. (eds.), *Soft Computing in Industrial Applications*,
Advances in Intelligent Systems and Computing 223, DOI: 10.1007/978-3-319-00930-8_17,

The lipid tissue drained from the broken fibrous tissue causes a thrombus. The thrombus prevents a blood flow. The tissue characterization and estimation of plaque constitution are thus very important to predict the breakdown of plaque.

As a tissue characterization technique for finding the fibrous and lipid tissues, an intravascular ultrasound (IVUS) method [1] is often used in a medical diagnosis. In the IVUS method, the high frequency ultrasound signals are transmitted from a probe attached to a catheter rotating in a blood vessel. The reflected ultrasound signals are received again by the probe. Those signals are called radio frequency (RF) signals.

The medical doctor makes a diagnosis of the tissue properties by evaluating a B-mode image, which is made from those RF signals. The tissue characterization from B-mode image, however, is difficult, because the B-mode image is very grainy.

For those reasons, many classification algorithms and signal processing methods for tissue characterization have been proposed. In the VH-IVUS method proposed by Nair et al. [2], the tissue characterization is made by using the classification tree based on a dictionary of the frequency spectrum features obtained from the RF signals. The iMap method proposed by Sathyanarayana et al. [3] also uses the dictionary of the frequency spectrum features.

In addition, Kubota et al. [4] and Uchino et al. [5] have newly proposed the tissue characterization method based on the frequency spectrum features by using multiple k-nearest neighbor method. The frequency spectrum of the RF signals is generally used in many conventional works, because the differences of tissue are reflected on the RF signals.

In the conventional methods, the frequency spectrum of only around the target point of tissue has been considered. However, the plaque tissues are spread spatially, and thus, it is important to focus on a spatial feature of the RF signals rather than considering only the target point of tissue.

In this work, we propose a new tissue characterization technique considering the spatial and statistical characteristics of RF signals. In the proposed method, a local region matrix, called a window matrix, is extracted from a matrix that is composed of RF signals observed at the entire circumference of a section of a coronary artery. And a 2-dimensional histogram of the frequency spectrum is then generated from the window matrix. In the tissue characterization, k-nearest neighbor method (k-NN), in which 2D histogram features are employed as the prototype vectors, is applied.

By using the 2D histogram, it is possible to make a tissue characterization considering the spread of the frequency spectrum on a space of RF signals. It is expected that the accuracy of tissue characterization is improved compared to the conventional methods in which the single frequency spectrum is used.

In this paper, we have made a comparative study in tissue characterization to demonstrate the effectiveness of the proposed method. The promising experimental results are given.

2 Methodology

This chapter describes briefly the intravascular ultrasound (IVUS) method, 2D histogram of the frequency spectrum, and k-nearest neighbor (k-NN) method.

2.1 IVUS Method

The ultrasound probe attached at the tip of the catheter is inserted into a coronary artery, and those RF signals are processed as follows. The ultrasound signal is transmitted and received at the probe while the catheter is rotating (Fig. 1a). The ultrasound signal observed by the probe is called a radio frequency (RF) signal. The intensity of the reflected RF signal from the tissue depends on the characteristics of the tissue and also on the location of the probe.

The RF signals are observed in all directions in the coronary artery. The absolute value of the sampled RF signal is first taken, and its envelop, and then finally its logarithmic value is calculated. This transformed signal is converted into 8-bit luminosity values. Those luminosity values in all radial directions are used to obtain a tomographic cross sectional image of a coronary artery. This image is called a B-mode image (Fig. 1b). A medical doctor diagnoses the condition of a coronary artery by seeing this B-mode image. However, this B-mode image is very grainy as shown in Fig. 1b.

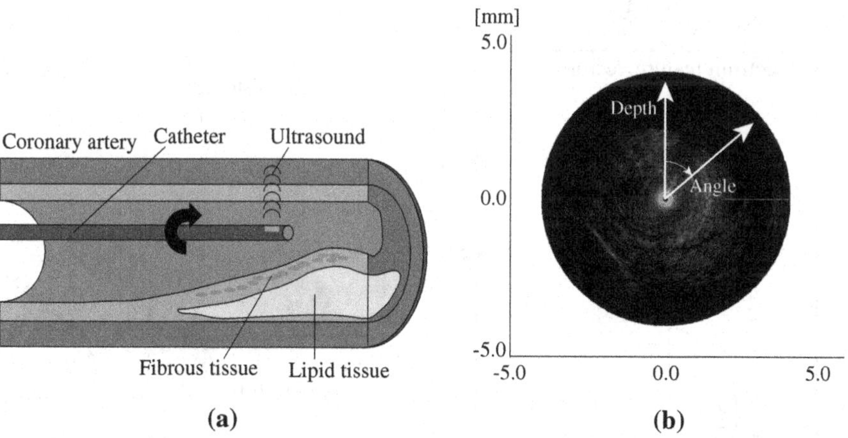

Fig. 1 **a** An ultrasound probe attached to the distal end of a catheter. The ultrasound signal is transmitted from the probe and the reflected signal from the tissue is observed again by the probe. **b** An example of the B-mode image obtained by the IVUS method. This is a real time ultrasound cross-sectional image of a blood vessel where a catheter probe is currently rotating

2.2 2D Histogram

An RF matrix, where RF signals at every angle are aligned in the column, is prepared in order to create the 2D histogram. In RF matrix, the row and the column correspond to the angle of each RF signal and the depth of the vessel, respectively (see Fig. 2).

The region around the target tissue point in which the tissue characterization is carried out is extracted from the RF matrix as shown in Fig. 2. In this paper, the extracted region is defined as "window." The size of the window is arbitrary.

The 2D histogram relevant to the frequency and to the spectrum intensity is created from the window. Specifically, The short time signals extracted from the window are processed with the fast Fourier transform (FFT). The frequency spectrum vectors are obtained for the windows. The 2D histogram is created from those frequency spectrum vectors. The 2D histogram is considered to be a feature representing the tissue at the target point.

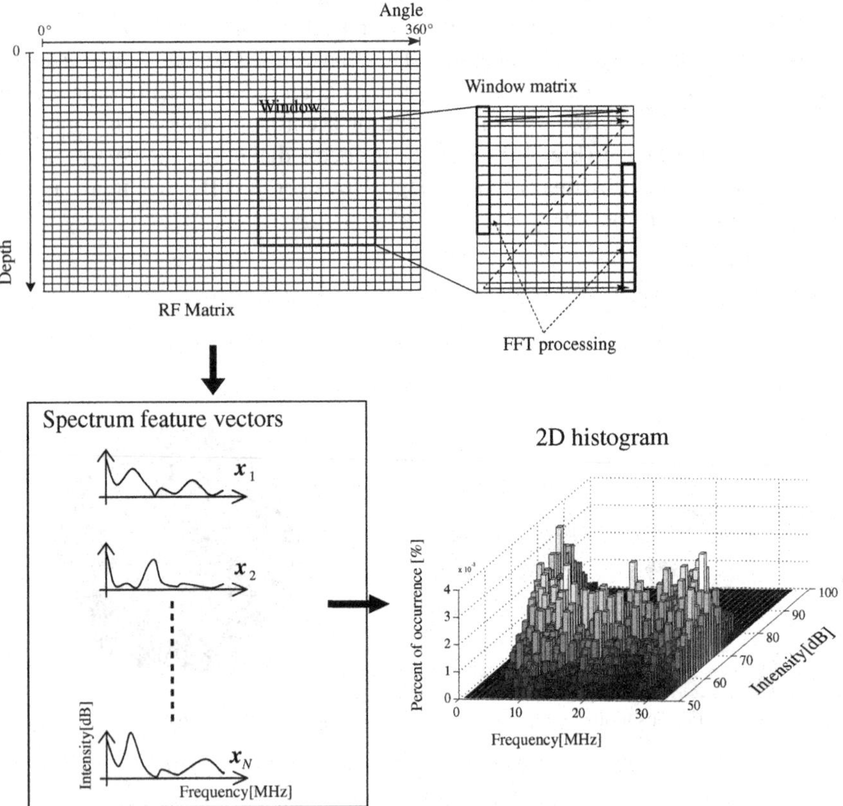

Fig. 2 A flow of creating a 2D histogram. First, the window matrix is extracted from RF matrix. Second, the spectrum feature vectors are calculated by Fast Fourier Transform for window matrix. Finally, the 2D histogram is created from the spectrum feature vectors

2.3 k-Nearest Neighbor Method

A k-NN (k-nearest neighbor method) makes a statistical classification based on the training prototype neighborhood vectors in a feature space [6]. The algorithm is briefly described in the following.

Suppose that the feature vectors $w_i(i = 1, 2, ..., N)$ are given, and let the class label ω_i of each feature vector be known. The feature vectors are used as the training prototype vectors.

When the input vector x, whose class label is unknown, is given to k-NN, the class label of the input vector is determined as follows:

$$l = \arg_{\omega} max \sum_{i=1}^{N} \delta(x; w_i|\omega_i = \omega), \tag{1}$$

$$\delta(\vec{x}) = \begin{cases} 1 & if\ \|\vec{w}_i - \vec{x}\| \le r(k), \\ 0 & otherwise, \end{cases} \tag{2}$$

where $r(k)$ represents the Euclidean distance between the input vector x and the k-th nearest prototype vector.

In determining the class label for the input vector, the prototype vectors within the k-th nearest neighbor are determined, according to the distance of Eq. (1). After that, the class label of the input vector is determined by a majority vote for the class labels of the k nearest neighbor prototype vectors by Eq. (2).

3 Tissue Characterization by Proposed Method

In the tissue characterization by the proposed method, the 2D histograms are employed as the prototype vectors w_i in the k-NN. The 2D histogram is transformed to $M_{Freq} \times M_{Int}$ dimensional vector, where M_{Freq} and M_{Int} are numbers of bins for the frequency and the spectrum intensity of the 2D histogram, respectively.

The class label of each prototype vector is known from the findings of a medical doctor examining the corresponding dyed tissue looking through a microscope.

In this work, the fibrous tissue, the fibrofatty tissue, the lipid tissue are classified. Specifically, the 2D histograms are created for each of the corresponding RF signal, i.e., for the fibrous tissue, for the fibrofatty tissue, and for the lipid tissue. Figure 3 shows the examples of 2D histogram for each tissue. The representative vectors that are used as the prototype vectors are selected manually or selected according to the simple rule in this work.

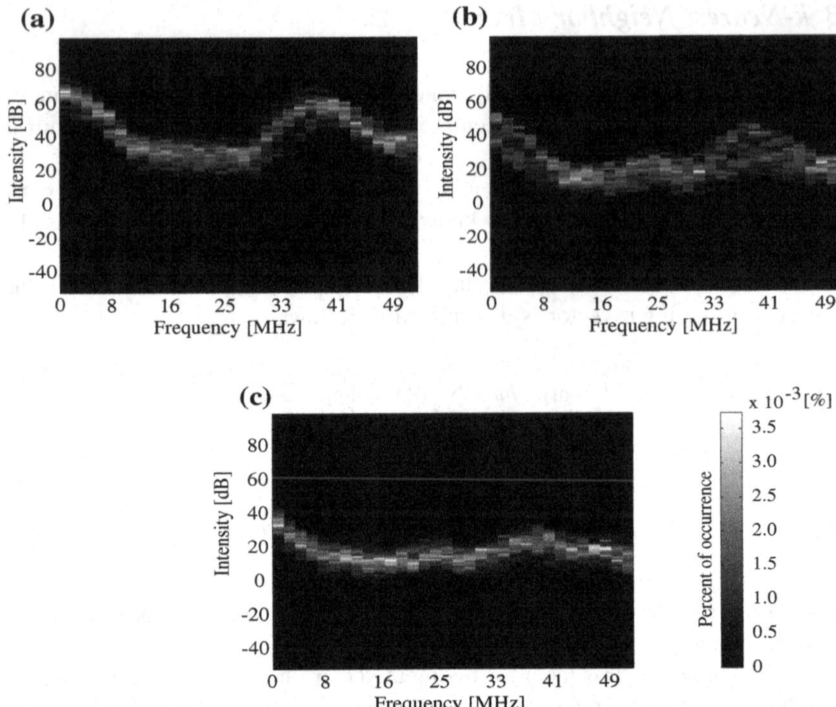

Fig. 3 Examples of 2D histogram for each tissue. **a** Fibrous tissue, **b** fibrofatty tissue, and **c** lipid tissue. Percent of occurrence is illustrated with gray scale color. High frequency is represented by white

4 Experiments

The performances of the proposed method are compared with those of the conventional k-NN method.

4.1 Experimental Settings

In the experiments, RF signals observed from two different sections of the coronary artery are prepared. One is used as the training data, and the other is used as the test data for classification.

The width of window for the Fourier transformation is 64 points for both methods. Every frequency spectrum is log-transformed. The width and height of window for 2D histogram are 21 and 41, respectively. As for the width of bin for 2D histogram, 10 MHz for the frequency axis and 10 dB for the spectral intensity axis.

The prototype vectors are selected manually from each tissue in the training stage. The numbers of selected prototype vectors from the fibrous, fibrofatty and lipid tissues are 800, 600 and 200, respectively.

4.2 Experimental Results

Figure 4 shows the results for the training data. Figure 4a shows the tissue composition given by a medical doctor through a microscope analysis. In other words, it is the desirable result of tissue characterization. Each tissue, i.e. fibrous tissue, fibrofatty tissue, and lipid tissue, is shown in gray scale color. Figure 4b and c show the tissue characterization results by the conventional and the proposed methods, respectively. In Fig. 4b, the conventional method gives false tissue characterization at many points. The false tissue characterization may lead to a doctor's misdiagnosis. In contrast, the result by the proposed method of Fig. 4c is almost equal to Fig. 4a, which is a desirable result.

Figure 5 shows the results for the test data. Figure 5a, b and c show the desirable result, the results by the conventional method and the results by the proposed method, respectively. In the results of the conventional method of Fig. 5b, it is difficult to specify the areas of fibrofatty and lipid tissues. However, the proposed method gives it better.

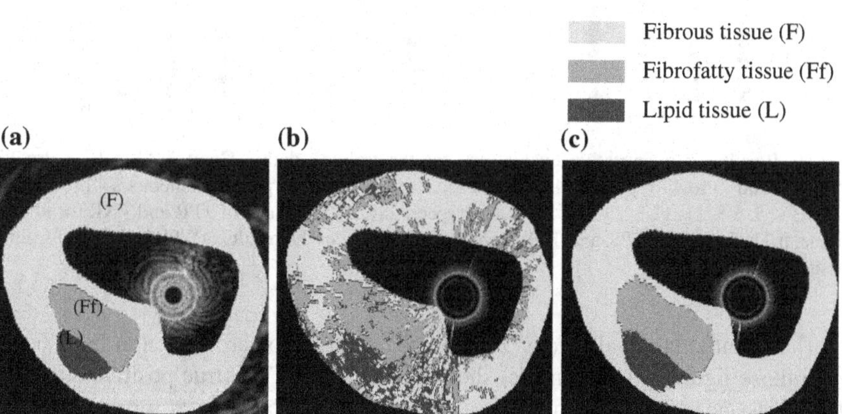

Fig. 4 Tissue characterization results by the conventional method and by the proposed method for the training data. **a** The tissue composition given by a medical doctor by examining the dyed tissue looking through a microscope. **b** The tissue characterization results by the conventional method. The numbers of prototype vectors for fibrous, fibrofatty and lipid tissues are 800, 600, and 200, respectively. The prototype vectors are selected at random. **c** The tissue characterization results by the proposed method. The numbers of prototype vectors for fibrous, fibrofatty and lipid tissues are 800, 600, and 200, respectively. The prototype vectors are selected at random

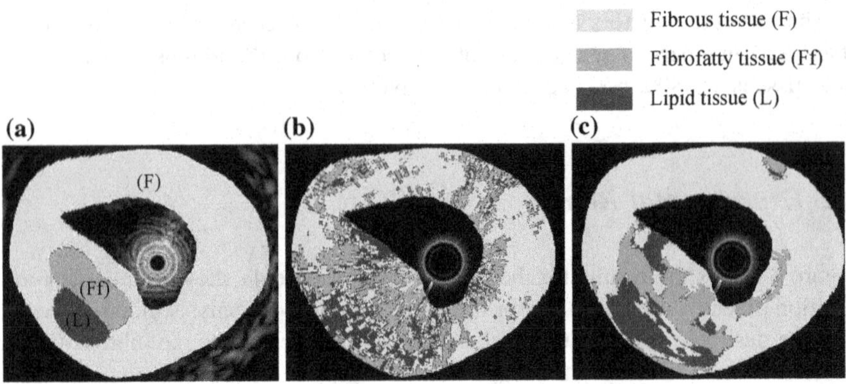

Fig. 5 Tissue characterization results by the conventional method and by the proposed method for the test data. **a** The tissue composition given by a medical doctor by examining the dyed tissue looking through a microscope. **b** The tissue characterization results by the conventional method. The numbers of prototype vectors for fibrous, fibrofatty and lipid tissues are 800, 600, and 200, respectively. The prototype vectors are selected at random. **c** The tissue characterization results by the proposed method. The numbers of prototype vectors for fibrous, fibrofatty and lipid tissues are 800, 600, and 200, respectively. The prototype vectors are selected at random

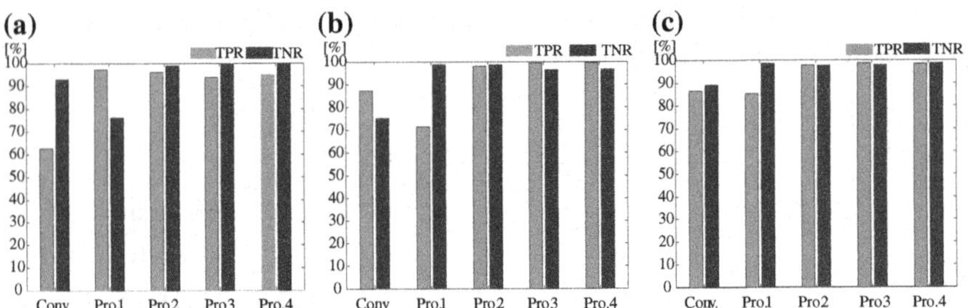

Fig. 6 Results of quantitative evaluations for the training data. “Conv.,” “Pro.1,” “Pro.2,” “Pro.3,” and “Pro.4” represent the conventional method and the proposed methods with window sizes 3 × 5, 5 × 11, 11 × 21, and 21 × 41, respectively. **a** The results of TPR and TNR for fibrous tissue. **b** The results of TPR and TNR for fibrofatty tissue. **c** The results of TPR and TNR for lipid tissue

The quantitative evaluations of tissue characterization are shown in Figs. 6 and 7. In those figure, (a), (b) and (c) show the results of TPR (true positive rate) and TNR (true negative rate) for fibrous, fibrofatty and lipid tissues, respectively. In each figure, “Conv.,” “Pro.1,” “Pro.2,” “Pro.3,” and “Pro.4” represent the conventional method and the proposed methods with window sizes 3 × 5, 5 × 11, 11 × 21, and 21 × 41, respectively. From those results, it is seen that the window size affects the accuracy of tissue characterization, and the proposed method with window size 21 × 41 gives the best result.

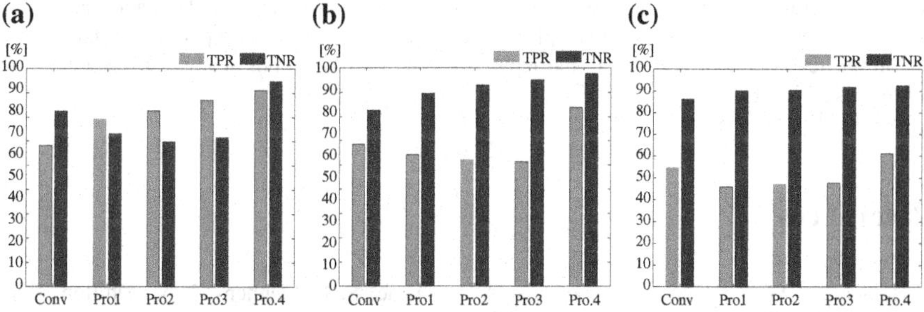

Fig. 7 Results of quantitative evaluations for the test data. "Conv.," "Pro.1," "Pro.2," "Pro.3," and "Pro.4" are the same as in Fig. 6. **a** The results of TPR and TNR for fibrous tissue. **b** The results of TPR and TNR for fibrofatty tissue. **c** The results of TPR and TNR for lipid tissue

5 Discussions

It is seen from the experimental results in Chap. 4 that the proposed method provides better accuracy of tissue characterization than the conventional method. As for this reason, it is considered that the distribution of frequency spectrum in the window is concerned. In Fig. 3a and c, i.e., the 2D histograms of fibrous and lipid tissues, the intensities around 40 MHz are biased toward low or high. By contrast, in Fig. 3b, i.e., the 2D histogram of fibrofatty tissue, the intensities around 40 MHz are dispersed from low to high. The differences in distribution of intensities yielded a good tissue characterization. In the conventional method, the difference in distribution of intensities has not been considered.

In addition, the window size and the bin range of histogram affect the results. For example, if the window size is small, the information around the tissue point of concern is not extracted enough. By contrast, if big, the histogram includes various features of tissue.

It is also important to adjust the bin range of histogram. If a small bin range is employed, tissue characterization becomes sensitive to noise. By contrast, if big, the tissue characterization becomes insensitive to noise.

The good results have been obtained by the proposed method. It is not enough, however, to be used right away in a medical practice.

6 Conclusions

We have proposed in this paper to use a histogram of frequency spectrum vector with window for tissue characterization of coronary plaque. In the experiments, the proposed method have given the better results than the conventional method. A further direction of this work will be to increase the number of the training data, and to think out the method to select automatically the best prototype vectors from the training data.

Acknowledgments Many thanks are due to Dr. T. Hiro for providing the IVUS data. Thanks are also due to Dr. G. Vachkov for his helpful discussions. This work was supported by the Grant-in-Aid for Scientific Research (B) of the Japan Society of Promotion of Science (JSPS) under the contract No.23300086.

References

1. Hodgson, J.M., Graham, S.P., Savakus, A.D., Dame, S.G., Stephens, D.N., Dhillon, P.S., Brands, D., Sheehan, H., Eberle, M.J.: Clinical percutaneous imaging of coronary anatomy using an over-the-wire ultrasound catheter system. Int. J. Card. Imaging **4**, 187–193 (1989)
2. Nair, A., Margolis, M.P., Kuban, B.D., Vince, D.G.: Automated coronary plaque characterisation with intravascular ultrasound backscatter: ex vivo validation. Euro. Intervention **3**, 113–120 (2007)
3. Sathyanarayana, S., Carlier, S., Li, W., Thomas, L.: Characterisation of atherosclerotic plaque by spectral similarity of radiofrequency intravascular ultrasound signals. Euro. Intervention **5**, 133–139 (2009)
4. Kubota, R., Uchino, E., Suetake, N.: Hierarchical k-nearest neighbor classification using feature and observation space information. IEICE Electron. Exp. **5**, 114–119 (2008)
5. Uchino, E., Suetake, N., Kubota, R., Koga, T., Hashimoto, G., Hiro, T.: An roc performance validation of hierarchical k-nearest neighbor classifier applied to tissue characterization using ivus-rf signal. In: International Workshop on Nonlinear Circuits and Signal Processing, pp. 333–336 (2009)
6. Dasarathy, B.: Nearest neighbor (NN) norms: nn pattern classification techniques. IEEE Computer Society Press tutorial. IEEE Computer Society Press, Washington (1991)

Face Recognition Using Convolutional Neural Network and Simple Logistic Classifier

Hurieh Khalajzadeh, Mohammad Mansouri and Mohammad Teshnehlab

Abstract In this paper, a hybrid system is presented in which a convolutional neural network (CNN) and a Logistic regression classifier (LRC) are combined. A CNN is trained to detect and recognize face images, and a LRC is used to classify the features learned by the convolutional network. Applying feature extraction using CNN to normalized data causes the system to cope with faces subject to pose and lighting variations. LRC which is a discriminative classifier is used to classify the extracted features of face images. Discriminant analysis is more efficient when the normality assumptions are satisfied. The comprehensive experiments completed on Yale face database shows improved classification rates in smaller amount of time.

1 Introduction

Effective methods in the extraction of features and classification methods for the extracted features are the key factors in many real-world pattern recognition and classification tasks [1]. Neural networks such as multilayer perceptron (MLP) are considered as one of the simplest classifiers that can learn from examples. An MLP can approximate any continuous function on a compact subset to any desired accuracy [2].

H. Khalajzadeh (✉) · M. Mansouri · M. Teshnehlab
K. N. Toosi University of Technology, Tehran, Iran
e-mail: hurieh.khalajzadeh@gmail.com

M. Mansouri
e-mail: mohammad.mansouri@ee.kntu.ac.ir

M. Teshnehlab
e-mail: teshnehlab@eetd.kntu.ac.ir

V. Snášel et al. (eds.), *Soft Computing in Industrial Applications*,
Advances in Intelligent Systems and Computing 223, DOI: 10.1007/978-3-319-00930-8_18,

The performance of such classifiers depends strongly on an appropriate preprocessing of the input data [3]. In the traditional models a hand designed feature extractor congregates relevant information from the input. Designing the feature extractor by hand requires a lot of heuristics and, most notably, a great deal of time [4]. Moreover, it is not apparent what illustrates an optimal preprocessing or if there even exists an optimal solution. Sometimes it is required the input dimension to be reduced as far as possible [3]. The main deficiency of MLP neural networks for high dimensional applications such as image or speech recognition is that they offered little or no invariance to translations, shifting, scaling, rotation, and local distortions of the inputs [4].

Convolutional neural networks [5] were proposed to address all problems of simple neural networks such as MLPs. CNNs are feed-forward networks with the ability of extracting topological properties from the unprocessed input image without any preprocessing needed. Thus, these networks integrate feature selection into the training process [3]. Furthermore, CNNs can recognize patterns with extreme variability, with robustness to distortions and simple geometric transformations like translation, scaling, rotation, squeezing, stroke width and noise [6, 7]. Different versions of convolutional neural networks are proposed in the literature.

In this paper, two types of discriminative learning methods: Logistic regression classifier and Convolutional Networks are combined. LRCs [8, 9] could recognize unseen data in same accuracy as popular discriminative learning methods such as support vector machines (SVMs) [10] in a very significant less time. To investigate its efficiency and learning capability, a training algorithm is developed using back propagation gradient descent technique. The proposed architecture was tested by training the network to recognize faces.

The rest of this paper is organized as follows. Section 2 discusses an introduction to CNNs and Logistic Regression method. The proposed combined structure base on CNN and LRC is outlined in Sect. 3. The Yale dataset [11] which is used in experiments is introduced in Sect. 4. Section 5 reports the simulation results completed on the introduced database. Finally, in Sect. 6 conclusive remarks are resumed.

2 Logistic Regression, Convolutional Neural Networks

In this section, we briefly describe the logistic regression method and the Convolutional neural networks. The focus is to inspect their internal structures to provide insights into their respective strengths and weaknesses on the present vision task. This analysis will lead us to propose a method that combines the strengths of the two methods.

2.1 Convolution Neural Networks

Yann LeCun and Yoshua Bengio introduced the concept of CNNs in 1995. A convolutional neural network is a feed-forward network with the ability of extracting topological properties from the input image. It extracts features from the raw image and then a classifier classifies extracted features. CNNs are invariance to distortions and simple geometric transformations like translation, scaling, rotation and squeezing.

Convolutional neural networks combine three architectural ideas to ensure some degree of shift, scale, and distortion invariance: local receptive fields, shared weights, and spatial or temporal sub-sampling [5]. The network is usually trained like a standard neural network by back propagation.

CNN layers alternate between convolution layers with feature map $C^i_{k,l}$ shown by (1),

$$C^i_{k,l} = g(I^i_{k,l} \otimes W_{k,l} + B_{k,l}) \tag{1}$$

And non-overlapping sub-sampling layers with feature map $S^i_{k,l}$ shown by (2),

$$S^i_{k,l} = g(I^i_{k,l} \downarrow W_{k,l} + Eb_{k,l}) \tag{2}$$

Where g(x) = tanh(x) is a sigmoidal activation function, B and b the biases, W and w the weights, $I^i_{k,l}$ the i'th input and $\downarrow$ the down-sampling symbol. E is a matrix whose elements are all one and $\otimes$ denotes a 2-dimensional convolution. Note that upper case letters symbolize matrices, while lower case letters present scalars [12].

A convolutional layer is used to extract features from local receptive fields in the preceding layer. In order to extract different types of local features, a convolutional layer is organized in planes of neurons called feature maps which are responsible to detect a specific feature. In a network with a 5×5 convolution kernel each unit has 25 inputs connected to a 5×5 area in the previous layer, which is the local receptive field. A trainable weight is assigned to each connection, but all units of one feature map share the same weights. This feature which allows reducing the number of trainable parameters is called weight sharing technique and is applied in all CNN layers. LeNet5 [5], a fundamental model of CNNs proposed by LeCun, has only 60,000 trainable parameters out of 345,308 connections. A reduction of the resolution of the feature maps is performed through the subsampling layers. In a network with a 2×2 subsampling filter such a layer comprises as many feature map numbers as the previous convolutional layer but with half the number of rows and columns.

In the rest of this section LeNet5 which is a particular convolutional neural network is described. LeNet5 takes a raw image of 32×32 pixels as input. It is composed of seven layers: three convolutional layers (C1, C3 and C5), two sub-sampling layers (S2 and S4), one fully connected layer (F6) and the output layer. The output layer is an Euclidean RBF layer of 10 units (for the 10 classes) [13]. These layers are connected as shown in Fig. 1.

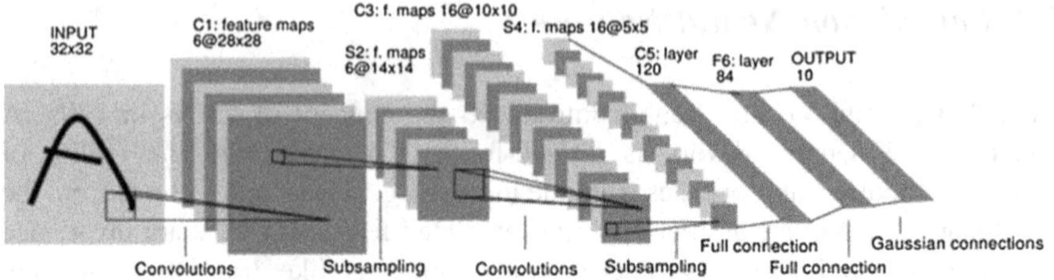

Fig. 1 LeNet5 architecture [5]

Table 1 The interconnection of the S2 layer to C3 layer [5]

	0	1	2	3	4	5	6	7	8	9	10	11	12	13	14	15
0	X				X	X	X			X	X	X	X		X	X
1	X	X				X	X	X			X	X	X	X		X
2	X	X	X				X	X	X			X		X	X	X
3		X	X	X			X	X	X	X			X		X	X
4			X	X	X			X	X	X	X		X	X		X
5				X	X	X			X	X	X	X		X	X	X

As shown in Table 1 the choice is made not to connect every feature map of S2 to every feature map of C3. Each unit of C3 is connected to several receptive fields at identical locations in a subset of feature maps of S2 [5, 13].

2.2 Logistic Regression Classifier

Logistic Regression is an approach to learning functions of the form $f : X \rightarrow Y$, or $P(Y|X)$ in the case where Y is discrete-valued, and $X = <X_1 \cdots X_n>$ is any vector containing discrete or continuous variables. In this section we will primarily consider the case where Y is a boolean variable, in order to simplify notation. More generally, Y can take on any of the discrete values $y = \{y_1 \cdots y_k\}$ which is used in experiments [9].

Logistic Regression assumes a parametric form for the distribution $P(Y|X)$, then directly estimates its parameters from the training data. The parametric model assumed by Logistic Regression in the case where Y is boolean is:

$$P(Y = 1|X) = \frac{1}{1 + exp(w_0 + \sum_{i=1}^{n} w_i X_i)} \tag{3}$$

and

$$P(Y = 0|X) = \frac{exp(w_0 + \sum_{i=1}^{n} w_i X_i)}{1 + exp(w_0 + \sum_{i=1}^{n} w_i X_i)} \tag{4}$$

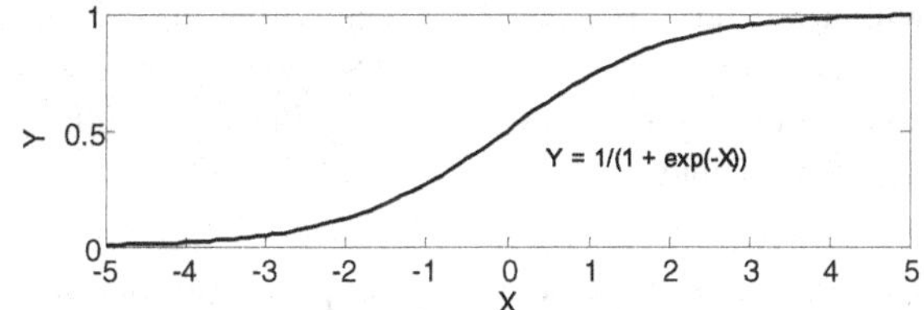

Fig. 2 Form of the logistic function. $P(Y|X)$ is assumed to follow this form [9]

Notice that Eq. (4) follows directly from Eq. (3), because the sum of these two probabilities must equal 1 [9].

One highly expedient property of this form for $P(Y|X)$ is that it leads to a simple linear expression for classification. To classify any given X we generally want to assign the value y_k that maximizes $P(Y = y_k|X)$. Put another way, we assign the label $Y = 0$ if the following condition holds [9]:

$$1 < \frac{P(Y = 0|X)}{P(Y = 1|X)}$$

Substituting from Eqs. (3) and (4), this becomes

$$1 < exp(w_o + \sum_{i=1}^{n} w_i X_i)$$

Form of the logistic function is shown in Fig. 2.

And taking the natural log of both sides, we have a linear classification rule that assigns label $Y = 0$ if X satisfies

$$0 < w_o + \sum\nolimits_{i=1}^{n} w_i X_i, \tag{5}$$

And assigns $Y = 1$ otherwise [9].

3 Proposed Method

In this paper, a CNN structure depicted in Fig. 3 is considered for feature extraction. The applied CNN is composed of four layers: input layer, two convolutional layers and one subsampling layer. Size of input layer is considered as 64×64, so the input images of the applied database were resized to 64×64 to be compatible with the proposed structure. The first convolutional layer has six feature maps, each of which has resolution of 58×58, with a receptive field of 7×7. The second layer, or the subsampling layer, contains six feature maps of size 29×29, with a receptive field of 2×2. The second convolutional layer has sixteen feature maps, each of which has resolution of 22×22, with a receptive field of 8×8. The second subsampling layer contains sixteen feature maps of size 11×11, with a receptive field of 2×2. The output layer is a fully connected layer with 15 feature maps with the size of 1×1, with a receptive field of 11×11. The interconnections between first Subsampling layer and second convolutional layer are shown in Table 2.

Table 2 Interconnection of first subsampling layer with the second convolutional layer

	0	1	2	3	4	5	6	7	8	9
0	X				X	X	X			
1	X	X				X	X	X		
2	X	X	X				X	X	X	
3		X	X	X				X	X	X
4			X	X	X				X	X
5				X	X	X				X

The primary weights and biases are considered in the range of -0.5 and 0.5.

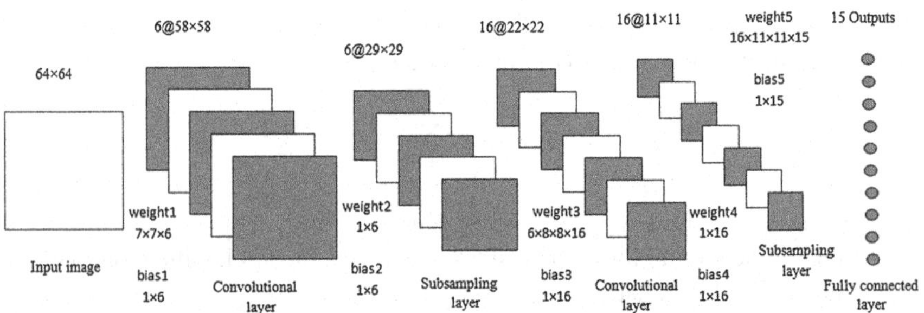

Fig. 3 CNN structure used for feature extraction

Due to computational complexity and time limitation of CNNs, we considered 500 epochs for the comparisons. At first, learning rate is considered the same as [7]:

$$\eta = \frac{\eta_0}{\frac{n}{N/2} + \frac{c_1}{max(1,(c_1 \frac{max(0,c_1(n-c_2N))}{(1-c_2)N}))}} \tag{6}$$

Where η = learning rate, η_0 = initial learning rate = 0.1, N = total training epochs, n = current training epoch, $c_1 = 50$ and $c_2 = 0.65$. Learning rate as a function of epoch number is shown in Fig. 4.

Using the mentioned learning rate, after passing approximately half of the maximum number of epochs, no more improvement was resulted for accuracy and error rate. Therefore, a new learning rate is created by a little alteration in the Eq. (6) which is shown in Eq. (7). In the proposed learning rate, after passing 0.65% of epochs a small value (0.001) is replaced. It causes an abrupt change in both accuracy and error rates. Keeping the learning rate as a constant value leads to increase exploitation and to yield the convergence of learning algorithm sooner.

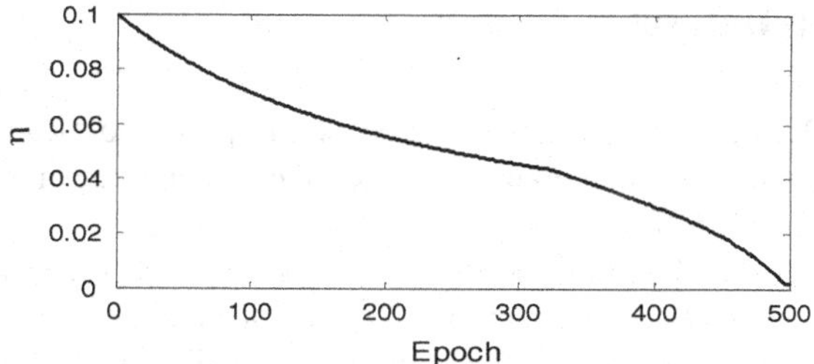

Fig. 4 Learning rate (η_1) [7]

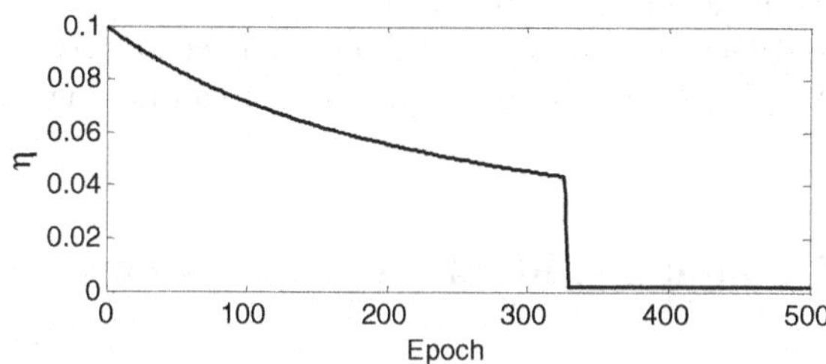

Fig. 5 Learning rate (η_2)

$$\eta = \frac{\eta_0}{\frac{n}{N/2} + \frac{c_1}{max(1,(c_1 \frac{max(0,c_1^2(n - c_2N))}{(1-c_2)N}))}} \tag{7}$$

The new learning rate is shown in Fig. 5. The comparisons based on the two learning rate are brought in the following.

Images are normalized to lie in the range -1 to 1 by removing the mean value and dividing them by their standard deviation as shown in (8).

$$I_n = \frac{I - \bar{I}}{\sigma} \tag{8}$$

Where I is the input image, $\bar{I}$ is the mean of all input image pixels, σ is the standard deviation and I_n is the normalized image.

By applying CNNs to a small database, number of trainable parameters surpasses the number of data. For this reason, weights in the network were updated sequentially after each pattern presentation, as opposed to batch update that weights are only updated once per pass through the training set.

A simple back-propagation gradient descent algorithm without any optimization is applied for training the network. Optimizing the network is postponed to the future works. Error function is assumed to be the squared error,

$$E = \frac{1}{2}e^2 = \frac{1}{2}(t - y)^2 \tag{9}$$

Where t is the target output, y is the actual network output, and e is the network error.

4 Dataset

To evaluate the performance of the proposed structure, various simulations are performed on a commonly used face database. Some samples of this database are shown in Fig. 6.

The Yale face database contains 165 face images of 15 individuals. There are 11 images per subject, one for each facial expression or configuration: center-light, glasses/no glasses, happy, normal, left-light, right-light, sad, sleepy, surprised and wink. Each image was digitized and presented by a 61×80 pixel array whose gray levels ranged between 0 and 255.

Table 3 summarizes what the number of data and classes in the database is. Also presented are train and test data partitioning for database images.

5 Simulation Results and Discussions

Firstly, classifying the extracted features is done using a winner takes all mechanism. There are 15 output neurons extracted for each image data. The biggest output neuron specified the class of the data. The recognition accuracy, training time and number of trainable parameters are summarized in Table 4. Also, train and test accuracy is shown in Fig. 7.

Several machine learning algorithms are applied on the features extracted from the database images using CNN. Machine learning algorithms are simulated using WEKA software. Results with these methods are reported and compared in Tables 5 and 6.

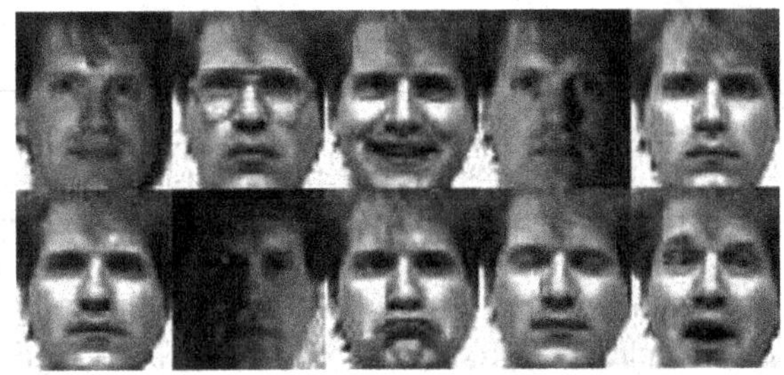

Fig. 6 Samples of Yale face database images

Table 3 Number of data and classes in database

Number of data	Number of train data	Number of test data	Classes	Train sample per class	Test sample per class
165	135	30	15	9	2

Table 4 Recognition accuracy, training time and number of parameters for Yale database

Number of parameters	Average training per epoch time (s)	Train accuracy (%)	Test accuracy (%)
35559	65	88.88	80

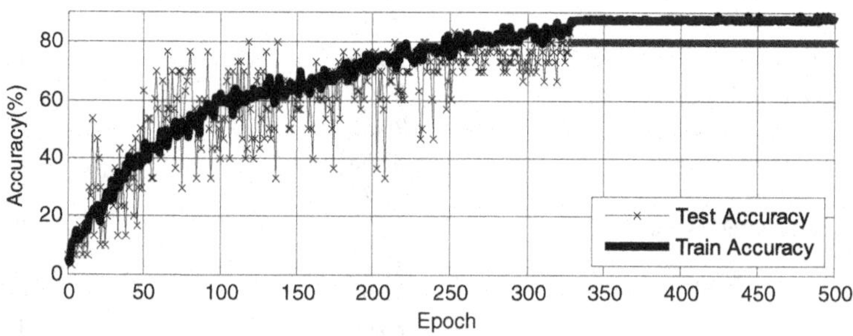

Fig. 7 Train and test data classification accuracy on Yale well-known face data set

Three different methods of Naïve Bayes classifier consist of NaiveBayes, NaiveBayesSimple and NaiveBayesUpdateable, functional classifiers consist of Logistic, MultiLayerPerceptron, RBFNetwork, SimpleLogistic and SOM which is shown the Support Vector Machine, lazy methods consist of IB1, IBk and LWL (Local Weighted Learning), Classification via regression methods and trees consist of BFTree, FT, LMT and SimpleCart are tested on feature vectors which are resulted using CNN. IBk method is repeated for k in the range of 1 to 10. The best accuracy is resulted by k = 7 which is shown in Table 5.

As it is shown, experimental results show that Simple Logistic algorithm could recognize the unseen face data with the highest classification accuracy of 86.06 in the least time in compare with other algorithms. LMT tree is also resulted the same classification accuracy in a bit more time. Classification accuracy and classification time on the Yale dataset for various classification methods are shown and compared in Fig. 8.

Table 5 Results of IBk algorithm

k	1	2	3	4	5	6	7	8	9	10
Accuracy(%)	76.36	79.39	81.21	81.81	81.81	83.03	83.63	83.03	83.03	83.03

Table 6 Comparison of different algorithms

	Accuracy (%)	Time (s)
NaiveBayes	75.15	0.05
NaiveBayesSimple	75.15	0.01
NaiveBayesUpdateable	75.15	0
Logistic	65.45	1.23
MultiLayerPerceptron	81.21	2.88
RBFNetwork	72.72	23
SimpleLogistic	86.06	1.22
SupportVectorMachine	83.03	9.79
IB1	76.36	0
IBk	83.63	0
LWL	78.78	0
ClassificationViaRegression	81.21	0.28
BFTree	75.15	0.21
FT	71.51	0.64
LMT	86.06	1.64
SimpleCart	75.75	0.22

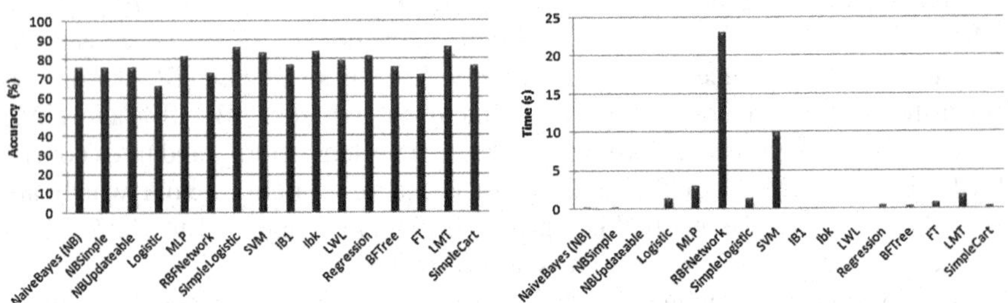

Fig. 8 Classification accuracy and time on the test set for various classification methods

6 Conclusion

In this paper, convolutional neural networks and simple logistic regression method are investigated with results on Yale face dataset. To combine the advantages of the two methods a two step learning process is proposed: first, training a convolutional neural network and view the first N – 1 layers as a feature extractor. Second, applying a simple logistic regression method on the features produced by the convolutional neural network. Proposed structure benefits from all CNN advantages such as feature extracting and robustness to distortions. The network was trained using back-propagation gradient descent algorithm. Applicability of these networks in recognizing face images is presented in this paper. Based on experiments on the Yale dataset, feature extraction using CNN which is applied to normalized data causes the system to cope with faces subject to pose and lighting variations. Moreover, simple logistic regression method classifies the extracted features with the highest classification accuracy and lowest classification time in compare with other machine learning algorithms.

References

1. Lo, S.B., Li, H., Wang, Y., Kinnard, L., Freedman, M.T.: A multiple circular path convolution neural network system for detection of mammographic masses. IEEE Trans. Med. Imaging 150–158 (2002)
2. Hu, Y.H., Hwang, J.: Handbook of neural network signal processing. CRC Press, Boca Raton (2002)
3. Gepperth, A.: Object detection and feature base learning by sparse convolutional neural networks. In: Lecture notes in artificial intelligence 4807, Springer Verlag Berlin, 221–231 (2006)
4. Bouchain, D.: Character Recognition Using Convolutional Neural Networks. In: Seminar Statistical Learning Theory, University of Ulm, Germany (2006/2007)
5. LeCun, Y., Bottou, L., Bengio, Y., Haffner, P.: Gradient-based learning applied to document recognition. In: Proceedings of the IEEE. vol. 86, pp. 2278–2324 (1998)
6. Ahranjany, S.S., Razzazi, F., Ghassemian, M.H.: A very high accuracy handwritten character recognition system for Farsi/Arabic digits using Convolutional Neural Networks. In: IEEE Fifth International Conference on Bio-Inspired Computing, Theories and Applications (BIC-TA) pp. 1585–1592 (2010)
7. Lawrence, S., Giles, C.L., Tsoi, A.C., Back, A.D.: Face recognition: a convolutional neural network approach. In: IEEE Trans. Neural Networks 98–113 (1997)
8. Palei, S.K., Das, S.K.: Logistic regression model for prediction of roof fall risks in bord and pillar workings in coal mines: An approach. Saf. Sci. (2009)
9. Mitchell, T.M.: Generative and discriminative classifiers: naive bayes and logistic regression. Machine Learning (2010)
10. LeCun, Y., Huang, F.-J.: Large-scale learning with SVM and convolutional nets for generic object categorization. In: Proceedings of the Computer Vision and Pattern Recognition Conference (CVPR'06). IEEE Press, Salt Lake City (2006)
11. Yale face database, http://cvc.yale.edu/projects/yalefaces/yalefaces.html
12. Lyons, M., Akamatsu, S., Kamachi, M., Gyoba, J.: Coding facial expressions with gabor wavelets. In: Third IEEE International Conference on Automatic Face and Gesture Recognition pp. 200–205 (1998)
13. Fasel, B.: Robust face analysis using convolutional neural networks. In: Proceedings of the 16th International Conference on Pattern Recognition, pp. 40–43 (2002)

Continuous Features Discretization for Anomaly Intrusion Detectors Generation

Amira Sayed A. Aziz, Ahmad Taher Azar, Aboul Ella Hassanien and Sanaa El-Ola Hanafy

Abstract Network security is a growing issue, with the evolution of computer systems and expansion of attacks. Biological systems have been inspiring scientists and designs for new adaptive solutions, such as genetic algorithms. In this paper, an approach that uses the genetic algorithm to generate anomaly network intrusion detectors is used. An algorithm is proposed using a discretization method for the continuous features selection of intrusion detection, to create some homogeneity between values, which have different data types. Then, the intrusion detection system is tested against the NSL-KDD data set using different distance methods. A comparison is held amongst the results, and it is shown by the end that this proposed approach has good results, and recommendations are given for future experiments.

A. S. A. Aziz (✉)
French University in Egypt (UFE), Shorouk City, Egypt
e-mail: amira.abdelaziz@egyptscience.net

A. S. A. Aziz · A. T. Azar
Scientific Research Group in Egypt (SRGE), Cairo, Egypt
e-mail: ahmed_t_azar@yahoo.com

A. T. Azar
Misr University for Science & Technology (MUST), 6th of October City, Egypt

A. E. Hassanien
Chairman of Scientific Research Group in Egypt (SRGE), Cairo, Egypt
e-mail: abo@egyptscience.net

S. E.-O. Hanafy
Faculty of Computers and Information, Cairo University, Cairo, Egypt

V. Snášel et al. (eds.), *Soft Computing in Industrial Applications*,
Advances in Intelligent Systems and Computing 223, DOI: 10.1007/978-3-319-00930-8_19,

1 Introduction

With the evolution of computer networks during the past few years, security is a crucial issue and a basic demand for computer systems. Attacks are expanding and evolving as well, making it important to come up with new and advanced solutions for network security. Intrusion Detection Systems (IDS) have been around us for a some time, as an essential mechanism to protect computer systems, where they identify malicious activities that occur in that protected system. Genetic Algorithms (GA) are a group of computational models inspired by natural selection [1, 2]. This solution works on a group of chromosomes-like data structure (a population) where they reproduce new individuals that would be more be fitting in the environment. These new generations are developed using selection and recombination functions such as crossover and mutation [3].

The GAs were first seen as optimization solutions, but now they are applied in a variety of systems, including the IDSs [4, 5]. The GA is used as a machine learning technique to generate artificial intelligence detection rules. The rules are usually in the if-then forms, where the conditions are values that represent normal samples or values to indicate an intrusion is in the act [3, 6]. For a Network-based IDS (NIDS), usually the network traffic is used to build a model and detect anomalous network activities. Many features can be extracted and used in a GA to generate the rules, and these features may be of different data types, and may have a wide range of values. So, this paper presents an approach that uses a discretization algorithm with continuous features to create homogeneity amongst features.

Discretization is simply a process of converting continuous attributes to discrete ones by partitioning them into intervals. Data sets used in intrusion detection systems are high-dimensional, hence discretization is needed as a preprocessing step before applying clustering, feature selection, training...etc processes on the data. Limited use and research was held concerning when to use which discretization algorithms in IDS. More details about discretization algorithms can be found in [7–9].

The rest of this paper is organized as follows: Sect. 2 gives a background of the different algorithms used in this approach. Section 3 gives a review on some of the previous work done in the area. Section 4 describes the proposed approach, Sect. 5 gives an overview of the experimental analysis and results. And finally in Sect. 6, conclusion and directions for future research are presented.

2 Background

2.1 Anomaly Intrusion Detection

Intrusion Detection Systems (IDSs) are security tools used to detect anomalous or malicious activity from inside and outside intruders [10]. An IDS can be host-based or network-based, which is the concern in this paper [11]. They are classified by many axes, one of them is the detection methodology that classifies them to signature-based and anomaly-based IDS. The former detects attacks by comparing the data to patterns stored in a signature database of known attacks. The later detects anomalies by defining a model of normal behaviour of the monitored system, then considers any behaviour lying outside the model as anomalous or suspicious activity. Signature-based IDS can detect well-known attacks with high accuracy but fails to detect or find unknown attacks. Anomaly-based IDS has the ability to detect new or unknown attacks but usually has high false positives rate (normal activities detected as anomalous). There are three types of anomaly detection techniques: statistical-based, knowledge-based, and machine learning-based [12]. IDS performance can be measured by two key aspects: the detection process efficiency and the involved cost of the operation [12].

2.2 Genetic Algorithms

Genetic Algorithm (GA) is an evolutionary computational technique that is used as a search algorithm, based on the concepts of natural selection and genetics. There are 3 meanings of search: 1- Search for stored data: where the problem is to retrieve some information stored in a computer memory efficiency. 2- Search for paths to goals: where one needs to find the best paths from an initial state to a goal. 3- Search for solutions: where one needs to find a solution or group of solutions in a large space of candidates. GA works on a population of individuals, where each individual is called a chromosome and is composed of a string of values called genes. The population goes through a process to find a solution or group of high quality solutions. The quality of an individual is measured by a fitness function that is dependant on the environment and application. The process starts with an initial population, that goes through transformation for a number of generations. During each generation, three major operations are applied sequentially to each individual: selection, crossover, and mutation until target is met [6, 13].

2.3 Negative Selection Approach

Artificial Immune Systems (AIS) are inspired by the nature's Human Immune System (HIS), which is an adaptive, tolerant, self-protecting, and dynamic defence system [14]. AIS is a set of algorithms that mimic the different functionalities of the HIS, and they can perform a range of tasks. The major algorithms are: negative selection, clonal selection, and immunity networks. The Negative Selection Approach (NSA) is based on the concept of self-nonself discrimination, by first creating a profile of the self (normal) behaviour and components. Then use this profile to rule out any behaviour that doesn't match with that profile. The training phase goes on the self samples. And then, the detectors are exposed to different samples, and if a detector matches a self as nonself then it's discarded. The final group of detectors (mature detectors) are released to start the detection process [15, 16].

2.4 Equal-Width Binning Algorithm

There are many algorithms for continuous features discretization for algorithmic purposes. These discretization algorithms are very important for machine learning, as it is required by some algorithms. But more importantly, the discretization increases the speed of induction algorithms. The discretization algorithms are classified in many ways: Local or Global, Supervised or Unsupervised Static or Dynamic, Top-down or Bottom-up, and Direct or Incremental [17–19].

Local methods apply partitions on localized regions of the instance space, where Global methods works on the entire instance space, that every feature is partitioned into number of regions independent of other attributes. Supervised methods make use of the class labels associated with the instances in the process, while unsupervised methods perform discretization regardless of class label. Static or Dynamic, where Static methods determine the number of bins for each feature independent of other features after performing one discretization pass of the data (performed before the classification). Dynamic methods determine the number of bins for all features simultaneously by a search through the data space (performed while the classifier is built). Top-down(Splitting) or Bottom-up (Merging), where Splitting methods start with an empty group of cut points, and build up during the discretization process. While in merging, the algorithm starts with a list of cut points, then discards unneeded ones during discretization by merging intervals. Direct or Incremental, where In direct methods, number of bins (intervals) is predefined either by user or using an algorithm. Incremental methods start with simple discretization that gets improved and refined until stopped by a condition (meeting a certain criterion).

3 Related Work

There have been several studies reported focusing on discretization algorithms [19, 20]. Dougherty et al. [17] Applied EWB, 1R, and Recursive Entropy Partitioning as preprocessing step before using C4.5 and Naive-Bayes classifiers on data. The data set they used was 16 data sets from the UC Irvine (UCI) Repository. C4.5 performance improved on 2 data sets using entropy discretization, but slightly decreased on some. At 95 % confidence level, Naive-Bayes with entropy discretization is better than C4.5 on 5 data sets and worse on 2 (with average accuracy

83.97 % vs. 82.25 % for C4.5). Clarke and Barton [21] applied Minimum Descriptive Length (MDL) to select number of intervals, and modified a version of the k2 method for one test, entropy based discretization for another test. They ran their experiment on NGHS and DISC from two epidemiological studies. The Dynamic Partitioning with MDL metric lead to more highly connected BBN than with only entropy partitioning. New proposed method lead to better representations of variable dependencies in both data sets. But generally, using entropy and MDL partitioning provided clarification and simplification in the BBN. Zhao and Zhou [22] suggested a rough set based heuristic method, enhanced in two ways: (1) decision information is used in candidate cut computation (SACC), and (2) an estimation of cut selection probability is defined to measure cut significance (ABSP). The experiment was applied on continuous UCI data sets. Their SACC was compared to an algorithm known as UACC, and ABSP was compared to some typical rough set based discretization algorithm. SACC performs better with less number of cuts, and ABSP slightly improves predictive accuracies.

Gupta et al. [18] applied K-means clustering with euclidean distance similarity metric, and Shared Nearest Neighbour (SNN). MDL was used for discretization with alpha=Beta=0.5 (ME-MDL), and was applied on 11 data sets from UCI repository. Comparing their algorithm results with ME-MDL results: in all data sets, when SNN or k-means clustering was used, the proposed algorithm gave better results. In heart data set, SNN clustering performed better than ME-MDL. In other data sets, k-means was better than SNN. Joita [23] proposed a discretization algorithm based on the k-means clustering algorithm, that avoids the O(n log n) time required for sorting. The algorithm was proposed to be tested in the future. Chen et al. [24] proposed an improved method of continuous attributes discretization by: (1) hierarchical clustering was applied to form initial division of the attribute, and (2) merging adjacent ranges based on entropy, taking into consideration not to affect level of consistency of the decision table. They used data of their provincial educational committee project for the experiment. The results were not listed, but mentioned to prove the validity of their algorithm. Ferreira and Figueirdo [25] used clustering with discretization for better results. Updated versions of the well-know Linde-Buzp-Gray (LBG) algorithm were proposed: U-LBG1 (used a variable number of bits) and U-LBG2 (used a fixed number of bits). For clustering,Relevance-Redundancy Feature Selection (RRFS) and Relevance Feature Selection (RFS) methods were used. Data sets from UCI, the five data sets of the NIP2003 FS challenge, and several micro-array gene expression data sets were used for their experiments (no normalization was applied on any of the used data sets). The proposed approaches allocated a small number of bits per feature. RRFS performs better than RS for eliminating redundant features.

4 Proposed Approach

4.1 Motivation

In [26] the algorithm was originally suggested with the application on real-valued features in the NSL-KDD data set. It was used with a variation parameter defining the upper and lower limits of the detectors values (conditions). It had very good results, but the real-valued features are not enough to detect all types of attacks, so the algorithm should expand to include features of different types. In [27], the algorithm was applied on the KDD data set, using a range of features to detect anomalies. The problem with using different features is that they have different data types ranges: binary, categorical, and continuous (real and integer). This may lead to problems while applying the algorithm. First of all, a wide range of values need to be covered in a way that can represent each region uniquely. Secondly, there should be some sort of homogeneity between features values to apply the GA. So, the use of some discretization algorithm for continuous features lead to the suggestion of the following approach.

4.2 Suggested Approach

Equal-width interval binning [17] is the simplest method for data discretization, where the range of values is divided into k equally sized bins, as k is a parameter supplied by the user as the required number of bins. The bin width is calculated as:

$$\delta = \frac{x_{max} - x_{min}}{k} \tag{1}$$

and the bin boundaries are set as: $x_{min}+i\delta$, $i=1,\ldots,k-1$.

The equal-width interval binning algorithm is a global, unsupervised, and static discretization algorithm. The suggested approach starts with binning the continuous features with a previously defined number of bins, Then, replace each feature value with its enclosing bin number. Finally, run the GA on the modified data set samples to generate the detectors (rules). Following the NSA concepts, this is applied on the normal samples through the training phase. The self samples are presented in the self space S. The process is shown in Algorithm I.

Algorithm 1 Proposed Algorithm

Run equal-width binning algorithm on continuous features.
Initialize population by selecting random individuals from the space *S*.
for The specified number of generations **do**
  for The size of the population **do**
    Select two individuals (with uniform probability) as $parent_1$ and $parent_2$.
    Apply crossover to produce a new individual (*child*).
    Apply mutation to child.
    Calculate the distance between *child* and $parent_1$ as d_1, and the distance between *child* and $parent_2$ as d_2.
    Calculate the fitness of *child*, $parent_1$, and $parent_2$ as f, f_1, and f_2 respectively.
    if $(d_1 < d_2)$ and $(f > f_1)$ **then**
      replace $parent_1$ with *child*
    else
      if $(d_2 <= d_1)$ and $(f > f_2)$ **then**
        Replace $parent_2$ with *child*.
      end if
    end if
  end for
end for
Extract the best (highly-fitted) individuals as your final solution.

The fitness - which was inspired from [28] - is measured by calculating the matching percentage between an individual and the normal samples, as:

$$fitness(x) = \frac{a}{A} \tag{2}$$

where a is the number of samples matching the individual by 100 % , and A is the total number of normal samples. Three different distance methods were tested (one at a time), to find the best results. The distances measured between a child X and a parent Y using the following formulas:

- The Euclidean distance as:

$$d(X,Y) = \sqrt{(x_1 - y_1)^2 + (x_2 - y_2)^2(x_n - y_n)^2} \tag{3}$$

- The Hamming distance, which defines the difference between 2 strings (usually binary) as the number of places in which the strings have different values [29]. So it's calculated as (where n is number of features):

$$d(X,Y) = \sum_{i=0}^{n} |(x_i - y_i)| \tag{4}$$

- The Minkowski Distance, which is similar to the Euclidean distance but uses the p-norm dimension as the power value instead. So, the formula goes as:

$$d(X,Y) = (\sum_{i=0}^{n}(|x_i - y_i|^p))^{1/p} \quad (5)$$

In the Minkowski distance case, p can be any value larger than 0 and up to infinity. It can be have real value between 0 and 1. If we are interested in finding the difference between objects, then we should aim for high p values. If we are interested in finding the how much the objects are similar, then we should go for low p values [30]. In our experiment, a small value of 0.5 was used, and a big values of 18 was used to compare results.

5 Experiment

5.1 Data Set

The NSL-KDD IDS data set [31] was proposed in [32] to solve some issues in the widely use KDD Cup 99 data set. These issues affect the performance of the systems that use the KDD data set and results in very poor evaluation of them. The resulted data set is having a reasonable size and is unbiased, and it's affordable to use in the experiments without having to select a small portion of the data. The data sets used in our experiment are:

- KDD Train+_20 Percent normal samples for training and generating the detectors.
- KDD Train+ and KDDTest+ for testing, where the difference between them is that the Test set include additional unknown attacks that are not included in the Train set.

5.2 Experiment Settings

In the proposed approach, the features were selected as in [17], which are shown in Table 1. Ports classification was performed manually, and they were classified into 9 categories as in [27]. The procedure was done manually because it is dependent on the network and system settings more than number ranges.

The values used for the GA parameters are summarized in Table 2:

Different values of population size and number of generations were used to compare the results to see which would lead to better results, and threshold value of 0.8 was used for the experiment.

Table 1 Features selected from NSL-KDD data set

Feature	Data type	No. of bins
duration	Integer	8
protocol_type	Categorial	N/A
service	Integer	9
land	Binary	N/A
urgent	Integer	1
host	Integer	3
num_failed_logins	Integer	3
logged_in	Binary	N/A
root_shell	Binary	N/A
su_attempted	Binary	N/A
num_file_creations	Integer	4
num_shells	Integer	2
is_host_login	Binary	N/A
is_guest_login	Binary	N/A
count	Integer	10
same_srv_rate	Real	3
diff_srv_rate	Real	3
srv_diff_host_rate	Real	3

Table 2 GA parameters

Population size	200, 400, 600
Number of generations	200, 500, 1,000, 2,000
Mutation rate	2/L, where L is the number of features
Crossover rate	1.0

5.3 Results

After running the algorithm on the train set normal samples, varieties of detectors (rules) were obtained, based on population size and number of generations. Running those detectors on the test set, the detection rates are shown in Fig. 1.

As shown in Fig. 1, the detection rates are all above 75 % - as the maximum detection rates realized are 81.93 % and 81.54 % obtained by the detectors generated by GA applying Euclidean and Minkowski (p=18) distances respectively. The overall average detection rates are 79.06, 78.62, 79.25, and 79.18 % obtained using Euclidean, Hamming, and Minkoswki distance functions respectively, and in most cases the Minkowski distance gave better performance. The rates obtained with detectors generated using population size 200 are generally better. To measure the IDS efficiency, true positives and true negatives rates (TPR and TNR respectively) are calculated and shown in Figs. 2 and 3.

In Fig. 2, TPRs lie between 60 % and 80 %, which means that at least 20 % of the attacks are detected as normal. Figure 3 indicated that the recognition of normal samples is high as the TNRs are mostly above 90 % with low FPRs—all less than

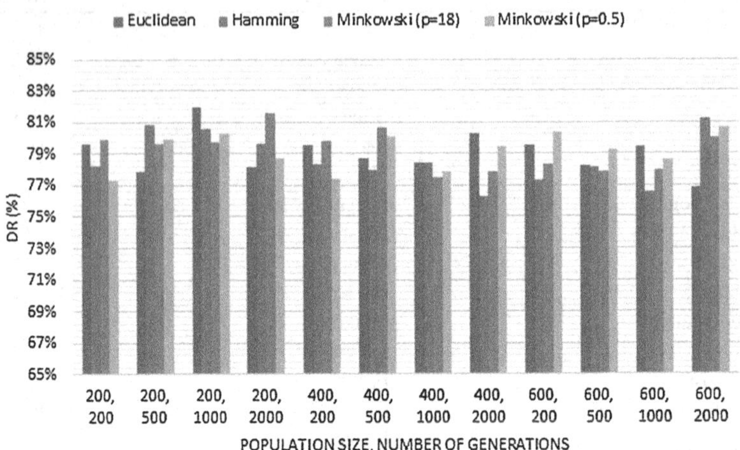

Fig. 1 Detection rates

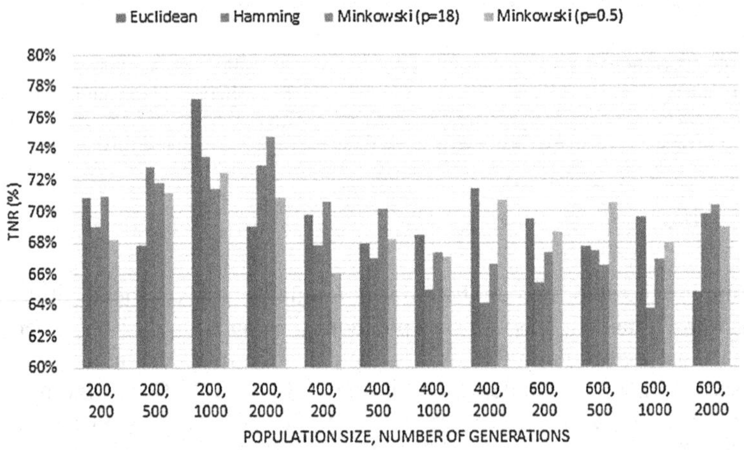

Fig. 2 True positives rates

9 %. Detectors generated using bigger populations give higher TNRs, while higher TPRs are obtained with detectors generated by smaller populations. Other experiments similar to this one use classifiers for multiple classification detection, but 2-classes detection (normal/anomaly) is applied in the algorithm. So, a real comparison can't be held in the moment until a multi-class classifier is applied for more precise results.

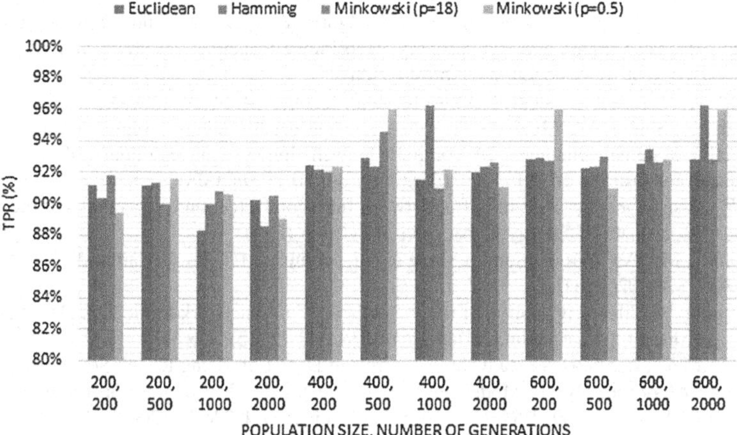

Fig. 3 True negatives rates

6 Conclusion and Future Work

In the paper, an algorithm is implemented to generate detectors that should be able to detect anomalous activities in the network. The data was pre-processed before using them in the algorithm, by discretizing the continuous features to create homogeneity between data values, and then replacing values with bin numbers. The results indicated that the equal-width interval binning algorithm that was used is very simple and has good results. As for the parameters of the GA, the detectors generated by GA with smaller population size gave better detection rates and true alarms than others generated using higher population sizes. The advantage of using smaller population sizes are: (1) less time-consuming while generating the detectors, and (2) less number of detectors are generated. The detectors generated using the Minkowski distance gave better results in most cases than those generated using the Euclidean distance (which is widely used) and the Hamming distance. Future work will be focused on applying other discretization algorithms that are more dynamic. Also, using classifiers to increase detection accuracy, and be able to define which type of anomalies have been detected.

References

1. Haupt, R.L., Haupt, S.E.: Practical Genetic Algorithms. 2nd edn. Wiley, New York (2004)
2. Polhlheim, H.: Genetic and Evolutionary Algorithms: Principles, Methods and Algorithms. http://www.geatbx.com/docu/index.html (2006)
3. Whitley, D.: A genetic algorithm tutorial. Stat. Comput. **4**, 65–85 (1994)
4. Owais, S., Snasel, V., Abraham, A.: Survey: using genetic algorithm approach in intrusion detection systems techniques, 7th computer information systems and industrial management applications. Ostrava **26–28**, 300–307 (2008). doi:10.1109/CISIM.2008.49
5. Li, W.: Using Genetic Algorithm for Network Intrusion Detection. Proceedings of the United States Department of Energy Cyber Security Group (2004)
6. Nitchell, M.: An Introduction to Genetic Algorithms. MIT Press Cambridge, MA, USA (1998). ISBN 0262631857
7. Sengupta, N., Sil, J.: Evaluation of rough set theory based network traffic data classifier using different discretization method. IJIEE **2**(3), 338–341 (2012)
8. Wa'el, M.M., Agiza, H.N., Radwan, E.: Intrusion Detection Using Rough Sets based Parallel Genetic Algorithm Hybrid Model. Proceedings of the World Congress on Engineering and Computer Science 2009 (WCECS 2009). San Francisco, USA, II, (2009)
9. Ertoz, L., Eilertson, E., Lazarevic, A., Tan, P.N., Kumar, V., Srivastava, J., Dokas, P.: MINDS—minnesota intrusion detection system. In: Data Mining—Next Generation Challenges and Future Directions. MIT Press, Cambridge (2004)
10. Aleksandar, L., Vipin, K., Jaideep, S.: Intrusion detection: a survey. In: Kumar, V. et al. (eds.) Managing Cyber Threats Issues, Approaches, and Challenges vol. 5, pp. 19–78 (2005)
11. Murali, A., Roa, M.: A survey on intrusion detection approaches. First International Conference on Information and Communication Technologies, ICICT 2005, pp. 233–240, Aug (2005)
12. Garcia-Teodora, P., Daz-Verdejo, J., Maci-Fernndez, G., Vzquez, E.: Anomaly-based network intrusion detection: Techniques, systems and challenges. Comput. Secur. **28**(1–2), 18–28 (2009)
13. Akbar, S., Chandulal, J.A., Rao, K.N., Kumar, S.: Troubleshooting techniques for intrusion detection system using genetic algorithm. Int. J. Wisdom Based Comput. **1**(3), 86–92 (2011)
14. Dasgupta, D.: Advances in artificial immune systems. IEEE Comput. Intell. Mag. **1**(4), 40–49 (2006)
15. Greensmith, J., Whitbrook, A., Aickelin, U.: Artificial immune systems. In: Gendreau, M., Potvin J.-Y. (eds.) Handbook of Metaheuristics, International Series in Operations Research and Management Science. Springer, Springer US. vol. 146, pp. 421–448 (2010)
16. Aickelin, U., Greensmith, J., Twycross, J.: Immune system approaches to intrusion detection—a review. In: Proceedings of the 3rd International Conference on Artificial Immune Systems (ICARIS), LNCS 3239, 316–329 (2004)
17. Dougherty, J., Kohavi, R., Sahami, M.: Supervised and Unsupervised Discretization of Continuous Features. Proceedings of the Twelfth Conference on Machine Learning 95(10), pp. 194–202 (1995)
18. Gupta, A., Mehrotra, K., Mohan, C.: A Clustering-based discretization of supervised learning. Stat. Probab. Lett., Elsevier, **80**(910), 816–824 (2010)
19. Liu, H., Hussain, F., Tan, C.L., Dash, M.: Discretization: an enabling technique. Data Min. Knowl. Disc. **6**(4), 393–423 (2002)
20. Kotsiantis, S., Kanellopoulos, D.: Discretization techniques: a recent survey. GESTS Int. Trans. Comput. Sci. Engin. **32**(1), 47–58 (2006)
21. Clarke, E.J., Barton, B.A.: Entropy and MDL discretization of continuous variables for Bayesian belief networks. Int. J. Intell. Syst. **15**(61), 61–92 (2000)
22. Zhao, J., Zhou, Y.: New heuristic method for data discretization based on rough set theory. J. China Univ. Post. Telecommun. **16**(6), 113–120 (2009)

23. Joita, D.: Unsupervised Static Discretization Methods in Data Mining, Revista Mega, Byte. vol. 9 (2010)
24. Chen, S., Tang, L., Liu, W., Li, Y.: A Improved Method of Discretization of Continuous. Attribute, 2011 2nd International Conference on Challenges in Environmental Science and Computer Engineering (CESCE 2011), Elsevier, 11(A), 213–217 (2011)
25. Ferreira, A.J., Figueiredo, M.A.: An unsupervised approach to feature discretization and selection. Pattern Recogn. **45**(9), 3048–3060 (2012)
26. Aziz, A.S.A., Salama, M., Hassanien, A.E., Hanafi, S.O.: Detectors Generation using Genetic Algorithm for a Negative Selection Inspired Anomaly Network Intrusion Detection System. Proceedings of the IEEE FedCSIS, Wroclaw, Poland, pp. 625–663, ISBN:978-83-60810-51-4 (2012)
27. Powers, S.T., He, J.: A hybrid artificial immune system and self-organizing map for network intrusion detection. Int. J. Comput. Inf. Sci., Elsevier, **178**(15), 3024–3042 (2008)
28. Goyal, A., Kumar, C.: GA-NIDS: A genetic algorithm based network intrusion detection system. A Project at Electrical Engineering and Computer Science, Northwestern University, Evanston, IL, http://www.cs.northwestern.edu/ago210/ganids/ (2007)
29. He, M.X., Peroukhov, S.V., Ricci, P.E.: Genetic code, hamming distance, and stochastic matrices. Bull. Math. Biol. **66**(5), 1405–1421 (2004). doi:10.1016/j.bulm.2004.01.002
30. Kotnarowski, M.: Measurement of distance between voters and political parties—different approaches and their consequences, 3rd ECPR Graduate Conference. Dublin (2010)
31. NSL-KDD data set, http://nsl.cs.unb.ca/NSL-KDD/
32. Tavallaee, M., Nagheri, E., Lu, W., Ghorbani, A.A.: A detailed analysis of the KDD Cup 99 data set. Proceedings of the 2009 IEEE Symposium Computational Intelligence for Security and Defense Applications, CISDA09 (2009)

Visualisation of High Dimensional Data by Use of Genetic Programming: Application to On-line Infrared Spectroscopy Based Process Monitoring

Tibor Kulcsar, Gabor Bereznai, Gabor Sarossy, Robert Auer and Janos Abonyi

Abstract In practical data mining and process monitoring problems high-dimensional data has to be analyzed. In most of the cases it is very informative to map and visualize the hidden structure of complex data in a low-dimensional space. Industrial applications require easily implementable, interpretable and accurate projection. Nonlinear functions (aggregates) are useful for this purpose. A pair of these functions realise feature selection and transformation but finding the proper model structure is a complex nonlinear optimisation problem. We present a Genetic Programming (GP) based algorithm to generate aggregates represented in a tree structure. Results show that the developed tool can be effectively used to build an on-line spectroscopy based process monitoring system; the two-dimensional mapping of high dimensional spectral database can represent different operating ranges of the process.

1 Introduction

Dimensionality reduction methods can be performed in two ways. *Feature selection methods* try to select a subset of the features of data which contain the most important characters of data objects. The well known exhaustive search method [1] examines all possible subsets and selects the subset with the largest feature selection criterion as the solution. This method guarantees to find the optimum solution, but if the number of the possible subsets is large, it becomes impractical. There have been many methods proposed to avoid the enormous computational cost

T. Kulcsar · J. Abonyi (✉)
Department of Process Engineering, University of Pannonia, Veszprem H-8200, Hungary
e-mail: janos@abonyilab.com

G. Bereznai · G. Sarossy · R. Auer
MOL Ltd. Duna Refinery, Szazhalombatta H-2440, Hungary

V. Snášel et al. (eds.), *Soft Computing in Industrial Applications*,
Advances in Intelligent Systems and Computing 223, DOI: 10.1007/978-3-319-00930-8_20,

(e.g. branch and bound search [2], floating search [3], Monte Carlo algorithms, and Genetic Programming [4]. In contrast with the feature selection the *feature extraction methods* do not select the most relevant attributes but they combine them into some new attributes. The number of these new attributes is generally more less than the number of the original attributes. So feature extraction methods take all attributes into account and they provide reduced representation by feature combination and/or transformation. The most commonly used linear dimensionality reduction methods are for example the Principal Component Analysis (PCA) [5], the Independent Component Analysis (ICA) [6] or the Linear Discriminant Analysis (LDA) [7]. However if the manifolds are nonlinearly embedded into the higher dimensional space linear methods provide unsatisfactory representation of data. In these cases the *nonlinear dimensionality reduction methods* may outperform the traditional linear techniques and they are able to give a good representation of data set in the low-dimensional data space. To unfold these nonlinearly embedded manifolds many nonlinear dimensionality reduction methods are based on the concept of geodesic distance and they build up graphs to carry out the visualization process (e.g. Isomap, Isotop, TRNMap). The bets known nonlinear dimensionality reduction methods are Kohonen's Self-Organizing Maps (SOM) [8], Sammon mapping [9], Locally Linear Embedding (LLE) [10], Laplacian Eigenmaps or Isomap [11]. The topic of feature extraction and features selection methods is an active research area recently, a lots of research papers introduce new algorithms or utilize them in different scientific fields.

Industrial applications require easily implementable, interpretable and accurate projections. Nonlinear functions (aggregates) are useful for this purpose. A pair of these functions realise feature selection and transformation. These low dimensional mappings can also be utilized to index the spectral database by giving small number of primary key variables, and sophisticated prediction and clustering algorithms [12–14] can be developed based on this indexing.

Finding the proper model structure is a complex nonlinear optimisation problem. We present a Genetic Programming (GP) based algorithm to generate nonlinear aggregates. This method is based on a *tree representation* based symbolic optimization technique developed by John Koza. This representation is extremely flexible, trees can represent computer programs, mathematical equations or complete models of process systems. This scheme has been already used for circuit design in electronics, algorithm development for quantum computers, and it is suitable for generating model structures: e.g. identification of kinetic orders, steady-state models, and differential equations. In [15] GP is applied to find simple nonlinear functions by minimising the distance preservation based Sammon stress function. The drawback of this approach is that since the models were not parametrized only simple mappings with approximative distance preserving properties were generated.

We developed a much more sophisticated approach. The functions generated by GP are parameterised and a nonlinear parameter optimisation step is embedded into the GP. Furthermore, instead of distance preserving measures the cost function is based on the neighborhood preserving properties of the mapping since this measure is much closer reflects the application of the visualizer high dimensional instance based models.

The paper is organized as follows: In Sect. 2 the problem of topology preserving mapping of high dimensional data is introduced. We present the GP algorithm that can be used to generate such mappings in Sect. 3. Finally in Sect. 4 an application example is shown. Results show that the developed tool can be effectively used to build an on-line spectroscopy based process monitoring system; the two-dimensional mapping of high dimensional spectral database can represent different operating ranges of the process.

2 Topological Mapping for Visualization of High Dimensional Data

The goal of *dimensionality reduction* is to map a set of observations from a high-dimensional space (D) into a low-dimensional space (d, $d \ll D$) preserving as much of the intrinsic structure of the data as possible. Let $\mathbf{X} = \{\mathbf{x}_1, \mathbf{x}_2, \ldots, \mathbf{x}_N\}$ be a set of the observed data, where $\mathbf{x}_i$ denotes the i-th observation ($\mathbf{x}_i = [x_{i,1}, x_{i,2}, \ldots, x_{i,D}]^T$). Each data object is characterized by D dimensions, so $x_{i,j}$ yields the j-th ($j = 1, 2, \ldots, D$) attribute of the i-th ($i = 1, 2, \ldots, N$) data object. Dimensionality reduction techniques transform data set $\mathbf{X}$ into a new data set $\mathbf{Y}$ with dimensionality d ($\mathbf{Y} = \{\mathbf{y}_1, \mathbf{y}_2, \ldots, \mathbf{y}_N\}$, $\mathbf{y}_i = [y_{i,1}, y_{i,2}, \ldots, y_{i,d}]^T$). In the reduced space many data analysis tasks (e.g. classification, clustering, image recognition) can be carried out faster than in the original data space.

As dimensional reduction methods are based on the preservation of the dissimilarities and/or the neighborhood relation of the objects, the numeral evaluation of the mappings aims to measure the realization of these principles. *The neighborhood preservation of the mappings and the local and global mapping qualities* can be measured by functions of trustworthiness and continuity. Kaski and Vienna pointed out that every visualization method has to make a tradeoff between gaining good trustworthiness and preserving the continuity of the mapping [16, 17]. A projection is said to be *trustworthy* [16, 18] when the nearest neighbors of a point in the reduced space are also close in the original vector space. Let N be the number of the objects to be mapped, $U_k(i)$ be the set of points that are in the k size neighborhood of the sample i in the visualization display but not in the original data space. The measure of trustworthiness of visualization can be calculated in the following way:

$$M_1(k) = 1 - \frac{2}{Nk(2N - 3k - 1)} \sum_{i=1}^{N} \sum_{j \in U_k(i)} (r(i,j) - k), \tag{1}$$

where $r(i,j)$ denotes the ranking of the objects in input space.

The projection onto a lower dimensional output space is said to be *continuous* [16, 18] when points near to each other in the original space are also nearby in the output space. The measure of continuity of visualization is calculated by the following equation:

$$M_2(k) = 1 - \frac{2}{Nk(2N-3k-1)} \sum_{i=1}^{N} \sum_{j \in V_k(i)} (s(i,j) - k), \tag{2}$$

where $s(i,j)$ is the rank of the data sample i from j in the output space, and $V_i(k)$ denotes the set of those data points that belong to the k-neighbors of data sample i in the original space, but not in the mapped space used for visualization. Both trustworthiness and continuity functions are function of the number of neighbors k. Usually, the qualitative measures of trustworthiness and continuity are calculated for $k = 1, 2, \ldots, k_{max}$, where k_{max} denotes the maximum number of the objects to be taken into account. At small values of parameter k the local reconstruction performance of the model can be tested, while at larger values of parameter k the global reconstruction is measured.

3 Visualisation by Use of Genetic Programming

Industrial applications require easily implementable, interpretable and accurate projections. Nonlinear functions (often referred as aggregates) are useful for this purpose. A pair of these functions realise feature selection and transformation. Such mapping is used for the visualisation and indexing of spectroscopic databases in the Topological Mapping using Aggregates (TOPNIR) modelling framework [19]. The two main forms of the aggregates are shown by Eqs. (3) and (4).

$$y_1 = a_{1,0} \frac{a_{1,1}x_{1,1} * a_{1,2}x_{1,2}}{a_{1,3}x_{1,3} * a_{1,4}x_{1,4}} \tag{3}$$

$$y_2 = a_{2,0} \frac{a_{2,1}x_{2,1} + a_{2,2}x_{2,2}}{a_{2,3}x_{2,3} * a_{2,4}x_{2,4}} \tag{4}$$

Finding the optimal set of features ($x_{i,j}$), the optimal model structure structure and parameter set of these functions is a complex nonlinear optimisation problem. These functions can be represented by trees (see Fig. 1) and genetic programming can be used to find the optimal model structure.

Because the algorithm of Genetic Programming is well-known, we will not present the details of the algorithm but focus here on the specific details.

Unlike common optimization methods, in which potential solutions are represented as numbers (usually vector of real numbers), the symbolic optimization algorithms represent the potential solutions by structured ordering of several

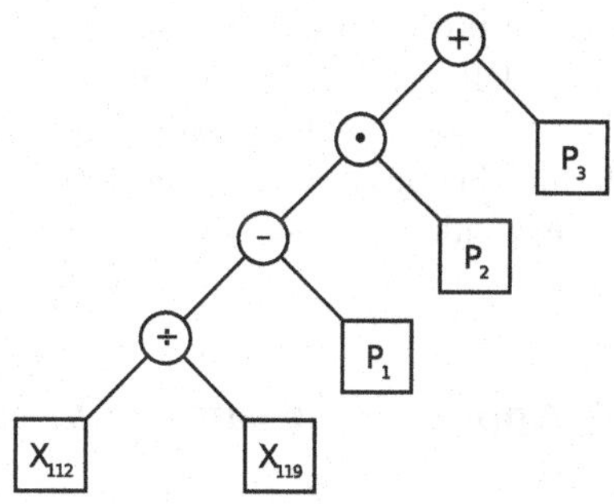

Fig. 1 Decomposition of a tree to function terms

symbols. One of the most popular method for representing structures is the binary tree. A population member in GP is a hierarchically structured tree consisting of functions and terminals. The functions and terminals are selected from a set of functions (operators) and a set of terminals. For example, the set of operators F can contain the basic arithmetic operations: $F = \{+, -, *, /\}$; however, it may also include other mathematical functions, Boolean operators, conditional operators or Automatically Defined Functions (ADFs). In this work we only used arithmetic operations. The set of terminals T contains the arguments for the functions. For example $T = \{x_1, \ldots x_n, p_j\}$ with x_i represents the elements of possible input variables and p_j represents the parameters. Now, a potential solution may be depicted as a rooted, labeled tree with ordered branches, using operations (internal nodes of the tree) from the function set and arguments (terminal nodes of the tree) from the terminal set.

Genetic Programming is an evolutionary algorithm. It works with a set of individuals (potential solutions), and these individuals form a generation. In every iteration the algorithm evaluates the individuals and selects the bets ones for reproduction according to their fitness values, generates new individuals by mutation, crossover and direct reproduction, and finally creates the new generation. The fitness function reflects the goodness of a potential solution which is proportional to the probability of the selection of the individual. In the current application the fitness function is based on the topology preserving property of the mapping:

$$fitness = M_1 M_2 = \frac{1}{N} \sum_{k=1}^{N} M_1(k) \sum_{k=1}^{N} M_2(k), \tag{5}$$

where N is the number of projected data-points.

The parameters of the functions (aggregates) have huge impact to the mapping performance. The evaluation of the fitness of the models are performed at optimal parameter values. Therefore a nonlinear parameter optimisation step is embedded into the GP. After GP generated the new population of model structures Sequential Quadratic Programming (SQP) calculates the optimal values of the parameters of these models.

The proposed approach has been implemented in MATLAB. The user should only define the high dimensional data that should be mapped, one aggregate function which optimal pair should be found by the optimisation, and the set of the terminal nodes (the set of the variables of the model and set of the internal nodes—mathematical operators.

4 Application Example: Visualization of Spectral Database

Near Infrared spectroscopy with Topological Mapping (TOPNIR) is widely used in oil industry to estimate product properties (e.g. aromatic components, cloud point, flash point, density etc.) of products and process streams [19]. TOPNIR performs a two dimensional mapping of the spectral space to visualise the operation regimes of the process. The TOPNIR algorithm utilises spectral databases for the prediction of product properties based on on-line measured infrared spectra by utilizing the well known k-nn algorithm [20]. There are 14 aggregates defined in the TOPWIN software used as a framework of the TOPNIR algorithm.The aggregates are equations that combine absorbances measured at significant wavelengths. In ideal case aggregates reflect product properties. Since these properties can be dependent on different rages of the spectra each aggregate built up several wavelengths to contain enough information related to a certain chemical property.

We used the topology preserving mapping based cost function to select the pairs of aggregates reflects the best of the hidden structure of the spectral database of an illustrative process at MOL Ltd. Duna Refinery. Figure 2 shows the mappings defined by these aggregates called Naro and Parox.

The model equations of these aggregates are the following:

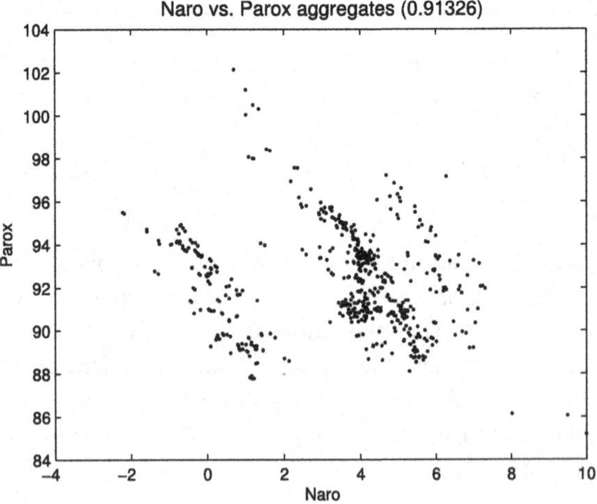

Fig. 2 Naro vs. parox aggregates (0.91326). $M_1 = 0.9379$, $M_2 = 0.9737$, $k = 10$

$$y_{Parox} = (x_{84}/(20 * x_{15} + x_{112}) - 0.0686) * 550 - 12.22 \tag{6}$$

$$y_{Naro} = ((x_{112}/x_{119}) - 1.2462) * 130 + 55 \tag{7}$$

As can be seen only four variables among the 195 $x_1 \cdots x_{195}$ are used by these aggregates, where x_i means the absorbance value at the i-th wavenumber in the the range (4,776–4,000) ($[\frac{1}{cm}]$). We tried to increase the be 0.913 performance of this mapping by adding more variables and the model and utilising the proposed genetic programming methodology.

Principal Component Analysis (PCA) utilises all the features as it extracts two independent linear combinations of these values (principal components) to map the variables .

$$y_j = PC_j = a_{j,0} + \sum_{i=1}^{n} a_{j,i} x_{j,i} \tag{8}$$

As can be seen, although all the features are used by this model its performance is not better than a simple nonlinear model of aggregates utilising four properly selected variables.

To obtain a much better model we applied GP to find the optimal pair of the Parox and also for the Naro aggregates. Based on our experiments we found that with the parameters given in Table. 1 the GP is able to find good solutions for various problems. Hence these parameters are the default parameters of the toolbox that have not been modified during the experiments presented in this paper.

The application of GP resulted the following two equations

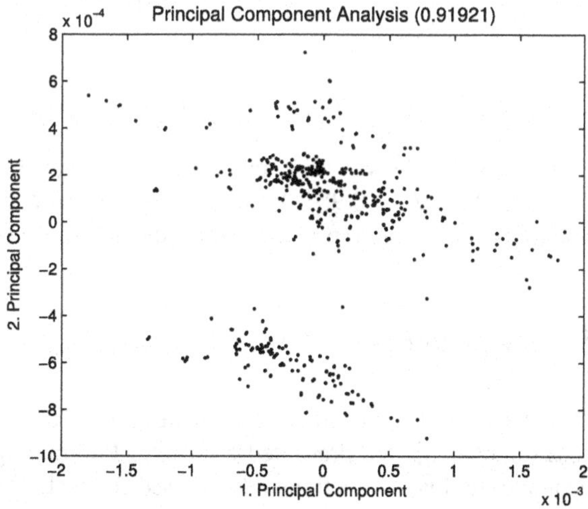

Fig. 3 Principal component analysis (0.91921). $M_1 = 0.9353$, $M_2 = 0.9828$, $k = 10$

Table 1 Parameters of GP in the application examples

Population size	50
Maximum number of evaluated individuals	2,500
Type of selection	Roulette-wheel
Type of mutation	Point-mutation
Type of crossover	One-point (2 parents)
Type of replacement	Elitist
Generation gap	0.667
Probability of crossover	0.5
Probability of mutation	0.5
Probability of changing terminal–non-terminal nodes (vica versa) during mutation	0.25

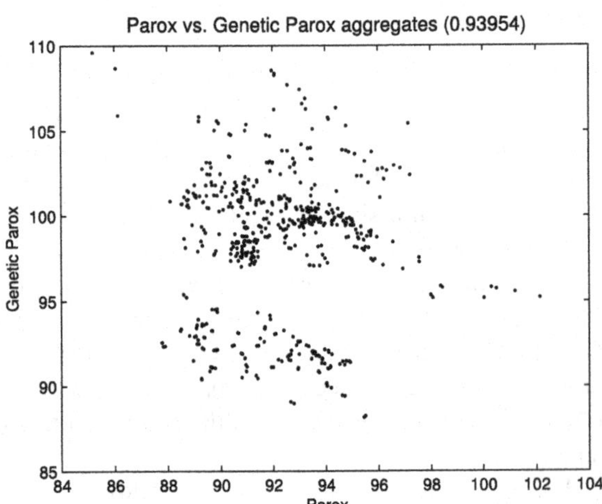

Fig. 4 Parox vs. genetic parox aggregates (0.93954). $M_1 = 0.955$, $M_2 = 0.9838$, $k = 10$

$$y_2^{Naro} = x_{108}/x_{143}/x_{153} - x_{185} \tag{9}$$

$$y_2^{Parox} = ((x_{87}/(x_{163} * x_{120}))/0.278332 \tag{10}$$

As Fig. 4 shows this mapping gives much better performance, 0.94 ($M_1 = 0.955$, $M_2 = 0.9838$) and the resulted map perfectly shows the different operating modes (summer and winter diesel) of the process.

5 Conclusions

Visualisation of high-dimensional data is important task in data mining and process monitoring. We presented a Genetic Programming (GP) based algorithm to generate nonlinear functions can be used for feature selection and transformation and applied to build an on-line spectroscopy based process monitoring system. We defined a novel cost function based on the topology preserving property of the mapping. The resulted tool was applied to design new aggregates for the TOPNIR modelling framework. An example based on industrial spactral database illustrated that the algorithm was able to generate compact and accurate mappings with better performance than PCA or classical aggregate based models.

Acknowledgments The financial support of the TAMOP-4.2.2/B-10/1-2010-0025 and the TAMOP-4.2.2.A-11/1/KONV-2012-0071 projects are gratefully acknowledged.

References

1. Jain, A., Zongker, D.: Feature selection: evaluation, application, and small sample performance. IEEE Trans. Pattern Anal. Mach. Intell. **192**, 153–158 (1997)
2. Narendra, P., Fukunaga, K.: A branch and bound algorithm for feature subset selection. IEEE Trans. Comput., **C-269**, 917–922 (1977)
3. Pudil, P., Novovičová, J., Kittler, J.: Floating search methods in feature selection. Pattern Recogn. Lett. **15**(1), 1119–1125 (1994)
4. Madr, J., Abonyi, J., Szeifert, F.: Genetic programming for the identification of nonlinear input-output models. Ind. Eng. Chem. Res., **44**(9), 3178–3186 (2005)
5. Jolliffe, T.: Principal Component Analysis. Springer, New York (1996)
6. Comon, P.: Independent component analysis: a new concept? Sig. Process. **36**(3), 287–317 (1994)
7. Fisher, R.A.: The use of multiple measurements in taxonomic problems. Annals Eugenics **7**, 179–188 (1936)
8. Kohonen, T.: Self-Organizing Maps. Springer, Berlin (2001)
9. Sammon, J.W.: A non-linear mapping for data structure analysis. IEEE Trans. Comput. **18**(5), 401–409 (1969)
10. Roweis, S., Saul, L.: Nonlinear dimensionality reduction by locally linear embedding. Science **290**, 2323–2326 (2000)
11. Tenenbaum, J.B., Silva, V., Langford, J.C.: A global geometric framework for nonlinear dimensionality reduction. Science **290**, 2319–2323 (2000)
12. Sonbul, Y.R.: Topological near infrared analysis modeling of petroleum refinery products (2005). US6.897.071 B2
13. Yang, J., Lee, I.: Common Clustering Algorithms. Comprehensive Chemometrics. Elsevier, Amsterdam, pp 577–618, (2009)
14. Erdil, E., Mimaroglu, S.: Combining multiple clusterings using similarity graph. Pattern Recogn. **44**(3), 694–703 (2011)
15. Chemaly, T.P., Aldrich, C.: Visualization of process data by use of evolutionary computation. Comput. Chem. Eng. **25**(9–10), 1341–1349 (2001)
16. Venna, J., Kaski, S.: Local multidimensional scaling with controlled tradeoff between trustworthiness and continuity. In: Proceedings of the Workshop on Self-organizing Maps, pp 695–702
17. Venna, J., Kaski, S.: Local multidimensional scaling. Neural Netw., **19**(6), 889–899 (2006)
18. Kaski, S., Nikkilä, J., Oja, M., Venna, J., Törönen, J., Castrén, E.: Trustworthiness and metrics in visualizing similarity of gene expression. BMC Bioinform., **4**(1), 48, (2003)
19. Descales, B., Lambert, D., Llinas, J.R., Martens, A., Osta, S., Sanchez, M., Bages, S.: Method for determining properties using near infra-red (nir), spectroscopy (2000). US6.070.128
20. Govindaraju, V., Wu, Y., Ianakiev, K.: Improved k-nearest neighbor classification. Pattern Recogn. **35**(1), 2311–2318 (2002)

Radial Basis Artificial Neural Network Models for Predicting Solubility Index of Roller Dried Goat Whole Milk Powder

Sumit Goyal and Gyanendra Kumar Goyal

Abstract In this work, Radial Basis (Exact Fit) and Radial Basis (Fewer Neurons) artificial neural network (ANN) models were developed to evaluate its capability in predicting the solubility index of roller dried goat whole milk powder. The ANN models were trained with a data file composed of variables: loose bulk density, packed bulk density, wettability and dispersibility, while solubility index was the output variable. The modeling results showed that there is an agreement between the experimental data and the predicted values, with coefficient of determination and Nash-Sutcliffe coefficient close to 1. Therefore, this method may be effective for rapid estimation of solubility index of roller dried goat whole milk powder.

1 Introduction

A study was planned for predicting the solubility index of roller dried goat whole milk powder by developing radial basis function (RBF) artificial neural network (ANN) models. In today's tough competition, a key issue that defines the success of a manufacturing organization is its ability to adapt easily to the changes of its business environment. It is very useful for a modern company to have a good estimate of how key indicators are going to behave in the future, a task that is fulfilled by forecasting. A competent predictive method can improve machine utilization, reduce inventories, achieve greater flexibility to changes and increase profits [1]. The contribution of goat milk to the economic and nutritional well being of humanity is undeniable in many developing countries, especially in the

S. Goyal (✉) · G. K. Goyal
National Dairy Research Institute, Karnal 132001, India
e-mail: thesumitgoyal@gmail.com

G. K. Goyal
e-mail: gkg5878@yahoo.com

V. Snášel et al. (eds.), *Soft Computing in Industrial Applications*,
Advances in Intelligent Systems and Computing 223, DOI: 10.1007/978-3-319-00930-8_21,

Mediterranean, Middle East, Eastern Europe and South American countries. Goat milk has played a very important role in health and nutrition of young and elderly people. It has been known for its beneficial and therapeutic effects on the people who have cow milk allergy. These nutritional, health and therapeutic benefits enlighten the potentials and values of goat milk and its specialty products. The chemical characteristics of goat milk can be used to manufacture a wide variety of products, including fluid beverage products (low fat, fortified, or flavoured) and ultra high temperature (UHT) milk; fermented products such as cheese, buttermilk or yogurt,; frozen products such as ice cream or frozen yogurt; butter, condensed/ dried products, sweets and candies. In addition, other specialty products such as hair, skin care and cosmetic products made from goat milk have recently gained further attention. Nevertheless, high quality products can only be produced from good quality goat milk. The quality milk should have the potential to tolerate technological treatment and be transformed into a product that satisfies the expectations of consumers in terms of nutritional, hygienic and sensory attributes. Taste is the main criteria used by consumers to make decisions to purchase and consume goat milk and its products [2]. In present era, the consumers are extremely conscious about quality of the foods they buy. Regulatory agencies are also very vigilant about quality and safety issues and insist on the manufacturers adhering to the label claims about quality and shelf life. Such discerning consumers, therefore, pose a far greater challenge in product development and marketing. The development of RBF-ANN models for predicting the solubility index of useful dairy product namely roller dried goat whole milk powder would be extremely beneficial to the manufactures, retailers, consumers and regulatory agencies from the quality, health and safety points of view.

2 Review of Literature

ANN has proved an efficient tool for predictive modelling concerning food products.

2.1 Butter

The seasonal variations of the fatty acids composition of butters over three seasons during a 12 month study in the protected designation of origin Parmigiano-Reggiano cheese area were studied. Fatty acids were analyzed by GC-FID, and then computed by ANN. Compared with spring and winter, butter manufactured from summer milk creams showed an optimal saturated/un-saturated fatty acids ratio (−8.89 and −5.79 %), lower levels of saturated fatty acids (−2.63 and

−1.68 %) and higher levels of mono-unsaturated (+5.50 and +3.45 %), poly-unsaturated fatty acids (+0.65 and +0.17 %), and rumenic acid (+0.55 and +3.41 %), while vaccenic acid had lower levels in spring and higher in winter (−2.94 and +2.91 %). ANN models were able to predict the season of production of milk creams, and classify butters obtained from spring and summer milk creams on the basis of the type of feeding regimens [3].

2.2 Cheese

Ni and Gunasekaran observed that a three-layer ANN model is able to predict more accurately than regression equations for the rheological properties of Swiss type cheeses on the basis of their composition [4]. The results of the experiments conducted by Jimenez-Marquez et al. [5] on prediction of moisture in cheese of commercial production using neural networks models can be used both for research to develop the base of knowledge on production variables and their complex interactions, as well as for the prediction of cheese moisture.

2.3 Processed Cheese

Linear Layer (Train) and Generalized Regression ANN models have been developed for predicting the shelf life of processed cheese stored at 7–8° C. The comparison of the two developed models showed that Generalized Regression model with spread constant as 10 got best simulated with less than 1 % root mean square error (RMSE). The study revealed that computational intelligence models are quite effective in predicting the shelf life of processed cheese [6]. Several other ANN models have been reported for processed cheese [7, 8].

2.4 Milk

The accuracy of milk production forecasts on dairy farms using a *ffann* (feed-forward ANN) with polynomial post-processing has been implemented. Historical milk production data was used to derive models that are able to predict milk production from farm inputs using a standard *ffann*, a *ffann* with polynomial post-processing and multiple linear regression. Forecasts obtained from the models were then compared with each other. Within the scope of the available data, it was found that the standard *ffann* did not improve on the multiple regression technique, but the *ffann* with polynomial post processing did [9].

2.5 Burfi

Radial basis (exact fit) model was proposed for estimating the shelf life of an extremely popular milk based sweetmeat namely burfi. The input variables were the experimental data of the product relating to moisture, titratable acidity, free fatty acids, tyrosine, and peroxide value; and the overall acceptability score was the output. Mean square error (MSE), RMSE, coefficient of determination (R^2) and Nash - Sutcliffe coefficient (E^2) were applied for comparing the prediction ability of the developed models. The observations indicated exceedingly well correlation between the actual data and predicted values, with high R^2 and E^2 values, establishing that the models were able to analyze non-linear multivariate data with very good performance and shorter calculation time. The developed model, which is very convenient, less expensive and fast, can be a good alternative to expensive, time consuming and cumbersome laboratory testing method for estimating the shelf life of the product [10].

2.6 ANN Modelling in Other Foodstuffs

ANNs have been used as a predictive modelling tool for several other foods, *viz.*, cherries [11], cakes [12], apple juice [13], chicken nuggets [14], Iranian flat bread [15], potato chips [16] and pistachio nuts [17].

The published literature shows that no work has been reported using ANN modelling for predictive analysis on goat milk powder. The present study would be of great significance to the dairy industry, academicians and researchers.

3 Method Material

For developing Radial Basis (Exact Fit) and Radial Basis (Fewer Neurons) models for predicting the solubility index of roller dried goat whole milk powder, several combinations were tried and tested to train the RBF-ANN models with spread constant ranging from 10 to 200. The dataset was randomly divided into two disjoint subsets namely, training set (having 78 % of the total observations) and testing set (22 % of the total observations). RBF-ANN consists of one layer of input nodes, one hidden radial-basis function layer and one output linear layer. The hidden layer contains n neurons. The hidden layer computes the vector distance (or radius) between the hidden layer weight vectors (which can be interpreted as the centers of the radial-basis functions of each neuron) and the input vectors. The resulting distances are multiplied by the hidden layer biases of each neuron and then a RBF (usually, a Gaussian function) is applied to the result [18]. The RBF-ANN topology has a special structure that has certain advantages over the more popular Feedforward ANN architecture, including faster training algorithms and more successful forecasting capabilities [1]. The input variables for RBF-ANN models were the data of the product pertaining to loose bulk density, packed bulk density, wettability and dispersibility, while solubility index was the output variable (Fig. 1).

In the present investigation, manual selection of spread variables (trial and error) was performed. The size of the deviation (also known as spread) determines how spiky the Gaussian functions are [19].

$$MSE = \left[\sum_{1}^{N}\left(\frac{Q_{\exp} - Q_{cal}}{n}\right)^2\right] \tag{1}$$

$$RMSE = \sqrt{\frac{1}{n}\left[\sum_{1}^{N}\left(\frac{Q_{\exp} - Q_{cal}}{Q_{\exp}}\right)^2\right]} \tag{2}$$

$$R^2 = 1 - \left[\sum_{1}^{N}\left(\frac{Q_{\exp} - Q_{cal}}{Q_{\exp}^{2}}\right)^2\right] \tag{3}$$

$$E^2 = 1 - \left[\sum_{1}^{N}\left(\frac{Q_{\exp} - Q_{cal}}{Q_{\exp} - \overline{Q_{\exp}}}\right)^2\right] \tag{4}$$

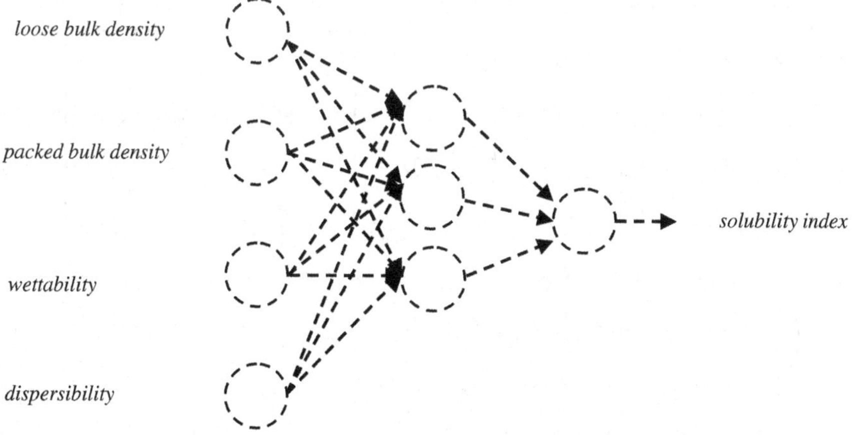

Fig. 1 Input and output variables of ANN model

where, Q_{exp} = Observed value; Q_{cal} = Predicted value; Q_{exp}=Mean predicted value; n= Number of observations in dataset. MSE (1); RMSE (2); R^2(3); and E^2 (4) were used with the aim to compare the prediction ability of the developed models. Neural Network Toolbox under MATLAB software was used for performing the experiments. Training pattern of ANN models is illustrated in Fig. 2.

4 Results and Discussion

The results of Radial Basis (Exact Fit) and Radial Basis (Fewer Neurons) models developed for predicting solubility index of roller dried goat whole milk powder are displayed in the Tables 1 and 2, respectively.

The Radial Basis (Exact Fit) and Radial Basis (Fewer Neurons) models got simulated very well, and gave high R^2 and E^2 values (Tables 1 and 2). The best results for radial basis model were with the spread constant 20→MSE 6.18519E-05; RMSE: 0.007864599; R^2: 0.992135401; E^2: 0.999938148. However, no difference was found between the results of the Radial Basis (Exact Fit) and Radial Basis (Fewer Neurons) models as both the models gave similar results with the same spread constants ranging from 10 to 200. Our observations are similar to the earlier findings of Sutrisno et al. [20], who developed ANN models with back-propagation algorithm to predict mangosteen quality during storage at the most appropriate pre-storage conditions which performed the longest storage period. In their experiments R^2 was found close to 1 (more than 0.99) for each parameter,

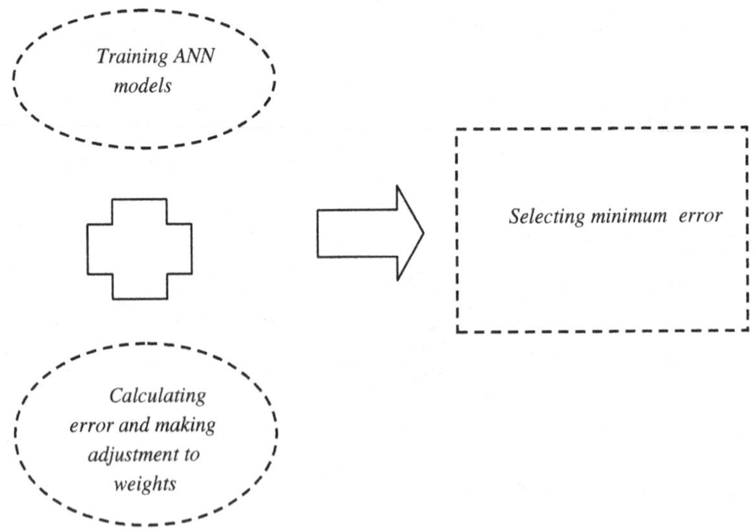

Fig. 2 Training pattern for ANN network

Table 1 Performance of radial basis (exact fit) model

Spread Constant	MSE	RMSE	R^2	E^2
10	9.09751E-05	0.009538085	0.990461915	0.999909025
20	**6.18519E-05**	**0.007864599**	**0.992135401**	**0.999938148**
30	6.23472E-05	0.007896026	0.992103974	0.999937653
40	6.9358E-05	0.008328147	0.991671853	0.999930642
50	7.61927E-05	0.00872884	0.99127116	0.999923807
60	8.0645E-05	0.008980256	0.991019744	0.999919355
70	8.30617E-05	0.009113821	0.990886179	0.999916938
80	8.45E-05	0.009192388	0.990807612	0.9999155
90	8.52238E-05	0.009231672	0.990768328	0.999914776
100	8.56595E-05	0.009255242	0.990744758	0.99991434
110	8.60964E-05	0.009278812	0.990721188	0.999913904
120	8.63882E-05	0.009294526	0.990705474	0.999913612
130	8.65343E-05	0.009302383	0.990697617	0.999913466
140	8.68269E-05	0.009318096	0.990681904	0.999913173
150	8.69734E-05	0.009325953	0.990674047	0.999913027
160	8.712E-05	0.00933381	0.99066619	0.99991288
170	8.74136E-05	0.009349523	0.990650477	0.999912586
180	8.75606E-05	0.00935738	0.99064262	0.999912439
190	8.78549E-05	0.009373093	0.990626907	0.999912145
200	8.80022E-05	0.00938095	0.99061905	0.999911998

Table 2 Performance of radial basis (fewer neurons) model

Spread Constant	MSE	RMSE	R^2	E^2
10	9.09751E-05	0.009538085	0.990461915	0.999909025
20	**6.18519E-05**	**0.007864599**	**0.992135401**	**0.999938148**
30	6.23472E-05	0.007896026	0.992103974	0.999937653
40	6.9358E-05	0.008328147	0.991671853	0.999930642
50	7.61927E-05	0.00872884	0.99127116	0.999923807
60	8.0645E-05	0.008980256	0.991019744	0.999919355
70	8.30617E-05	0.009113821	0.990886179	0.999916938
80	8.45E-05	0.009192388	0.990807612	0.9999155
90	8.52238E-05	0.009231672	0.990768328	0.999914776
100	8.56595E-05	0.009255242	0.990744758	0.99991434
110	8.60964E-05	0.009278812	0.990721188	0.999913904
120	8.63882E-05	0.009294526	0.990705474	0.999913612
130	8.65343E-05	0.009302383	0.990697617	0.999913466
140	8.68269E-05	0.009318096	0.990681904	0.999913173
150	8.69734E-05	0.009325953	0.990674047	0.999913027
160	8.712E-05	0.00933381	0.99066619	0.99991288
170	8.74136E-05	0.009349523	0.990650477	0.999912586
180	8.75606E-05	0.00935738	0.99064262	0.999912439
190	8.78549E-05	0.009373093	0.990626907	0.999912145
200	8.80022E-05	0.00938095	0.99061905	0.999911998

indicating that the model was good to memorize data. Fernandez et al. [21] studied the weekly milk production in goat flocks and clustering of goat flocks by using self organizing maps for prediction, establishing the effectiveness of ANN modelling in animal science applications. Another study showed that ANN modelling is a successful alternative to statistical regression analysis for predicting amino acid levels in feed ingredients [22]. The experimental results indicate that RBF-ANN modelling could potentially be used to predict the solubility index of roller dried goat whole milk powder.

5 Conclusion

The possibility of using radial basis function artificial neural network (RBF-ANN) model as an alternative to expensive, time consuming and cumbersome laboratory testing method for predicting the solubility index of roller dried goat whole milk powder has been successfully explored. The methodology is particularly useful for dairy industry, since meaningful prediction of milk powder quality using RBF-ANN modelling reduces costs and time of experimentation; thereby increasing income of the dairy industry. The RBF-ANN models predicted the solubility index of roller dried goat whole milk powder with reasonable accuracy with coefficient of determination and Nash - Sutcliffe coefficient close to 1. From the study, it is concluded that RBF-ANN models are a promising tool for predicting the solubility index of the product.

References

1. Doganis, P., Alexandridis, A., Patrinos, P., Sarimveis, H.: Time series sales forecasting for short shelf-life food products based on artificial neural networks and evolutionary computing. J. Food Engg., **75**, 196–204 (2006)
2. Ribeiro, A.C., Ribeiro, S.D.A.: Specialty products made from goat milk. Small Ruminant Res. **89**, 225–233 (2010)
3. Gori, A., Chiara, C., Selenia, M., Nocetti, M., Fabbri, A., Caboni, M.F., Losi, G.: Prediction of seasonal variation of butters by computing the fatty acids composition with artificial neural networks. Euro. J. Lip. Sci. Tech. **113**(11), 1412–1419 (2011)
4. Ni, H., Gunasekaran, S.: Food quality predication with neural networks. Food Tech. **52**(10), 60–65 (1998)
5. Jimenez-Marquez, S.A., Thibault, J., Lacroix. C.: Prediction of moisture in cheese of commercial production using neurocomputing models. Int. Dairy J. **15**, 1156–1174 (2005)
6. Goyal, S., Goyal, G.K.: Radial basis (exact fit) and linear layer (Design) ANN models for shelf life prediction of processed cheese. Int. J. u- e- Service Sci. Tech., **5**(1), 63–69 (2012)
7. Goyal, S., Goyal, G.K.: Supervised machine learning feedforward backpropagation models for predicting shelf life of processed cheese. J. Engg., **1**(2), 25–28 (2012)
8. Goyal, S., Goyal, G.K.: Analyzing shelf life of processed cheese by soft computing. Sci. J. of Ani. Sci., **1**(3), 119–125 (2012)

9. Sanzogni, L., Kerr, D.: Milk production estimates using feed forward artificial neural networks. Comp. Electro. Agri. **32**(1), 21–30 (2001)
10. Goyal, S., Goyal, G.K.: Radial basis (exact fit) artificial neural network technique for estimating shelf life of burfi. Adv. Comp. Sci. App., **1**(2), 93–96 (2012)
11. Guyer, D., Yang, X.: Use of genetic artificial neural networks and spectral imaging for defect detection on cherries. Comp. Electro. Agri. **29**(3), 179–194 (2000)
12. Goyal S., Goyal, G.K.: Central nervous system based computing models for shelf life prediction of soft mouth melting milk cakes. Int. J. Info. Tech. Comp. Sci., **4**(4), 33–39 (2012)
13. Raharitsifa, N., Ratti, C.: Foam-mat freeze-drying of apple juice part 1: experimental data and ANN simulations. J. Food Process Engg. **33**, 268–283 (2010)
14. Qiao, J., Wang, N., Ngadi, M.O., Kazemi, S.: Predicting mechanical properties of fried chicken nuggets using image processing and neural network techniques. J. Food Engg. **79**(3), 1065–1070 (2007)
15. Omid, M., Akram, A., Golmohammadi, A.: Modeling thermal conductivity of Iranian flat bread using artificial neural networks. Int. J. Food Prop. **14**(4), 708–720 (2011)
16. Serpen, A., Gökmen, V.: Modeling of acrylamide formation and browning ratio in potato chips by artificial neural network. Mol. Nut. Food Res. **51**(4), 383–389 (2007)
17. Omid, M., Baharlooei, A., Ahmadi, H.: Modeling drying kinetics of pistachio nuts with multilayer feed-forward neural network. Drying Tech. Int. J. **27**(10), 1069–1077 (2009)
18. Mateo. F., Gadea. R., Medina. Á., Mateo. R., Jiménez, M.: Predictive assessment of ochratoxin a accumulation in grape juice based-medium by aspergillus carbonarius using neural networks. J. App. Microbio., **107**(3), 915–927 (2009)
19. Loukas, Y.L.: Radial basis function networks in host-guest interactions: instant and accurate formation constant calculations. Anal. Chimica Acta **417**(2), 221–229 (2000)
20. Sutrisno., Edris, I.M., Sugiyono, P.: Quality prediction of mangosteen during storage using artificial neural network. In: International Agricultural Engineering Conference, Bangkok, Thailand, 7–10 Dec 2009
21. Fernandez, C., Soria, E., Martin, J.D., Serrano, A.J.: Neural networks for animal science applications: two case studies. Exp. Sys. Applic. **31**, 444–450 (2006)
22. Cravener, T., Roush, W.: Improving neural network prediction of amino acid ledients. Poult. Sci. **78**, 983–991 (1999)

Online Prediction of Wear on Rolls of a Bar Rolling Mill Based on Semi-Analytical Equations and Artificial Neural Networks

Yukio Shigaki and Marcos Antonio Cunha

Abstract This paper presents a computer model for online prediction of the wear contour of grooved rolls in the round-oval-round pass rolling process based on semi-analytical equations and artificial neural networks (ANN). This wear may adversely affect the shape quality of final product and is a result of complex interactions of many variables in the rolling process. The temperature of the material, amount of rolled material, water cooling system efficiency, diameters of the rolls, rolling speed and rolling load are some of these factors that play important role when assessing the wear of the rolls. A first ANN learns the average electrical current for thousands of hot rolled billets, and is done for ideal conditions, with new rolls. A second ANN calculates empirical coefficients in order to define the spread of the workpiece and then its contour is calculated accurately. This second ANN has inputs of differences on ideal and real electrical currents (and, thus, the rolling load variation) generated from the first ANN, temperature, water cooling pressure, speed of the rolls, diameters of the rolls, etc. Then the coefficients γ and $\boldsymbol{\kappa}$ (for wear profile) are calculated and input in semi-analytical equations to define the wear and its contour, as an online prediction. The works of Shinokura and Takai (1984) and Byon and Lee (2007) apply constant values for these two coefficients, limiting its application for other operational data variation during the rolling process. The model presented in this work uses an ANN to adapt γ and $\boldsymbol{\kappa}$ to cope with this variation. More than 50,000 billets were monitored and their operational data collected. The model was tested and the results agree well in real operational situations.

Y. Shigaki (✉)
CEFET-MG, Federal Center of Technological Education of Minas Gerais, Belo Horizonte, Brazil
e-mail: yukio@des.cefetmg.br

M. A. Cunha
GERDAU, Divinopolis mill, Brazil
e-mail: marcos.cunha@gerdau.com.br

V. Snášel et al. (eds.), *Soft Computing in Industrial Applications*,
Advances in Intelligent Systems and Computing 223, DOI: 10.1007/978-3-319-00930-8_22,

1 Introduction

A hot rod rolling process has several rolling mills in tandem, and each mill has grooved rolls that changes and reduces the cross sectional area of the billet into rod until its final form. A round-oval-round pass rolling means that the billet is formed into a round cross section, and next is rolled to an oval form, and then into a round cross section, and so on. Tons of rods are produced by this process and the surface and dimensional quality are deteriorated with the wear of the grooves of the rolls (Fig. 1).

It is not a very easy task to determine exactly when the amount of roll's wear seems to affect adversely on the rod's quality, and today its verification is done manually on the rod being rolled at high temperature, around 1000 °C. It can be seen that it is a risky operation. When this problem is detected, the process is stopped and the rolls are changed, and the old ones are sent to maintenance, interrupting the production of bars.

The wear and its profile in the rolls of hot rod rolling were studied by many researchers. For flat rolling products Oike et al. [1] proposed a wear model dependent on roll load, length of the arc of contact, strip width, thickness reduction, roll diameter, exit strip length and some empirical coefficients. Archard [2] developed a relationship between the amount of material lost by wear and the contact length, force and hardness of the roll's material. Sachs et al. showed that the rolling load is the most important factor when assessing the wear of rolls. Shinokura and Takai [3] studied several types of cross-sections for bar rolling and proposed an equation for the spread of the material of the billet inside the channel, as it has direct impact on the profile of the wear of the channels.

Lee et al. [4] propose a new analytical model to predict the surface profile of the wear contour of grooved roll in the oval–round (or round–oval) pass rolling process computed by using a linear interpolation of the radius of curvature of an incoming billet and the radius of roll groove in the roll axis direction. The results of the billet surface provided by this analytic model is consistent with that obtained experimentally for certain rolling conditions such as the shape of the channel and the distance between the cylinders. The proposed model is very accurate for predicting the cross-sectional area of the rod, compared to the processing time

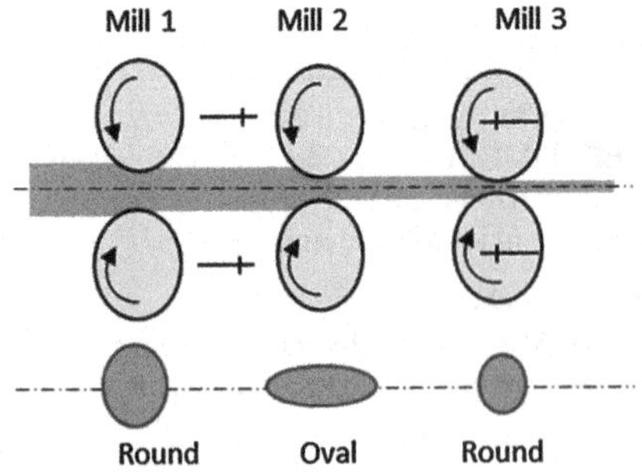

Fig. 1 Bar rolling tandem mills and cross sections

required using the finite element method. But further studies are needed to ensure the accuracy of the model for most cases, including changing rolling speed, channel geometry and material.

Kim et al. [5, 6] presented a study using neural network to preserve a uniform cross-sectional area of the output bar production line, considering the wear of the rolls. To predict the profile of wear on all passes in the process of hot rolling, they proposed a modification on the Archard's wear model considering the hardness of the cylinder. Then, the depth of wear of the cylinder was calculated at each stage of deformation in the contact region using the results of finite element simulation. This demonstrated that the proposed wear model could be used effectively in quantitative prediction of the wear profile of the rolls for oval and round sections. Furthermore, it was found that the use of neural networks produces results online to maintain uniform cross sectional area during the hot rolling process. These results can be used as a guide to adjust the distance between the rolls in the gap.

Byon and Lee [7] have proposed a semi-analytical model which predicts the contour of wear of the grooved rolls for oval section to round section in the hot rolling process. In this model the contours of the wear is assumed to be a second order polynomial function that is determined by applying a linear interpolation to the radius of curvature of the workpiece at the entrance of the channel of the roll and a weighting function which takes into account the rolling load, length of contact arc with the rolled rod, the hardness of the roll and the tonnage rolled. A system for measuring the amount of wear of the channel of the rolls was developed and contour of wear was measured using a plastic resin. This flows into the device created without restriction, filling the eroded area. The results show that the proposed model in this study, in general, have reasonable accuracy in predicting the wear of the channel contour of the roll, but the model does not consider, for example, the effect of increasing the roll speed on wear.

The calculated wear has shown to be linearly dependent on the length of the area of material in contact with the roll and the tonnage rolled, and demonstrated a nonlinear behavior with respect to rolling load, and a behavior inversely proportional to the hardness of the material of the roll. In a similar study, Byon et al. [8] have shown that the prediction of wear contour is in agreement with those obtained experimentally.

In another article, Byon and Lee [9] investigated the relationship between the variation of the gap between the rolls and the wear occurring in the channels. They made several experiments to find out what should be done for adjusting the distance between the rolls to maintain the profile of the rolled material with the geometric shape desired, depending on the wear of the grooves of the rolls. Based on experiments, they have proposed a model that predicts the adjustment of the gap when the groove already have a certain amount of wear. In this study the changes that occur in some process variables during bar rolling in a real situation were not taken into account. Among these variables, the actual temperature of the material changes, but in the study it is kept constant, and, therefore, the relationship should change. Other variables which change during the rolling process are the temperature of cooling water of the rolls, the pressure and flow of coolant.

In the article by Dong and Zheng [10], they studied the influence of alloying materials of the bar on spread, and found values 20–30 % higher for alloyed bars than in ordinary carbon steel bars. Then a greater spread, have a greater area of contact with the roller bar, thus causing a greater wear of the rolls.

Some models for determining the wear of rolls use empirical coefficients as one of variables of the process. Thus, the online application of these models with the suggested empirical coefficients by the above mentioned authors during the production process of the mill won't give precise results in the calculation of the wear, since the process parameters change significantly with time, and these changes alter the speed of the wear of rolls.

This study aims to analyze and correlate the wear occurred with changes in process parameters such as temperature of the workpiece, roll's speed, pressure and flow rate of cooling, rolling load, the hardness of the material, the mean temperature of cooling water and rolled tonnage, by two artificial neural networks (ANN).

A first ANN learns the average electrical current for thousands of hot rolled billets, and is done for ideal conditions, with new rolls. A second ANN calculates empirical coefficients in order to define the spread of the workpiece and then its contour is calculated accurately. This second ANN has inputs of differences on ideal and real electrical currents (and, thus, the rolling load variation) generated from the first ANN, temperature, water cooling pressure, speed of the rolls, diameters of the rolls, etc. Then the coefficients γ and $\boldsymbol{\kappa}$ (for wear profile) are calculated and input in semi-analytical equations to define the wear and its contour, as an online prediction. The works of Shinokura and Takai [3] and Byon and Lee [7] apply constant values for these two coefficients, limiting its application for other operational data variation during the rolling process. The model presented in this work uses an ANN to adapt γ and $\boldsymbol{\kappa}$ to deal with this variation. More than 50,000 billets were monitored and their operational variables collected.

2 Semi-Analytical Models for Spread and Wear

In this section is presented briefly the model of Shinokura and Takai for calculating the spread and the model of Byon and Lee to calculate the amount of wear of rolls for round-oval rolling sequence.

Equation (1) calculates the maximum spread.

$$W_{\max} = w_i \left(1 + \gamma \frac{\sqrt{R(H_{is} - H_{os})}}{w_i + 0.5H_i} \frac{A_h}{A_o} \right) \tag{1}$$

$$H_{is} = \frac{A_o - A_s}{B_c} \tag{2}$$

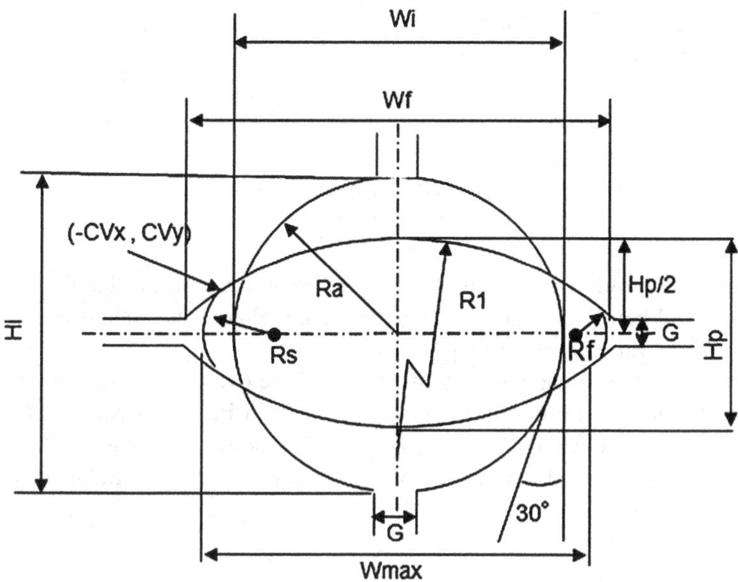

Fig. 2 A round workpiece entering into an oval groove

$$H_{os} = \frac{A_o - A_s A_h}{B_c} \tag{3}$$

When the maximum spread is known, Eqs. (2) and (3) can be used.

$$R_s = R_a W_t + R_f (1 - W_t) \tag{4}$$

$$W_t = \frac{W_f - W_{\max}}{W_f - W_i} \tag{5}$$

When the maximum spread is equal to the total width of the oval groove, Eq. (4) can be used to calculate R_f.

Where,

W_{max} Spread max
W_i Width of the entering workpiece
W_f Width of the oval groove
R_1 Radius of the oval groove
R_a Radius of the round profile
R_s Radius of the entering profile after spread
R_f Radius in the situation when $Wmax = Wf$ or the radius for maximum spread
H_i Height of the entering profile

H_p Height of the oval profile
γ,κ Empirical factors
G Roll gap

$$R_f = \frac{R_1 H_p - \left(W_f^2 + H_p^2\right)/4}{2R_1 - W_f} \tag{6}$$

If the entering profile with radius R_a is not deformed, this means that there wasn't total spread. When W_{max} equals W_i this means that there was no deformation, and $W_t = 1$. Then, Eq. (2) shows that $R_s = R_a$.

If the maximum spread equals to the width of the oval groove after rolled, then $W_{\max} = W_f$, and W_t in the Eq. (3) is null. And R_s in Eq. (2) is equal to R_f.

Sometimes R_s is calculated using Eq. (1). The points of contact (CV_X, CV_y), where the workpiece and the oval groove make contact, can be estimated from the Eqs. (7), (8) and (9).

$$\beta = \cos^{-1}\left(\frac{R_1 - H_p/2}{R_1 - R_s}\right) \tag{7}$$

$$CV_x = -R_1 \sin\beta \tag{8}$$

$$CV_y = -R_1 \cos\beta \tag{9}$$

Byon and Lee [7] proposed that the wear contour takes a parabolic form with the origin, O. ρ_{oval} is calculated by linear interpolation of R_1 and R_a, according to

$$\rho_{oval} = R_1 J_w + R_a(1 - j_W) \tag{10}$$

Figure 2 depicts the variables.

J_w is a weighting function at a given pass and is expressed as

$$J_W = 1 - \kappa\left(\frac{F_r^2 L_c N_B}{H_S}\right) \tag{11}$$

where

F_r	Rolling load at a pass
$Lc(= A_0 L_0/A_p)$	Length that the roll is in contact with workpiece during rolling
A_0, L_0	Cross sectional area and length of billet, respectively
A_p	Represents the cross sectional area and length of workpiece at a pass
H_s	Shore hardness of roll
κ	Correction coefficient

The parameter bo is computed as follows

$$\text{bo} = CV_y + R_1 - \frac{H_p}{2} - \sqrt{\rho_{oval}^2 - CV_x^2} \tag{12}$$

The roll wear contour, *f(x)* for the oval roll groove is given by

$$f(x) = \sqrt{\rho_{oval}^2\ x^2} + \text{ bo } \quad \text{for } \quad -CVx < x < CVx \tag{13}$$

Analogous equations are presented for a round groove. More details may be found in the excellent work by Byon and Lee [7].

3 Experimental Data

The wear contour was measured for 50,000 billets rolled in 22 stands tandem rolling mills. Each reading was done during the maintenance intervals with a dial indicator, and all operational parameters were collected (Fig. 3).

Figure 4 shows a sample of reading for groove 1 of the 11th stand. Each billet is 12 m long.

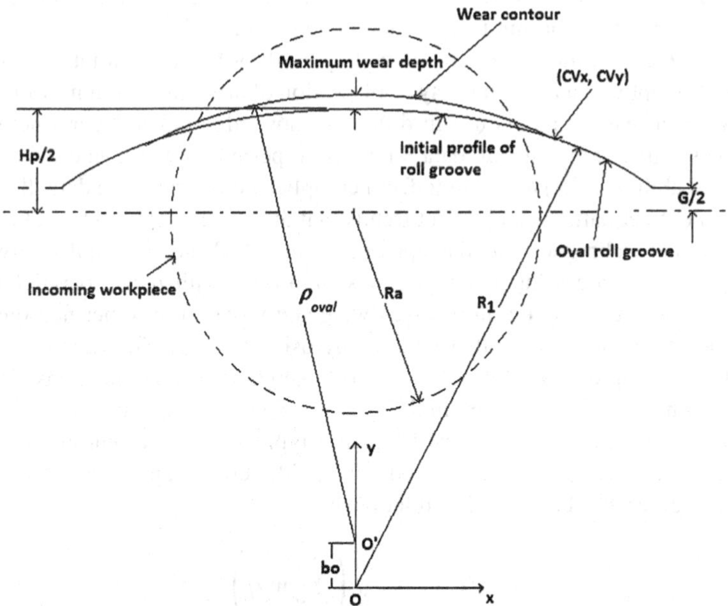

Fig. 3 Variables used to calculate the roll wear contour for an oval groove

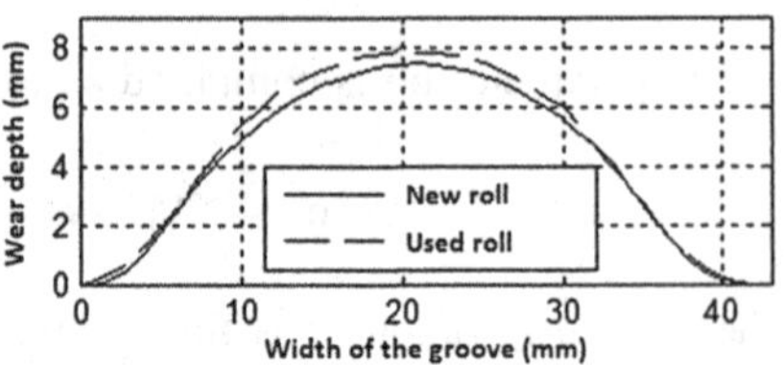

Fig. 4 Wear depth x width of the groove

Another important data is the electrical current that was measured for each billet with new rolls, and it was noted that the current level arises slightly from the start until the end of rolling process, as its temperature decreases.

4 Neural Network Model

Artificial neural networks (ANN) have been developed as generalizations of mathematical models of biological nervous systems and the basic processing elements of neural networks are called artificial neurons (or simply neurons or nodes). In a simplified mathematical model of the neuron, the effects of the synapses are represented by connection weights that modulate the effect of the associated input signals, and the nonlinear characteristic exhibited by neurons is represented by a transfer function. The neuron impulse is then computed as the weighted sum of the input signals, transformed by the transfer function, and the learning capability of an artificial neuron is achieved by adjusting the weights in accordance to the chosen learning algorithm, according to Sydenham and Thorn [11].

ANN have been a modeling scheme widely used for the multivariate non-linear analysis. The optimal nature and easy implantation of artificial neural networks has been the main reasons for its use in different knowledge fields. Neural networks can process great amounts of data in a short period of time, and with their remarkable ability to derive meaning from complicated or imprecise data, they can be used to extract patterns and detect trends that are too complex to be noticed by other computer techniques due to adaptive learning. ANN are essentially networks of many simple processing units (neurons or nodes) with dense parallel interconnections where each neuron receives weighted inputs from other neurons and communicates its outputs to other neurons by using an activation function. Thus, information is represented by massive cross-weighted interconnections. Neural networks might be single or multilayered, Kewalramani and Gupta [12]. According to Sydenham and Thorn [11], in a typical artificial neuron the signal flow from inputs $x_1, \ldots, x_n$ is considered to be unidirectional and the neuron output signal O is given by the following relationship:

$$0 = f(\text{ net }) = f\left(\sum_{j=1}^{n} w_j x_j\right) \tag{14}$$

where w_j is the weight vector, and the function f(net) is referred to as an activation (transfer) function. The variable net is defined as a scalar product of the weight and input vectors:

$$\text{net} = w^T x = w_1 x_1 + \ldots + w_n x_n \tag{15}$$

where T is the transpose of a matrix. In the simplest case, the output value O is computed as:

$$0 = f(\text{ net }) = \begin{cases} 1 & \text{if } \ w^T x \geq \theta \\ 0 & \text{otherwise} \end{cases} \tag{16}$$

where θ is the threshold level (this type of node is called a linear threshold unit).

A model with two ANNs were developed. The first ANN was trained with operational data for new rolls, and the target is the mean electrical current. The second ANN was trained with the wear data collected, with all operational data. The first ANN delivers the "ideal" current, and it is compared in the second ANN with the "real" current. This difference is taken into account as a kind of measure of the influence of the wear in the rolls. The targets of this second ANN are the empirical coefficients γ and $\boldsymbol{\kappa}$. With these values it is possible to calculate the amount of wear and its contour.

The backpropagation training algorithm was employed for both ANNs.

5 Results and Discussion

The ANN model was trained with real production data and adjusted to predict the rod profile, taking into account the variability of the operational parameters and the wear of the grooves.

The target of the ANN model are the empirical coefficients γ and $\boldsymbol{\kappa}$. With these values in hand it is possible to apply the equations of Shinokura and Takai for spread and the equations of Byon and Lee to calculate the amount of wear and its profile.

The model was implemented on the tandem mill and comparisons between the real profile and the prediction of the ANN were made.

Figure 5 shows a comparison between the model trained and new real data for stand number 11 and 9422 workpieces.

Table 1 presents a comparison of results for the presented new model with the prediction obtained with Shinokura/Takai/Byon/Lee's empirical coefficients and also the Archard's model.

Table 1 Comparison of results of the present model with two other methods

Number of billets	Measured wear (mm)	Results of presented new model (variable empirical coeff.) Variable **κ**	**Error**	Shinokura/Takai/ Byon/Lee's model (fixed empirical coeff.) 19,6E-12	**Error**	Achard's model (fixed empirical coeff.) 800,0E-12	**Error**
655	0.09	0.08	0.01	0.05	0.04	0.04	0.05
1347	0.16	0.14	0.02	0.10	0.06	0.08	0.08
2189	0.22	0.21	0.01	0.16	0.06	0.14	0.08
2970	0.28	0.27	0.01	0.22	0.06	0.19	0.09
3619	0.30	0.29	0.01	0.27	0.03	0.23	0.07
4449	0.33	0.315	0.02	0.34	0.01	0.28	0.05
5333	0.38	0.37	0.01	0.40	0.02	0.33	0.05
5990	0.40	0.41	0.01	0.46	0.06	0.38	0.02
7388	0.52	0.52	0.00	0.58	0.06	0.46	0.06
7986	0.58	0.58	0.00	0.63	0.05	0.50	0.08
8561	0.61	0.62	0.01	0.68	0.07	0.54	0.07
9422	0.65	0.66	0.01	0.76	0.11	0.59	0.06
		Average error	**0.01**	Average error	**0.05**	Average error	**0.06**

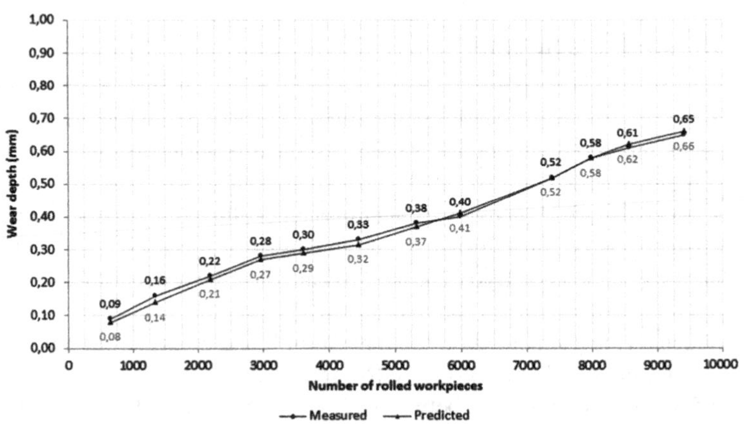

Fig. 5 Wear depth comparison

6 Conclusions

This work has developed ANN and semi-analytical models in order to calculate the wear and its profile in the grooves for rolls of bar hot rolling mills.

Preliminary results compared with the prediction of the model for new data has shown that it works very well. A comparison of this model with two other methods in Table 1 shows a better precision. In the near future this model will be prepared to automatically adjust the roll gap, and the human intervention, very risky as shown above, should not be necessary anymore.

References

1. Oike, Y., Okubo, I., Hirano, H., Umeda, K.: Tetsu-to-Hagane **63**, S222 (in Japanese) (1977)
2. Archard, J.F.: Contacts and rubbing of flat surfaces. J. Appl. Phys. **24**, 981–988 (1953)
3. Shinokura, T., Takai, K.A.: A new method for calculating spread in rod rolling. J. Appl. Metalworking **2**, 94 (1982)
4. Lee, Y., Choi, S., Kim, Y.H.: Mathematical model and experimental validation of surface profile of workpiece in round-oval-round pass sequence. J. Mater. Process. Technol. **108**, 4465–4470 (2000)
5. Kim, D.H., Kim, B.M., Lee, Y.: Application of ANN for the dimensional accuracy of workpiece in hot rod rolling process. J. Mater. Process. Technol. **130–131**, 214–218 (2002)
6. Kim, D.H., Kim, B.M., Lee, Y.: Adjustment of roll gap for the dimension accuracy of bar in hot bar rolling process. Int. J. KSPE **4**(1), 56–62 (2003)
7. Byon, S.M., Lee, Y.: Experimental and semi-analytical study of wear contour of roll groove and its applications to rod mill. ISIJ Int. **47**(47), 1006–1015 (2007)
8. Byon, S.M., Lee, Y.: A study of roll gap adjustment due to roll wear in groove rolling: experiment and modeling. In: Proceedings of the Institution of Mechanical Engineers, Part B: J. Eng. Manuf. **222**(7), (2008)
9. Byon, S.M., Lee, Y.: Experimental study for roll gap adjustment due to roll wear in single-stand rolling and multi-stand rolling test. J. Mech. Sci. Technol. **22**, 937–945 (2008)
10. Zheng, J., Dong, Y.: International Conference on Mechanic automation and control engineering (MACE), 26–28, (2010)
11. Sydenham, P.H.; Thorn, R.: Handbook of measuring system design. John Wiley & Sons (2005)
12. Kewalramani, M.A., Gupta, R.: Concrete compressive strength prediction using ultrasonic pulse velocity through artificial neural networks. Autom. Constr. **15**, 374–379 (2006)

Part II
Hybrid Machine Learning for Non-stationary and Complex Data

Real-Time Analysis of Non-stationary and Complex Network Related Data for Injection Attempts Detection

Michał Choraś and Rafał Kozik

Abstract The growing use of cloud services, increased number of users, novel mobile operating systems and changes in network infrastructures that connect devices create novel challenges for cyber security. In order to counter arising threats, network security mechanisms and protection schemes also evolve and use sophisticated sensors and methods. The drawback is that the more sensors (probes) are applied and the more information they acquire, the volume of data to process grows significantly. In this paper, we present real-time network data analysis mechanism. We also show the results for SQL Injection Attacks detection.

1 Rationale

Recently there is an increasing number of security incidents reported all over the world. The national CERTs (e.g. CERT Poland [1]) report that number of attacks in 2011 has increased significantly when compared to 2010. In annual reports they explain that most of network events submitted by automated feeds concern bot nets, spam, malicious URLs and Brute Force attacks.

The increased number of incidents is strongly related to the fact that recently there is also an increasing number of mobile devices users that form the population of connect-from-anywhere terminals that regularly test the traditional boundaries of network security. Also the so called BYOD (bring your own device [2, 3]) movement exposes the traditional security of many enterprises to novel and emerging threats. Many of nowadays malwares like ZITMO (Zeus In The Mobile)

M. Choraś
ITTI Ltd., Poznań, Poland

M. Choraś (✉) · R. Kozik
Institute of Telecommunications, UT&LS Bydgoszcz, Bydgoszcz, Poland
e-mail: mchoras@itti.com.pl

V. Snášel et al. (eds.), *Soft Computing in Industrial Applications*,
Advances in Intelligent Systems and Computing 223, DOI: 10.1007/978-3-319-00930-8_23,

do not aim at mobile device itself anymore but on gathering the information about the users and gaining the access to remote services like bank web services. This significantly expands cyber space network security perimeter.

There is also a significant number of reported incidents that are connected with huge widespread adoption of social media. Today, users are provide the content driving the growth at the same. This trend has a significant impact on accelerated spread of different kinds of malwares and viruses. As reported by SophosLabs [2] the number of malware pieces they have analyzed has been doubled since 2010.

Also as more and more cloud services and SaaS have been adapted by small and medium enterprises a big challenge for network security arises, since crucial for companies data started to be stored, maintained and transported by third party infrastructure where traditional points of inspection cannot be deployed. According to CISCO 2011 report [3] this trend is connected with the criminals that see the potential to get more return on their investment with cloud attacks, since they only need to hack one to hack them all.

Other well known problems like attacks on the web applications to extract data or to distribute malicious code still remain unsolved. Cybercriminals continuously steal data and distribute their malicious code via legitimate web servers they have compromised. Also the emerging technologies such as HTML5 bring new cyber threats that web services providers have to deal with.

In order to counter all the mentioned arising threats, network security mechanisms and protection schemes also evolve and use sophisticated sensors and methods. The drawback is that the more sensors (probes) are applied and the more information they acquire, the volume of data to process grows significantly.

Therefore, in this paper, we present real-time network data analysis mechanism and we prove its effectiveness for SQL Injection Attacks detection. The paper is structured as follows: SQLIA attacks and detection tools are shortly presented in Sect. 2. Our own solution for SQL Injection attempts detection based on the evolutionary algorithm is presented in Sect. 3. The experimental setup and results are provided in Sects. 4 and 5. Conclusions are given thereafter.

2 Current SQLIA Attack Detection Methods and Tools

One of the most important network threat is SQLIA (SQL Injection Attack) which ranks as top threat in the OWASP list [4]. SQL injection and other similar exploits are the results of interfacing a scripting language by directly passing information through another language and are ultimately caused by insufficient input validation. SQL Injection Attacks (SQLIA) refer to a code-injection attacks category in which part of the users input is treated as SQL code. Such code, if executed on the database, may change, erase, or expose sensitive data stored in the database.

One of the most significant examples of SQL Injection Attacks include:

- hacking the Royal Navys website and recovering user names and passwords of the sites administrators (November 2010) [5];
- stealing information related to almost 100,000 accounts of subscribers registered on ISP news and review site DSLReports.com (April 2011) [6];
- exploiting SQL injection vulnerabilities of approximately 500,000 web pages (April–August 2008) [7].

Several publications provide surveys, as well as analysis evaluating and comparing injection detection and prevention techniques. For example, more than twenty detective and preventive techniques are examined in [10]. In the publication, authors identified various types of SQLIAs and investigated ability to stop SQL injection provided by the most commonly used, current techniques. Similar approaches are presented in [11] and [12], where prevention techniques and security tools for the detection of SQL injection attacks were investigated.

The set of tools used in this paper for detecting the SQL Injection attacks consists of both an algorithms proposed by authors and known (state of the art) solutions and tools. The tools evaluated in our tests are:

1. Apache Scalp. It is an analyzer of Apache server access log file. It is able to detect several types of attacks targeted on web application. The detection is a signature-based one. The signatures have form of regular expressions that are borrowed from PHP-IDS project.
2. Snort. It the most widely deployed IDS system that uses set of rules that are used for detecting web application attacks. However most of the available rules are intended to detect very specific type of attacks that usually exploit very specific web-based application vulnerabilities.
3. ICD (Idealized Character Distribution [13]). The method is similar to the one proposed by Kruegel in [13]. The proposed character distribution model for describing the genuine traffic generated to web application. The Idealized Character Distribution (ICD) is obtained during the training phase from perfectly normal requests send to web application. The IDC is calculated as mean value of all character distributions. During the detection phase the probability that the character distribution of a query is an actual sample drawn from its ICD is evaluated. For that purpose Chi-Square metric is used.
4. SQL_ADS based on the Genetic Algorithm (proposed by authors [14] and described in Sect. 3.

3 Genetic Algorithm Description

In order to detect the anomalies in SQL queries a novel method is proposed. It exploits genetic algorithm, where the individuals in the population explore the log file that is generated by the SQL database. Each individual aims at delivering an generic rule (which is a regular expression) that will describe visited log line. It is important for the algorithm to have an set of genuine SQL queries during the learning phase.

The algorithm is divided into the following steps:

- Initialization. Each individual and line from log file is assigned. Each newly selected individual is compared to the previously selected in order to avoid duplicates.
- Adaptation phase. Each individual explores the fixed number of lines in the log file (the number is predefined and adjusted to obtain reasonable processing time of this phase).
- Fitness evaluation. Each individual fitness is evaluated. The global population fitness as well as rule level of specificity are taken into consideration, because we want to obtain set of rules that describe the lines in the log file.
- Cross over. Randomly selected two individuals are crossed over using algorithm for string alignment. If the newly created rule is too specific or too general it is dropped in order to keep low false positives and false negatives.

In order to obtain the regular expression from two strings a modified version of the Neddleman and Wunch algorithm is proposed [15]. The authors used this algorithm to find the best match between two DNA sequences which can diverge over time (e.g. by insertion or deletion) for different organisms. In order to find correspondence between those two sequences, it is allowed to modify the sequences by inserting the gaps. However, for each gap (and for mismatch) there is an penalty and award for genuine matches.

For Needleman and Wunsch algorithm the most important is to find the best alignment between two sequences (the one with highest award). From anomaly detection point of view the parts where gaps are inserted are also important, because they are the points of injections. These parts are described with regular expressions using guidelines proposed in [16]. Therefore, the obtained result can be represented with the following regular expression: "SELECT [a-z,]+ FROM patient WHERE name like [a-zA-z]+".

Needleman and Wunsch first suggested that in order to find the match with highest award a dynamic programming (DP) approach can be adapted. More details explaing how this is implemented can be found in [15].

The fitness function, that is used to evaluate each individual, takes into account the particular regular expression effectiveness (number of times it fires), the level of specificity of such rule and the overall effectiveness of the whole population. The fitness function is described by Eq. 1, where I indicates the particular individual regular expression, $E_{population}$ indicates the fitness of the whole population, E_f effectiveness of regular expression (number of times the rule fires), and E_s indicates the level of specificity. The α, β, and γ are constants that normalize the overall score and balance the each coefficient importance.

$$E(I) = \alpha * E_{population} + \beta * E_f(I) + \gamma * E_s(I) \tag{1}$$

$$E_{population} = \sum_{I \in Population} E_f(I) \tag{2}$$

The level of specificity indicates balance between number of matches and number of gaps.

4 Experiments

In this section our evaluation methodology is described. The SQL Injection Attacks are conducted on php-based web service with state of the art tools for services penetration and SQL injection. The traffic generated by attacking tools are combined together with normal traffic (genuine queries) in order to estimate the effectiveness of the proposed methods. The genuine queries are both man-made and generated by web crawlers as well.

The web service used for penetration test is so called LAMP (Apache + MySQL + PHP) server with MySQL back-end. It is one of the most common worldwide used servers and therefore it was used for validation purposes. The server was deployed on Linux Ubuntu operation system. For penetration tests examples services developed in PHP scripts and shipped by default with the server are validated.

Attack injection methodology is based on the known SQL injection methods, namely: boolean-based blind, time-based blind, error-based, UNION query and stacked queries. For that purpose sqlmap tool is used. It is an open source penetration and testing tool that allows the user to automate the process of validating the tested services against the SQL injection flaws.

In order to avoid double-counting the same attack patterns during the evaluation process, we decided to gather first the malicious SQL queries generated by sqlmap (several hundreds of different injection trials). After that genuine traffic (generated by crawlers and during the normal web service usage) is gathered. Such prepared data is used during the evaluation test that results are presented in Sect. 5.

5 Results

The conducted experiments were aimed at estimating the effectiveness of different tools commonly used for injection attack detection. Namely these are:

- Apache-Scalp (HTTP access log),
- Snort (HTTP packet content),
- ICD (HTTP access log),
- proposed SQL_ADS (SQL DB log).

Table 1 Effectiveness of injection attack detection (shown separately for genuine and malicious requests)

	SQL_ADS (%)	SNORT (%)	ICD (%)	SCALP (%)
Attack	87.8	66.3	97.9	50.9
Genuine	97.7	80.5	94.5	96.1
Weighted Avg.	96.2	78.3	95.0	89.0

It must be noticed that both Apache-Scalp and Snort tools do not require any learning phase, since the signatures of anomaly (having symptoms of SQL Injection) SQL queries and malicious HTTP request are provided together with theses tools. The signatures are developed by security experts in form of regular expressions.

The ICD and SQL_ADS require dedicated learning phase and focus only on genuine HTTP and SQL queries. Method used for evaluation engages classic 10-fold algorithm. As it is shown in Table 1, the proposed SQL_ADS algorithm slightly outperforms other state-of-the-art approaches when it comes to modelling the genuine queries. For queries having the symptoms of attack, the SQL_ADS is about 10 % worse when compared with ICD.

Another experiment aimed at investigating whenever combining above methods together can additionally improve overall effectiveness of injection attack detection. For that purpose 10-fold approach is used. The informations obtained from SQL_ADS, ICD and SNORT is used to build classifier for attack detection. Following classifiers were considered during this experiment:

- PART,
- NB (Naive Bayes),
- REPTree,
- J48,
- RIDOR.

The effectiveness of above classifiers is shown in Table 2. It can be noticed that overall weighted average effectiveness has increased (from 96 to 99 %) when we combine the proposed methods for injection attack detection.

Table 3 shows that without SQL_ADS the effectiveness of attack detection is worse, but it is still about 2 % better than the strongest classifier alone (in this case ICD).

Table 2 Effectiveness of different classifiers (with SQL_ADS)

	Attack	Genuine	Weig. Avg.
PART	0.967	0.995	0.991
NB	0.982	0.967	0.97
REPTree	0.955	0.997	0.99
J48	0.967	0.995	0.991
RIDOR	0.961	0.996	0.991

Table 3 Effectiveness of different classifiers (without SQL_ADS)

	Attack	Genuine	Weig. Avg.
PART	0.902	0.981	0.969
NB	0.881	0.983	0.967
REPTree	0.887	0.984	0.969
J48	0.902	0.981	0.969
RIDOR	0.887	0.978	0.963

6 Conclusions

In this paper an innovative correlation-base approach for injection attack detection was proposed. The described algorithm aims at efficient processing of large volume data that is generated by web applications. The advantage of the proposed solutions is that it allows for reusing existing efficient detectors for injection attack detection (e.g. SNORT, SCALP, character distribution approaches, etc.). Our experiments show that combining several weak injection attack detectors and engaging the machine learning techniques can lead to overall effectiveness improvement. In this paper we also proposed an novel evolutionary algorithm for modelling the genuine traffic with regular expressions. Presented results show that proposed algorithm, when combined with other approaches, can increase effectiveness of injection attack detection. The experiments show that proposed approach can achieve high effectiveness and can outperform other state of the art approaches like SNORT and SCALP.

Acknowledgments This work was partially supported by Applied Research Programme (PBS) of the National Centre for Research and Development (NCBR) funds allocated for the Research Project number PBS1/A3/14/2012 (SECOR)).

References

1. CERT Polska Annual Report. http://www.cert.pl/PDF/Report_CP_2011.pdf (2011)
2. SOPHOS homepage http://www.sophos.com
3. Cisco Annual Report (2011)
4. OWASP Top 10 2010, The Ten Most Critical Web Application Security Risks (2010)
5. Royal Navy Website Attacked by Romanian Hacker. http://www.bbc.co.uk/news/technology-11711478 (2008)
6. Mills, E.: DSL Reports Says Member Information Stolen (2011)
7. Keizer, G.: Huge Web Hack Attack Infects 500,000 pages (2008)
8. Rao, T.K., Kum, G.Y., Reddy, E.K., Sharma, M.: Major issues of web applications: a case study of SQL injection. J. Curr. Comput. Sci. Technol. **2**(1), 16–20 (2012)
9. Halfond, W., Orso, A.: AMNESIA: analysis and monitoring for neutralizing SQL-injection attacks. Proceedings of the 20th IEEEACM International Conference on Automated Software Engineering (2005)
10. Tajpour, A., JorJor Zade Shooshtari, M.: Evaluation of SQL injection detection and prevention techniques. In: CICSyN 2010 Second International Conference on Computational Intelligence, Communication Systems and, Networks (2010)

11. Amirtahmasebi, K., Jalalinia, S.R., Khadem, S.: A survey of SQL injection defense mechanisms. In: ICITST International Conference for Internet Technology and Secured, Transactions (2009)
12. Elia, I.A., Fonseca, J., Vieira, M.: Comparing SQL injection detection tools using attack injection: an experimental study. In: 2010 IEEE 21st International Symposium on Software, Reliability Engineering (2010)
13. Kruegel, C., Toth, T., Kirda, E.: Service specific anomaly detection for network intrusion detection. In: Proceedings of ACM Symposium on Applied, Computing, pp. 201–208 (2002)
14. Choraś, M., Kozik, R., Puchalski, D., Holubowicz, W.: Correlation approach for SQL injection attacks detection. In: Herrero, A. et al. (eds.) Advances in Intelligent Systems and Computing, vol. 189, pp. 177–186. Springer, Heidelberg (2012)
15. Needleman, S.B., Wunsch, C.D.: A general method applicable to the search for similarities in the amino acid sequence of two proteins. J. Mol. Biol. (1970)
16. Conrad, E.: Detecting Spam with Genetic Regular Expressions. SANS Institute InfoSec Reading Room (2007)

Recommending People to Follow Using Asymmetric Factor Models with Social Graphs

Tianle Ma, Yujiu Yang, Liangwei Wang and Bo Yuan

Abstract Traditional recommendation techniques often rely on the user-item rating matrix, which explicitly represents a user's preference among items. Recent studies on recommendations in the scenario of social networks still largely follow this principle. However, the challenge of recommending people to follow in social networks has yet to be studied thoroughly. In this paper, by using the utility instead of ratings and randomly sampling the negative cases in the recommendation log to create a balanced training dataset, we apply the popular matrix factorization techniques to predict whether a user will follow the person recommended or not. The asymmetric factor models are built with an extended item set incorporating the social graph information, which greatly improves the prediction accuracy. Other factors such as sequential patterns, CTR bias, and temporal dynamics are also exploited, which produce promising results on Task 1 of KDD Cup 2012.

T. Ma (✉) · Y. Yang · B. Yuan
Intelligent Computing Lab, Graduate School at Shenzhen, Tsinghua University, 518055 Shenzhen, P. R. China
e-mail: matianle1988@gmail.com

Y. Yang
e-mail: yang.yujiu@sz.tsinghua.edu.cn

B. Yuan
e-mail: yuanb@sz.tsinghua.edu.cn

L. Wang
Huawei Noah's Ark Lab, Huawei Technologies Co., Ltd., 518055 Shenzhen, P. R. China
e-mail: wangliangwei@huawei.com

V. Snášel et al. (eds.), *Soft Computing in Industrial Applications*, Advances in Intelligent Systems and Computing 223, DOI: 10.1007/978-3-319-00930-8_24,

1 Introduction

Online social networking services have become tremendously popular in recent years, such as Facebook, Twitter, and Tencent-Weibo. These social media sites are generating huge amount of social data every day. For example, currently, there are more than 200 million registered users on Tencent-Weibo, generating 40 million messages each day. This scale benefits the Tencent-Weibo users but it can also flood users with huge volumes of information and hence puts them at the risk of information overload [1].

To cope with the issue of information overload, existing recommender systems generally follow three strategies: collaborative filtering, content-based recommendation, and hybrid recommendation [2]. Social recommender systems are recommender systems that target the social media domain. And recommending people to follow has become one of the hot topics in social recommender systems. In KDD Cup 2012, Task 1 [1] is a prediction task that involves predicting whether or not a user will follow an item that has been recommended to the user. In this paper, we compared several approaches to recommending items, which can be persons, organizations, or groups, for users to follow. We found that, by incorporating social graph information into the asymmetric factor models (AFM), the recommendation accuracy of traditional matrix factorization techniques can be significantly improved. We also closely investigated issues such as how to incorporate sequence patterns, clickthrough rates (CTR) bias and temporal dynamics into our model, sampling schemes for the recommendation logs and training schemes for the factor models.

The rest of this paper is organized as follows. Section 2 introduces the related work and highlights the contributions of our work. The problem of recommending people to follow is formulated in Sect. 3 together with the details of AFM. Some analysis and discussion on related issues are presented in Sect. 4. Experimental results are given in Sect. 5 and this paper is concluded in Sect. 6.

2 Related Work

The motivation of recommender systems (RS) is to automatically suggest items to each user that he or she may find appealing (see [3] for an overview). The recommender has the task to predict the rating for user u on a non-rated item i or to generally recommend some items for the given user u based on the ratings that already exist [4].

As one of the most successful approaches to recommender systems, collaborative filtering approaches predict users' interests by mining user rating history data [5–8] and are most effective when users have expressed enough number of ratings. However, it deals poorly with the so called *cold start* problems.

Over the last two decades many CF algorithms have been proposed such as matrix factorization techniques, neighborhood-based approaches and restricted Boltzmann machine [9]. In general, there are three main categories of CF techniques [10]: memory-based, model-based, and hybrid CF algorithms that combine CF with other recommendation techniques. The memory-based CF [11] explores the user-item rating matrix and makes recommendations based on the ratings of item *i* by a set of users whose rating profiles are most similar to that of user *u*. While these approaches are easy to implement, their performance may be compromised when data are sparse. Hybrid CF algorithms, such as the content-boosted CF algorithm [12], are found helpful to address the sparsity problem. However, these hybrid approaches can result in increased complexity and expense for implementation [10].

In most traditional recommender systems, only the information in the user-item rating matrix is exploited for recommendations, ignoring completely the social relationships among users [9–11, 13–15]. The results of experiments in [16] and other similar work have confirmed that social networks can provide an independent source of information, which can be exploited to improve the quality of recommendations.

Recently, memory-based approaches have been proposed for recommendation in social rating network [17, 18]. These methods use the transitivity of trust and propagate trust to indirect neighbors in social network. Model-based approaches have also been applied to social rating networks [19–21] by exploiting matrix factorization techniques to learn latent features for users and items from observed ratings. However, due to the sensitive nature of social data, most of the related research studies are only based on the Epinions dataset [18–21] or dataset crawled from https://Flixster.com.

In this paper, we focused on the people recommendation prediction task. More specifically, we investigated Task 1 in KDD Cup 2012. In our problem, there is an implicit feedback in terms of whether a person follows the recommended item or not. The difference between the implicit feedback and the explicit feedback is that explicit feedbacks such as ratings involve more initiatives since a user only needs to click the 'accept' button to follow someone. In contrast, a person has to take some initiative to rate a movie that has been watched.

In summary, the main contributions of this paper are as follows:

- By adopting the more general concept of utility [3] instead of ratings and randomly sampling the negative cases in the recommendation log to make a balanced training dataset, we successfully apply the popular matrix factorization techniques to predict whether a user will follow an item recommended.
- To address the *cold start* issue, which is serious in our task as half of the users in the test dataset are not in the training dataset, we apply the asymmetrical factor models representing a user by items that have been accepted and followed.
- We incorporate social graphs into the asymmetric factor models with an extended item set to greatly improve the prediction accuracy.

- Factors such as sequential patterns, clickthrough rate bias, and the temporal aspects are also taken into consideration to further improve the performance.

3 People Recommendation with AFM and Social Graphs

Recommender systems using social network information are often aimed at optimizing RMSE on observed ratings [17–21]. Compared to neighborhood [22, 23] and random walk [18] methods, matrix factorization methods were found to be the most accurate model in the context of social network information [19–21]. As there are no ratings in our problem setting, we cannot directly apply the above recommendation approaches and some modifications are necessary.

3.1 Preliminaries

To investigate the task of recommending people to follow, we focused on the recent competition, Task 1 in KDD Cup 2012 [1]. In this task, the recommendation log is included in training datasets (*rec_train_log.txt*) and test datasets (*rec_test_log.txt*). The format of both files is:$(UserId)\backslash t(ItemId)\backslash t(Result)\backslash t(Unix - timestamp)$.

The values of *'Result'* field are 1 or -1, where 1 represents the user *UserId* accepts the recommendation of item *ItemId* and follows it, and -1 represents the user rejects the recommended item. The true values of *'Result'* field are provided in the *rec_train_log.txt*, whereas the true values of *'Result'* field in *rec_test_log.txt* are withheld which need to be predicted.

In addition to the recommendation log dataset, the user sns dataset contains each user's following history. In our paper, we focused on the CF methods by exploiting the recommendation log and social graph information.

Let $U = \{u_1, \cdots, u_n\}$ and $I = \{i_1, \cdots, i_m\}$ be the sets of all users and all possible items that can be recommended, respectively. The space of U is large with more than two million users. The item set contains 6,095 items, which will be extended to around twenty thousand in our algorithm. Let r be a utility function [3] that measures the usefulness of item i to user $u(r : U \times I \to R)$ where R is a totally ordered set (e.g., non-negative integers or real numbers within a certain range). For each user $u \in U$, we want to choose such an item $i \in I$ that maximizes the user's utility:

$$i \in \underset{i \in I}{\operatorname{argmax}}\ r_{ui},\ u \in U$$

Here, utility can be an arbitrary function that indicates how a particular user likes this item.

3.2 Basic Matrix Factorization Models

Matrix factorization models [15] map users and items to a joint latent factor space of dimensionality f, such that user-item interactions are modeled as inner products. Each item i is associated with a vector $q_i \in \mathbb{R}^f$, and each user is associated with a vector $p_u \in \mathbb{R}^f$. The resulting dot product, $q_i^T p_u$, captures the interaction between user u and item i—the user's overall interest in the item's characteristics.

In our case, the result value is 1 if a user followed an item and -1 otherwise. Although there is no rating, this result value can reflect a user's overall interest in the item's characteristics or the utility of the item to the user. In our method, we used 0 to replace -1, and treated these values as ratings. Then we calculated each user's ratings for all the items recommended in the test dataset and ranked them. The top three items were outputted as the final recommendation results.

As defined earlier, r_{ui} is the utility of item i to the user u, which represents the user's interest in the item, and its estimate $\tilde{r}_{ui}$ is defined as follows:

$$\tilde{r}_{ui} = q_i^T p_u. \tag{1}$$

To learn the factor vectors (p_u and q_i), we minimizes the regularized squared error on the training set [15] :

$$min_{q^*,p^*} \sum\nolimits_{(u,i)\in\kappa} (r_{ui} - q_i^T p_u)^2 + \lambda(||q_i||^2 + ||p_u||^2). \tag{2}$$

Here κ is the set of (u, i) pairs in the training set (the recommendation log of users' following history). The constant λ controls the extent of regularization and is usually determined by cross-validation.

Equation (1) tries to capture the interactions between users and items that produce different utility values. However, large portions of utility values may be due to effects associated with either users or items, known as biases and intercepts, instead of any meaningful interactions.

Thus, we added a first-order approximation of the bias into the utility r_{ui} [15]:

$$\tilde{r}_{ui} = q_i^T p_u + b_i + b_u. \tag{3}$$

In Eq. (2), the utility value is broken into three components: item bias, user bias, and user-item interaction, allowing each component to explain only the part of a signal relevant to it. The system learns by minimizing the squared error function [15]:

$$max_{q^*,p^*} \sum\nolimits_{(u,i)\in\kappa} (r_{ui}) - b_u - b_i - q_i^T p_u)^2 + \lambda(||q_i||^2 + ||p_u||^2 + b_u^2 + b_i^2). \tag{4}$$

3.3 Asymmetric Factor Models

While a user is represented by the feature in the plain SVD model, the asymmetric factor model [5] represents a user by the items accepted or followed. In other words, AFM only parameterizes item features.

In our problem, the size of the provided item dataset (around 6 K items) is too small compared to the number of users (around two million). As a result, we extended the item dataset to around 20 K items using the social graph information. There are two datasets containing social graph information: *user_sns.txt* and *user_action.txt* and users with the most followers were added into the item dataset.

Given $N(u)$ as the set of items which are followed by user u, a virtual user feature p_u is given by: $p_u = |N(u)|^{-1/2} \sum_{i \in N(u)} q_i$ [5, 9]. This representation offers several benefits such as integrating new data and new user without retraining the whole model [24]. In AFM, the estimated matching value of user u and item i is:

$$\tilde{r}_{ui} = q_i^T \cdot \left(|N(u)|^{-1/2} \sum\nolimits_{j \in N(u)} q_j\right) + b_i + b_u. \quad (5)$$

3.4 Sampling Scheme

It is not unusual that many recommendations were not accepted by the user and the corresponding results are negative in the training dataset (*rec_log_train.txt*), resulting in an imbalanced dataset. Meanwhile, a negative result itself is not a strong indication that a user did not want to accept the recommendation and had no interest in the item at all. Instead, chances are that the user may follow the item the next time when the item is recommended to him/her. As a result, the original training dataset was re-sampled before training our asymmetric factor models. All records with positive results were retained while the negative cases were randomly down sampled to be roughly the same size as the positive cases.

4 Performance Considerations

In addition to the asymmetric factor model, better performance can be possibly achieved by taking more factors into consideration, such as sequential patterns, CTR bias, and temporal dynamics.

4.1 Sequence Patterns

The recommendation log is a time series dataset and there are some important sequential features. As mentioned above, there were only two different utility values in the training dataset that can affect the mean average precision (MAP). Using sequential information, we proposed to adjust the utility values as follows:

- If the result field is 1, the utility value was also set to 1;
- If the result field is -1 and the user did not accept any other items in the same time, the utility value was set to a small value between 0 and 1 (0.2 in this paper);
- If the result field is -1 and the user did accept one or more items in the same time, the utility value was set to 0.

The above scheme is based on a common sense: simply knowing that a user did not follow the item recommended in the log does not guarantee that he or she had no interest in the item. Consequently, it is inappropriate to set the corresponding utility value to 0. Instead, a small value such as 0.2 was chosen to represent this uncertainty. However, if a user accepted some other items in a very short time period, the possibility that the user might be really of no interest in the item could be reasonably high. As a result, we set the utility value of this rejected item to 0.

4.2 CTR Bias

The clickthrough rate (CTR) in our context was defined as the percentage of acceptance of different items of all the recommendations among certain user groups. First, we divided all users in the dataset into disjoint groups according to their age and gender. We found that different groups had different CTRs denoted as c_{ui}.

To emphasize this type of group characteristics, we add a coefficient to each user, denoted by α_u. The bias of the CTR term is defined as follows:

$$b_{ui}^{ctr} = \alpha_u \cdot c_{ui}. \tag{6}$$

In Eq. (6), each c_{ui} was pre-computed while the coefficients α_u were learned along with other parameters using stochastic gradient descent training scheme.

The estimate of the utility value between user u and item i becomes:

$$\tilde{r}_{ui} = q_i^T \cdot (|N(u)|^{-1/2} \sum\nolimits_{j \in N(u)} q_j) + b_i + b_u + \alpha_u \cdot c_{ui}. \tag{7}$$

4.3 Temporal Dynamics

So far, the presented models have not considered the temporal dynamics. However, one factor with time-drifting nature must be considered: a user's inclination to accept recommendations vary across different time periods of a week. According to the statistics that we have collected, most users were more likely to accept recommendations during weekends and nonworking time. A possible explanation may be that most people usually do not have enough time to spend on social networks during working hours.

As a result, we uniformly divided a week into 168 time slots (1 h per slot). In each slot, a user may have a general inclination to accept or reject a recommended item. Let each user has a new hidden layer relating to time slots $p_{ut}(t_{ui})$. Similarly, each time slot has its own feature $q_t(t_{ui})$. Both $p_{ut}(t_{ui})$ and $q_t(t_{ui})$ have the same number of hidden layers. As a result, this temporal effect can be modeled as follows:

$$b_{ut}(t_{ui}) = p_{ut}(t_{ui}) \cdot q_t(t_{ui}). \tag{8}$$

The new estimate of the utility value of item i to user u becomes:

$$\tilde{r}_{ui} = q_i^T \cdot \left(|N(u)|^{-1/2} \sum\nolimits_{j \in N(u)} q_j\right) + b_i + b_u + b_{ut}(t_{ui}). \tag{9}$$

4.4 The Combination Model

Taking both the CTR bias and the time-drifting effect into consideration, a more accurate estimate of the matching value of user u and item i is:

$$\tilde{r}_{ui} = q_i^T \cdot \left(|N(u)|^{-1/2} \sum\nolimits_{j \in N(u)} q_j\right) + b_i + b_u + \alpha_u \cdot c_{ui} + p_{ut}(t_{ui}) \cdot q_t(t_{ui}). \tag{10}$$

The system learns by minimizing the regularized squared error function:

$$\begin{aligned} & min_{q^*, b^*, \alpha^*, q_t^*, p_{ut}^*} \sum\nolimits_{(u,i) \in \kappa} (r_{ui} - \tilde{r}_{ui})^2 + \lambda(||q_i||^2 + ||\alpha_u||^2 + b_u^2 + b_i^2 + \\ & \sum\nolimits_{j \in N(u)} ||q_j||^2) + \lambda_t(||q_t(t_{ui}))||^2 + ||p_{ut}(t_{ui})||^2). \end{aligned} \tag{11}$$

The $\tilde{r}_{ui}$ in Eq. (11) is the same as in Eq. (10). Since the time-drifting term is different from other terms, we used a different regulation parameter λ_t.

5 Experiments

We conducted a series of experiments on the datasets of Task 1 in KDD Cup 2012 (see [1] for details). In this section, we report our experimental results and compare the results with different methods.

5.1 Experimental Setup

In the provided datasets, there are 2,320,895 users with 50,655,143 following relations. To evaluate the performance of our asymmetrical factor models, we considered several other algorithms:

- **Common-Follow Algorithm:** This algorithm is based on the intuition that a user will follow an item if his/her friends also follow it. Actually, many social network sites such as Facebook use similar methods to recommend people to connect with. In our work, when an item was recommended to a user, we calculated the percentage of his/her followees who also followed the same item. For example, if the percentage of user u's followees who followed item i is p, the utility of i to u was defined as $r_{ui} = \log_2(p+1)$.
- **Common-Retweet Algorithm:** This algorithm is similar to the Common-Follow Algorithm. The only difference is that we calculated the percentage of retweet times of item i from the followees of the user u. For example, if the percentage of retweet times of item i from the followees of the user u is p, then the utility of i to u was defined as $r_{ui} = \log_2(p+1)$.
- **Hot-Degree Algorithm:** We calculated the click rates of each of item, and recommended the hottest items to the user.

5.2 Results and Analysis

In Table 1, as the public and private leaderboards have a temporal order, we can see that the results were always better for public leaderboard than private leaderboard since it is temporally more close to the training dataset.

The performance of the Common-Follow Algorithm and the Common-Retweet Algorithm was very poor. In the meantime, the Hot-Degreee Algorithm performed relatively well, although it did not handle the interaction between users and items. This is acceptable because most of us tend to follow the general trend.

The results of the basic SVD techniques, with or without the user and item bias, were not good compared to the Hot-Degree Algorithm, although more elaborated matrix factorization techniques may improve the results [15].

By contrast, the asymmetrical factor models with an extended item set significantly improved the results. The reason is that, by extending the item set, we incorporated additional useful information in the social graph into the models. The adoption of sequential patterns, temporal dynamics, and CTR bias also contributed to the improvement of the results.

Note that the model parameters, such as the number of hidden layers, the regulation parameters, the iteration steps, and the number of iterations can also affect the final results.

Table 1 Results of different recommendation algorithms

Algorithms	Public leaderboard	Private leaderboard
Common-Follow	0.23611	0.23541
Common-Retweet	0.22182	0.21882
Hot-Degree	0.33383	0.32831
Basic SVD	0.30548	0.28697
SVD + CTR bias	0.31446	0.29664
AFM with extended item set	0.36983	0.36016
AFM + sequential patterns	0.37009	0.36035
AFM + sequential patterns + Temporal Dynamics	0.37028	0.36055
AFM + sequential patterns + CTR bias	0.37076	0.36093
AFM + sequential patterns + Temporal Dynamics + CTR bias	0.37132	0.36143

6 Conclusions

In this paper, we proposed to incorporate social graph information into asymmetrical factor models by extending the item dataset, in order to make accurate predictions on whether a user will follow an item or not. One of the major differences between traditional recommender systems and social recommender systems is that there are no ratings in social networks. To cope with this issue, we used the utility of an item to a user instead of ratings and developed novel latent factor models.

Experimental results show that social graph information can significantly improve the performance of recommender systems, compared to a number of standard recommendation algorithms. In the meantime, additional factors were also taken into consideration, such as sequential patterns, CTR bias, and temporal dynamics, which further boosted the predication accuracy.

As to future work, in our paper, the model was optimized in terms of RMSE. However, the evaluation metric in Task 1 of KDD Cup 2012 is MAP. Although RMSE and MAP are highly positively correlated, we can try to adapt our model to optimize MAP directly in the future. Also, we can build a directed weighted social graph that can be used to build a more elaborated model.

Acknowledgments This work was supported by the National Natural Science Foundation of China (No. 60905030) and Upgrading Plan Project of Shenzhen Key Laboratory (No. CXB201005250038A). The authors are also grateful to several colleagues who have provided constructive feedbacks to our work. Besides, we would like to gratefully acknowledge the organizers of KDD Cup 2012 as well as Tencent Inc. for making the datasets available.

References

1. Niu, Y., Wang, Y., Sun, G., Yue, A., Dalessandro, B., Perlich, C., Hamner, B.: The Tencent Dataset and KDD-Cup'12. In KDD-Cup, Workshop (2012)
2. Jannach, D., Zanker, M., Felfernig, A., Friedrich, G.: Recommender Systems: An Introduction. Cambridge University Press, Cambridge (2010)
3. Adomavicius, G., Tuzhilin, A.: Toward the next generation of recommender systems: a survey of the state-of-the-art and possible extensions. IEEE Trans. Knowl. Data Eng. **17**(6), 734–749 (2005)
4. Yang, X., Steck, H., Guo, Y., Liu, Y.: On Top-k Recommendation using Social Networks. In: Proceedings of 6th ACM Conference on Recommender Systems. ACM, Dublin (2012)
5. Paterek, A.: Improving regularized singular value decomposition for collaborative filtering. In: Proceedings of KDD Cup and Workshop 2007, pp. 5–8. ACM, San Jose (2007)
6. Keshavan, R.H., Montanari, A., Oh, S.: Matrix completion from noisy entries. J. Mach. Learn. Res. **11**, 2057–2078 (2010)
7. Pan, R., Zhou, Y., Cao, B., Liu, N.N., Lukose, R., Scholz, M., Yang, Q.: One-class collaborative filtering. In: Proceedings of the 8th IEEE International Conference on Data Mining, pp. 502–511. IEEE, Pisa (2008)
8. Srebro, N., Jaakkola, T.: Weighted low-rank approximations. In: Proceedings of the 20th International Conference on Machine Learning, pp. 720–727. AAAI Press, Washington, DC (2003)
9. Jahrer, M., Töscher, A., Legenstein, R.: Combining predictions for accurate recommender systems. In: Proceedings of the 16th ACM SIGKDD International Conference on Knowledge Discovery and Data Mining, pp. 693–702. ACM, Washington, DC (2010)
10. Su, X., Khoshgoftaar T.M.: A Survey of Collaborative Filtering Techniques. Adv. Artif. Intell. (2009)
11. Goldberg, D., Nichols, D., Oki, B.M., Terry, D.: Using collaborative filtering to weave an information tapestry. Commun. ACM **35**(12), 61–70 (1992)
12. Melville, P., Mooney, R.J., Nagarajan, R.: Content-boosted collaborative filtering for improved recommendations. In: Proceedings of the 18th National Conference on Artificial Intelligence and 14th Conference on Innovative Application of Artificial Intelligence, pp. 187–192. AAAI Press; MIT Press; Edmonton (2002)
13. Koren, Y.: Collaborative filtering with temporal dynamics. Commun. ACM **53**(4), 89–97 (2010)
14. Koren, Y.: Factorization meets the neighborhood: a multifaceted collaborative filtering model. In: Proceedings of the 14th ACM SIGKDD International Conference on Knowledge Discovery and Data Mining, pp. 426–434. ACM, Las Vegas, Nevada (2008)
15. Koren, Y., Bell, R., Volinsky, C.: Matrix factorization techniques for recommender systems. IEEE Comput. **42**(8), 30–37 (2009)
16. Singla, P., Richardson, M.: Yes, there is a correlation: from social networks to personal behavior on the web. In: Proceedings of the 17th International Conference on World Wide Web, pp. 655–664. ACM, Beijing (2008)
17. Massa, P., Avesani, P.: Trust-aware recommender systems. In: Proceedings of the 2007 ACM Conference on Recommender Systems, pp. 17–24. ACM, Minneapolis (2007)
18. Jamali, M., Ester, M.: TrustWalker: a random walk model for combining trust-based and item-based recommendation. In: Proceedings of the 15th ACM SIGKDD International Conference on Knowledge Discovery and Data Mining, pp. 397–406. ACM, Paris (2009)
19. Ma, H., King, I., Lyu, M.R.: Learning to recommend with social trust ensemble. In: Proceedings of the 32nd Annual International ACM SIGIR Conference on Research and Development in Information Retrieval, pp. 203–210. ACM, Boston (2009)
20. Jamali, M., Ester, M.: A matrix factorization technique with trust propagation for recommendation in social networks. In: Proceedings of the 2010 ACM Conference on Recommender Systems, pp. 135–142. ACM, Barcelona (2010)

21. Ma, H., Yang, H., Lyu, M.R., King, I.: Sorec: social recommendation using probabilistic matrix factorization. In: Proceedings of the 17th ACM Conference on Information and Knowledge Management, pp. 931–940. ACM, Napa Valley (2008)
22. Golbeck, J.A.: Computing and Applying Trust in Web-based Social Networks. University of Maryland at College Park, College Park (2005)
23. Jamali, M., Ester, M.: Using a trust network to improve Top-N recommendation. In: Proceedings of the 2009 ACM Conference on Recommender Systems, pp. 181–188. ACM, New York (2009)
24. Koren, Y.: Factor in the neighbors: scalable and accurate collaborative filtering. ACM Trans. Knowl. Discov. Data **4**(1), 1–24 (2010)

Part III
Emerging Theories and Applications in Transportation Science

Air Travel Demand Fuzzy Modelling: Trip Generation and Trip Distribution

Milica Kalić, Jovana Kuljanin and Slavica Dožić

Abstract This chapter describes the fuzzy logic approach to modelling of trip generation and trip distribution on country and country-pair levels. Different economic (GDP per capita of origin country, imports by destination countries) and social factors, as well as other ones (number of emigrants in destination country and destination country attractiveness) are considered. The case study of Serbia, illustrating possibilities of models, is given. Results of this research provide empirical evidence relating to successful use of fuzzy logic as a non-traditional technique.

1 Introduction

Air travel demand modelling and forecasting are key parts of the transportation planning process. There are different dimensions of air travel demand models depending on the purpose of study. From the aggregation level point of view, it is possible to categorize it into following levels: country level, country-pair, city, airport, city-pair, airport-pair and route. With regard to the process of modelling, there are the following model types: trip generation (trip emission and trip attraction), trip distribution, route choice and trip assignment. Depending on data, models may be based on aggregate or disaggregate data. Likewise, an air travel demand model can be more or less complex and sophisticated, causal or non-causal (time series, projection and trend extrapolation) as well as deterministic or stochastic.

M. Kalić (✉) · J. Kuljanin · S. Dožić
University of Belgrade, Faculty of Transport and Traffic Engineering, Belgrade, Serbia
e-mail: m.kalic@sf.bg.ac.rs

J. Kuljanin
e-mail: j.kuljanin@sf.bg.ac.rs

S. Dožić
e-mail: s.dozic@sf.bg.ac.rs

V. Snášel et al. (eds.), *Soft Computing in Industrial Applications*,
Advances in Intelligent Systems and Computing 223, DOI: 10.1007/978-3-319-00930-8_25,

This chapter presents country and country-pair levels of air travel demand models. With regard to different economic and social factors, two fuzzy models are developed in order to predict the total number of air passengers of a country and passenger flows between an origin country and destination countries. Trip generation and trip distribution problems are solved using fuzzy logic system. Fuzzy logic is shown to be a very promising approach to modelling the air travel demand process. It is very important to emphasize that fuzzy logic system is a universal approximator which means nonlinear mapping of an input data vector into a scalar output. In the proposed models, fuzzy logic system maps crisp inputs into crisp outputs.

The models incorporate experience from earlier works in this area. Models developed in this chapter are based on publicly available data for Serbia, as the origin country. Results of this research provide empirical evidence relating to successful use of fuzzy logic as a non-traditional technique.

This chapter has five sections. After the introduction and review of literature and previous experiences (Sect. 2), the conceptual framework of two fuzzy models is presented (Sect. 3). Also, in Sect. 3, the chapter examines different economic factors which may have influence on trip generation (concerning country level—aggregate model) and trip distribution (concerning links between origin country and selected destination countries—country-pair aggregate level). The fuzzy models based on the data related to Serbia are presented in Sect. 4. Section 5 offers some concluding remarks and further research.

2 Literature Review

Air transport demand forecast problem has been considered by many authors. Carson, Cenesizoglu and Parker considered the forecast of aggregate air travel demand and tried to give the answer whether it is better to use data at the national level or airport specific data [1]. Hsiao and Hansen developed a city-pair air demand model based on supply characteristics on routes and regional demand-side variables and applied it to the air transportation system of the United States [2]. Profillidis proposed econometric model and fuzzy linear regression for forecasting airport demand for the case of a tourist airport on Rhodes [3]. Alekseev and Seixas compared classical and neural modelling by introducing hybrid neural model for demand forecast (Brazil case study) [4]. Auto-regressive and exponential smoothing models are used by Samagaio and Wolters to develop independent forecasts for passenger numbers for the Lisbon Airport [5]. Many papers have shown that the fuzzy logic can be successfully applied in sequential procedures of passenger demand forecasting [6–9]. Those papers considered each phase in the transportation planning process (trip generation, trip distribution, modal split and route choice) separately. Kalić and Tošić developed a trip distribution model in air transportation under irregular conditions, when business and recreational trips were almost totally absent (Belgrade Airport case study between 1991 and 2000)

[8]. The basic assumption in this chapter was that the main impact on trip distribution was the number of people emigrating out of the country. Kalić, Dožić and Babić [10] developed a three-stage sequential air travel demand model that covered the last decade which was characterized by distinct political and market conditions (improved political situation, partially open air market, creation of competition, market penetration by low cost carriers, etc.), as well as increase in the standard of living. The stages are linked in such a way so that the output of each stage represents the input for the following one. The first stage—trip generation was estimated by a multiple linear regression analysis that related the variation in traffic at Belgrade airport to the variation of two independent variables (GDP per capita and foreign tourist arrivals). Trip distribution for the years 2012 and 2015 was done according to the historical data and expert opinion which considered the existing competitors and market trends. In the third stage, airline choice was considered in order to predict the flow share of the incumbent airline taking into account tariff differences and weekly frequencies. Following these researches [8, 10], this chapter investigates a similar problem, but on a different aggregation level as well as with a different approach. Namely, trip generation has been considered on the country level and trip distribution between origin country and destination countries (country-pair level). In this chapter, two separate fuzzy models (trip generation and trip distribution) are developed.

3 Conceptual Framework of Models

3.1 Theoretical Background

The air travel demand has traditionally been affected by the complex interaction of a vast number of factors that can roughly fall into two key sets. The first set reflects economic activity of a specific country and encompasses the variables such as Gross Domestic Products (GDP), level of trade growth, foreign direct investments (FDI), employment and unemployment rate, etc. Each of these factors will be further discussed in order to gain a better insight into how they affect the air travel demand. The second set of factors to consider arises from a change in supply. Those are primarily factors which emphasize supply of air travel that represent one of the main factors that influence demand. Of the general factors on the side of supply, penetration of low-cost carriers, declining real prices of air fares, speed of air travel and convenience of air travel are perhaps the most important ones.

Gross Domestic Product. GDP has always been a significant driver of air travel demand, and there is no doubt that it will certainly be so in the future. This factor captures the entire economic activity of a country on the most general level. It can be very often converted to the level of personal income when the focus of research is on the individuals who use air services (disaggregate level). GDP growth was used as a major indicator in modelling and trend projection of the air travel demand on country level. Doganis indicates income elasticity, the term which refers to the relation between a change in demand and a change in income (this is arrived at by dividing the percentage change in demand generated by an income change by the percentage change in personal income) [11]. Gillen et al. report that the elasticity of demand for air travel with respect to income varies from 0.8 to 2.6, depending on what country is being investigated [12]. In well-developed countries with matured economy and air transport system, response of air travel demand to change in GDP is weaker compared to developing countries where a small change in GDP causes a higher change in demand. It must be underlined that the relation between GDP (per capita) growth and air travel demand growth has usually been overestimated due to failure to take into account other variables, such as price of air fares, policy changes (liberalization of bilateral agreements, changing business models of carriers etc.) and network development.

Trade. Domestic trade together with international trade represent a fundamental factor of economic growth. The level of international trade directly influences the demand of business travel. According to many authors, what really contributes to international travel is the amount of trade in merchandise and service. Oum et al. argue that elasticity in this case is 0.83 (drop in trade by 10 % leads to a drop in international air travel demand of 8.3 %) [13]. It must be indicated that recent development of international supply chains boosted an increase in international air travel demand.

Foreign Direct Investments. In addition to the positive impact of GDP and domestic and international trade on air travel demand, FDI is another significant force which underlies the generation of air travel demand. Its magnitude is measured by the number of investments which comes from outside the county. According to Gillen, foreign direct investment represents a rough measure of the degree of globalization, and as more investments take place air passenger demand increase [12].

According to Doganis, all factors influencing air travel demand that are mentioned above are divided into two groups: factors that influence demand in all markets and more particular factors that may influence demand on specific routes but may be totally absent on others [11]. Such division of factors influencing air travel demand seems to be very reasonable and meaningful, since some routes tend to be quite leisure-related while others are still business-related. Historical events and cultural links are very important for understanding behaviour of a particular market. Thus, ethnic war and unrest during the 1990s which hit the region of the Balkans led to strong population movements towards North America, Canada, Australia and some countries of the European Union. During the period of recovery, this migration has caused strong residual ethnic links between communities (e.g. friends and relatives visiting their friends or families abroad—VFR travel). Cultural ties may also influence the trading activities and therefore they may have direct effect on business travel. Social environment and culture of a country highly determine the attitudes of its inhabitants towards tourist and free time preferences, as well as expenditure distribution. Migration of labor flows is still running and contributing to seasonal peaks.

Two models using fuzzy logic have been developed in this chapter. The first model refers to trip generation and can generally be applied to any country. The second model represents a trip distribution model which investigates strong flows between origin country and selected countries. This model can also be used for any general condition.

3.2 Trip Generation

This model emphasizes the role of an economic factor, such as GDP. As it was mentioned before, GDP was identified to have significant influence on a country's overall economic activity. Therefore, GDP can be employed as an explanatory variable in any robust model which tends to describe air travel demand on the aggregate level (in this case—country level). Consequently, GDP per capita will also share positive correlation with air travel demand. As expected, higher disposable income means more money that people can spend on travel by air for a variety of purposes. Growth rate of GDP per capita is highly important when dealing with the issue of modelling the air travel demand in developing countries. Thus, the model developed is highly suitable for countries of South East Europe (particularly countries of the Balkans region) where a single 1 % variation in income level will induce approximately 3 % change in demand.

A global description of the trip generation model is as follows: total passenger flow from origin country is determined according to the GDP per capita by using fuzzy logic.

3.3 Trip Distribution

The model evolved takes into account major flows between an origin country and selected countries. In order to properly describe flows between two countries, it is of vital importance to consider mutual activities that run between the two countries such as level of trade (export and import of goods), migration of labor flows, historical relations, level of tourist attraction, etc. Therefore, the flow between certain countries is primarily based on tourism. Still others are highly affected by the economic activity of given countries representing strong business flow. Some flows are fully conditioned by population migration for work and settlement. It is evident that countries such as the United States, Canada, Australia and western part of Europe are still appealing to emigrants from different parts of the world. Hence, relatively strong flows have been generated by movements of these immigrants to their home countries, especially in the period of holidays. In order to distinguish between and cover the characteristics of variety of purpose for travel, the model implies three explanatory variables: import value of goods, number of emigrants from selected country and attraction of countries of final destination.

This model is largely applicable to countries in the transition process presuming conversion from a planned economy to a market-oriented economy.

A global description of the trip distribution model developed in this chapter is as follows: passenger flows from origin country to selected countries are determined according to the number of emigrants, imports and attraction by using fuzzy logic.

4 Models' Application: Case Study of Serbia

In this section both models are illustrated using data related to Serbia.

The first model refers to determination of total passenger flow from Serbia (two airports: Belgrade and Niš). In order to describe how the total passenger flow from one country (trip generation) is changing, it is necessary to incorporate factors that affect the demand. As it was mentioned before, the most important factor recognized for this study is GDP per capita (constant 2000 USD) [14]. It is necessary to say that an increase in GDP in one year will cause an increase in passenger flow in the following year. In that respect, GDP for a specific year is considered in relation to the passenger flow in following year (Fig. 1). It can be observed that there was an exception in 2009, which can be explained by the effect of the world economic financial crisis.

Fuzzy logic system is used in order to make a short term forecast of total passenger flow from Serbia taking into account the previously mentioned factor. The experiments regarding the number and the shape of the input and output fuzzy

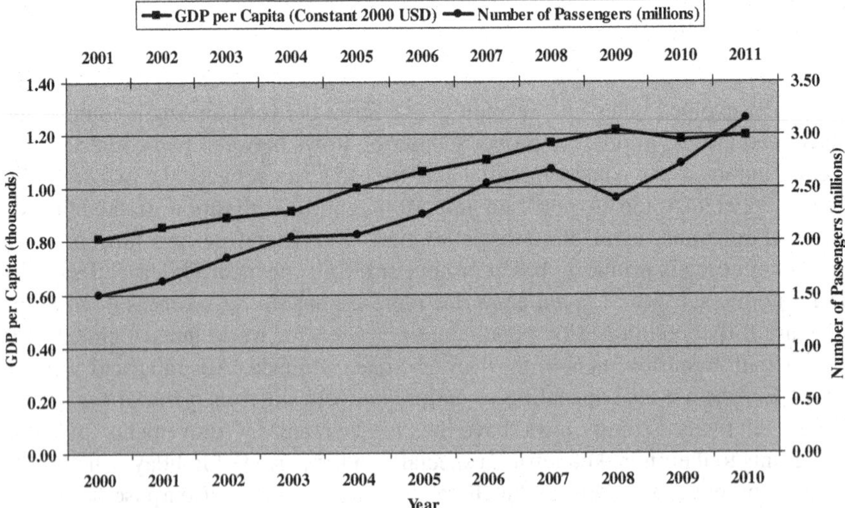

Fig. 1 GDP per capita and number of passengers (2000–2011)

sets were carried out and it was decided to use triangle and trapezoid sets as following. The membership functions of fuzzy sets Very Very Low, Very Low, Low, Low Medium and Medium are related to GDP, while the membership functions of fuzzy sets Approximately 1, Approximately 2, Approximately 2.5 and Approximately 3 are related to number of passengers (NP).

The output variable of the fuzzy system is the total flow from Serbia. The results are obtained by applying MAX-MIN fuzzy reasoning and defuzzification by centre of gravity, as given in Table 1. Also, the results obtained by linear regression are derived from the Eq. (1):

$$NP = -948796 + 3095.454^{*}GDP \tag{1}$$

Regression statistics are following: R^2 = 0.84, F = 49.22, t Stat = −2.05, p-value = 0.069 for intercept coefficient, t Stat = 7.01, p-value = 6.21*10−5 for GDP coefficient.

It can be seen from Table 1 that trip generation model based on fuzzy logic produced very satisfying results (average relative error is 0.06) in comparison to results obtained by regression (average relative error is 0.08). The world economic financial crisis caused a deviation in 2009. Total passenger flow from/to Serbia in 2012 obtained by this model is around 2.93 million passengers.

In order to determine passenger flows between Serbia and 14 selected countries that comprise more than 80 % of the total flows from/to Serbia, another fuzzy logic system is developed. Number of emigrants (EMG), imports by countries in millions of RSD (IMP) and attraction of destination (ATT) are used as input variables.

Number of emigrants reflects a strong connection between two countries and also a strong passenger flow, while imports indicate economic and business relation between the observed countries. A very important characteristic of a certain route is the attraction of origin and/or destination country. For this purpose,

Table 1 Input data and fuzzy logic and regression outputs (2000–2011)

Year	GDP per capita (USD)	Number of passengers (NP) (mil)	Estimated NP by fuzzy logic (mil)	Estimated NP by regression (mil)
2000	0.81	1.28	–	–
2001	0.85	1.50	1.39	1.69
2002	0.89	1.62	1.43	1.80
2003	0.92	1.85	1.96	1.88
2004	1.00	2.05	2.00	2.16
2005	1.06	2.06	2.00	2.33
2006	1.10	2.26	2.29	2.47
2007	1.17	2.54	2.51	2.66
2008	1.22	2.67	2.67	2.82
2009	1.18	2.40	3.11	2.70
2010	1.20	2.72	2.78	2.75
2011	1.19	3.15	3.08	2.73

Table 2 Input data and fuzzy logic output (FL Flow) for selected countries

	EMG (000)	IMP	ATT	Flow (000)	FL flow (000)
Austria	300	512	$2 + 1.5 + 2 = 5.5$	203	209
Denmark	7	91	$0 + 0 + 0 = 0$	35	33.5
France	120	482	$4 + 1 + 2 = 7$	129	120
Greece	10	228	$6 + 0 + 0 = 6$	77	87.6
Holland	20	256	$0 + 0 + 2 = 2$	47	37.1
Italy	102	1,432	$3 + 1 + 0 = 4$	138	150
Germany	845	1,768	$1 + 2 + 6 = 9$	538	447
United Kingdom	17	197	$1 + 1 + 4 = 6$	136	103
FRY Macedonia	1	272	$0 + 0.5 + 0 = 0.5$	47	37.7
Russian Federation	25	2,157	$0.5 + 2 + 0 = 2.5$	133	150
Switzerland	120	202	$0 + 0 + 6 = 6$	177	109
Bosnia and Herzegovina	1	556	$0 + 0.5 + 0 = 0.5$	24	31.3
Turkey	13	325	$6 + 2 + 0 = 8$	107	95.6
Montenegro	1	165	$6 + 4 + 0 = 10$	391	429

a quantitative indicator of attraction, which describes route taking into account tourist attraction of the destination country, historical relationship between two countries, and existence of the hub airport in the destination country, is derived. Tourist attraction means that there are some tourist resorts, historic monuments of culture etc. which can attract passengers during the season or during the whole year. Historical relationship refers to different political events such as for example generation of new states (the state of Serbia and Montenegro split into two states). Hub airport means that connecting passengers usually fly to that hub, where they can continue their trip flying to the chosen destination(s). Therefore, the formula derived for this purpose (2) represents the sum of all three factors with respect to the order in which they are mentioned above. Values for factor Tourism ranges from 0 to 6, while Historical (and other) links varies from 0 to 4. Variable Hub can take discrete value from the set of {0, 2, 4, 6}.

$$\text{ATTRACTION OF DESTINATION} = \text{Tourism} + \text{Historical (other) links} + \text{Hub} \tag{2}$$

Input data for 14 selected countries for the year 2010 can be seen in Table 2. Imports data (IMP) are taken from [15].

It can be seen from Table 2 that Denmark has the lowest possible value of attraction (0). That is expected, because of the weak relationship between Serbia and Denmark: there are no historical relationships between these countries, there are no hub airports in Denmark, and this country is not very attractive as a tourist destination for Serbian people. FRY Macedonia and Bosnia and Herzegovina have the value of attraction of 0.5 each, because those countries are neighboring countries (for Serbia) and they are very close to Serbia and more attractive for road transport (car or bus). Value of attraction for Holland is 2, which results from travelling through Amsterdam Schiphol airport (hub airport). Some economic, political and business relation between the Russian Federation and Serbia results in 2.5 value of attraction. Value of attraction for Italy is 4 and it is the result of partially Italian hub, tourist attraction (resorts and historic monuments of culture) and economic connections. A low cost carrier (Niki), historical relationship and tourist attraction (resorts and historic monuments of culture) set the value of attraction for Austria to 5.5. Greece, United Kingdom and Switzerland have the same value of attraction, i.e. 6, but for different reasons. For Greece, it is the result of very attractive tourist offers for summer seasons and partially of historic monuments of culture. With regard to the United Kingdom, it is about the low cost carrier (Wizz Air) and hub airports (London airports) which offer a lot of possibilities to passengers, as well as large distance between these two countries. Finally, Switzerland can also offer different possibilities to passengers through its hub airport (Zurich). Value of attraction for France is 7, because of the hub (Charles de Gaulle Airport), large distance, and tourist attraction (Paris, Côte d'Azzure, etc.). Turkey is attractive as a tourist destination (summer resorts, as well as historic monuments of culture (Istanbul)) and its value of attraction is 8. Germany is very interesting because of its hubs (Frankfurt, Munich), low cost carriers (Germanwings, Wizz Air), and strong economic and political links that set value of attraction to be 9. And finally, Montenegro has the highest possible value for attraction (10). The fact that Serbia and Montenegro comprised a single state in the past (very strong historical relationship) as well as summer resorts on the Adriatic coast make Montenegro the most attractive country for Serbian people.

The membership functions of fuzzy sets Very small, Small, Medium, Large and Very large is related to EMG, while the membership functions of fuzzy sets Low, Medium and High is related to IMP. ATT is described using fuzzy sets Low, Medium and Huge attraction. These fuzzy sets have been defined on the basis of the authors' experience and estimates. The membership functions of fuzzy sets Small, Medium and Large are related to passenger flow between two countries (Fuzzy Logic Flow—FL Flow). The fuzzy rule base consists of 45 rules. Some of them are presented below:

Rule 1: If EMG is Very small and IMP is Low and ATT is Low, then FL Flow is Small,

else

Rule 2: If EMG is Very small and IMP is Low and ATT is Medium, then FL Flow is Small,

else

...

Rule 21: If EMG is Medium and IMP is Low and ATT is Huge, then FL Flow is Medium,

else

...

Rule 44: If EMG is Very large and IMP is High and ATT is Medium, then FL Flow is Large,

else

Rule 45: If EMG is Very large and IMP is High and ATT is Huge, then FL Flow is Large.

The output variable of the fuzzy system is the passenger flow from Serbia to selected countries. The results are obtained by applying MAX-MIN fuzzy reasoning and defuzzification by centre of gravity. A very close correspondence can be observed (Table 2) if the real (Flow) and estimated value of passenger flows (FL Flow) are compared.

5 Conclusion

This chapter provides a detailed description of trip generation and trip distribution models using fuzzy logic. The benefits from the fuzzy logic will be more accurately assessed as the number of successful applications of the fuzzy logic in air travel demand modeling increases.

The models developed are particularly appropriate when dealing with the problem of air travel demand in countries passing through the process of transition. After several years of isolation, Serbia is now in the middle of this process which can be very long and difficult to handle. Since this process requires time in order to achieve a stable economic climate, challenges to further research would primarily be to explore economic environment and factors that can contribute to better understanding of air travel demand. Also, it would be interesting to explore the capabilities of the fuzzy approach from a less aggregated point of view as well as the elasticity of the model with respect to variables assessed by experts.

Acknowledgments This research has been supported by the Ministry of Education, Science and Technological Development, Republic of Serbia, as a part of the project TR36033 (2011–2014).

References

1. Carson, R.T., Cenesizoglu, T., Parker, R.: Forecasting (aggregate) demand for US commercial air travel. Int. J. Forecast. **27**, 923–941 (2011)
2. Hsiao, C.-Y., Hansen, M.: A passenger demand model for air transportation in a hub-and-spoke network. Transp. Res. Part E **47**, 1112–1125 (2011)
3. Profillidis, V.A.: Econometric and fuzzy models for the forecast of demand in the airport of Rhodes. J. Air Transp. Manag. **6**, 95–100 (2000)
4. Alekseev, K.P.G., Seixas, J.M.: A multivariante neural forecasting modeling for air transport—preprocessed by decomposition: a Brazilian application. J. Air Transp. Manag. **15**, 212–216 (2009)
5. Samagaio, A., Wolters, M.: Comparative analysis of government forecasts for the Lisbon airport. J. Air Transp. Manag. **16**, 213–217 (2010)
6. Teodorović, D., Kalić, M.: A fuzzy route choice model for air transportation networks. Transp. Plann. Technol. **19**(2), 109–120 (1995)
7. Kalić, M., Teodorović, D.: Modal split modelling using fuzzy logic. Proceedings of the Conference "Modelling and Management in Transportation", pp. 91–96. Poznan-Cracow, Poland (1999)

8. Kalić, M., Tošić, V.: Soft demand analysis: belgrade case study. Proceedings of the 8th Meeting of the Euro Working Group Transportation EWGT and Workshop IFPR on Management of Industrial Logistic Systems "Rome Jubilee 2000 Conference", pp. 271–275. Rome, Italy (2000)
9. Kalić, M., Teodorović, D.: Trip distribution modelling using fuzzy logic and a genetic algorithm. Transp. Plann. Technol. **26**(3), 213–238 (2003)
10. Kalić, M., Dožić, S., Babić, D.: Predicting Air Travel Demand Using Soft Computing: Belgrade Airport Case Study. Compendium of papers, Euro Working Group on Transportation, Paris (2012)
11. Doganis, R.: Flying Off Course. 2nd edn. Routledge, London (1992)
12. Gillen, D.: The future for interurban passenger transport bringing citizens closer together. Discussion Paper No. 2009–15. 18th International Transport Research, Symposium (2009)
13. Oum, T., Fu, X., Zhang, A.: Air transport liberalization and its impact on airline competition and air passenger traffic. Final report, OECD International Forum, Germany (2009)
14. World Bank Database. http://www.worldbank.org (October, 2012)
15. Statistical Office of the Republic of Serbia. (October, 2012). http://webrzs.stat.gov.rs/WebSite/

Design of Priority Transportation Corridor Under Uncertainty

Leonardo Caggiani and Michele Ottomanelli

Abstract Network design is one of the crucial activity in transportation engineering whose goal is to determine an optimal solution to traffic network layout with respect to given objectives and technical and/or economic constraints. In most of the practical problem the input data are not always precisely known as well as the information is not available regarding certain input parameters that are part of a mathematical model. Also constraints can be stated in approximate or ambiguous way. Thus, starting data and/or the problem constraints can be affected by uncertainty. Uncertain values can be represented using of fuzzy values/constraints and then handled in the framework of fuzzy optimization theory. In this paper we present a fuzzy linear programming method to solve the optimal signal timing problem on congested urban. The problem is formulated as a fixed point optimization subject to fuzzy constraints. The method has been applied to a test network for the case of priority corridors that are used for improve transit and emergency services. A deep sensitivity analysis of the signal setting parameters is then provided. The method is compared to classical linear programming approach with crisp constraints.

1 Introduction

In order to evaluate the performances of a transportation network or to define and suggest improvements to network layout, traffic managers and engineers use traffic simulation models. Generally these models can be divided into two principal categories: demand/supply interaction simulation models and supply design models.

L. Caggiani (✉) · M. Ottomanelli
Politecnico di Bari—D.I.C.A.T.E.Ch, Bari, Italy
e-mail: l.caggiani@poliba.it

M. Ottomanelli
e-mail: m.ottomanelli@poliba.it

V. Snášel et al. (eds.), *Soft Computing in Industrial Applications*, 291
Advances in Intelligent Systems and Computing 223, DOI: 10.1007/978-3-319-00930-8_26,

The simulation models are tools to predict users' behaviour (supply and activity systems are given, such as facilities, services and prices).

The supply design models are based on simulation models and allow to define supply and activity systems that best satisfy the prefixed objectives. In particular the models able to define the transportation network layout are named Network Design models [1]. Starting from an existing transportation network, these models allow to define an optimal network layout by means of objective functions subject to a set of constraints. The variables of these optimization problems can be continuous or discrete and they can be classified into three categories: layout variables (i.e. links capacity, links direction), supply performance variables (i.e. signals parameters) and price variables (i.e. transport fares).

The objectives of the design problem can be social (minimization of users costs) or operators objective (maximization of gains).

The constraints can be external (available budget, pollution), technical (i.e. link flows-capacity ratio) or consistency constraints among demand, flows and costs.

In literature different models, with single or multiple objective function [2, 4, 5], and algorithms have been presented to solve Network Design Problems (NDP). These optimization problems usually assume rigid ranges or minimum-maximum thresholds constraints, analytically defined with inequalities. Actually, both analysts available data and problem constraints can be affected by uncertainty; moreover within an interval, some values may better satisfy the purpose of the analysts (i.e. in a budget interval constraints the lower values are preferable to the others). These uncertain values can be better handled using soft computing approaches in particular through the use of fuzzy values/constraints [6, 7]. Recently, great attention has been given to paradigms developed in the theoretical framework of Fuzzy Set, in which Fuzzy Logic and Possibility Theory are mathematical tools for solving transportation problems [8] like fuzzy optimization [12–14]. Few authors have studied the opportunity to consider also this knowledge together with NDP [8, 10–12, 15].

In this paper we present a fuzzy non-linear programming to solve the equilibrium Network Design Problem in urban transportation in order to reduce the travel time on priority corridor [16]. The problem is formulated as a fixed point optimization subjected to fuzzy constraints. The proposed method has been applied to a test network. The preliminary results show that the proposed approach is very promising since it behaves as a multi-criteria optimization as well.

2 Statement of the Problem

The general formulation of the transportation supply design problem can be stated as:

$$\mathbf{x}^* = \arg \underset{\mathbf{x}}{opt}\, w(\mathbf{x}, \mathbf{f}^*) \tag{1}$$

subject to:

$$\mathbf{f}^* = \boldsymbol{\Delta}(\mathbf{x})\mathbf{P}(\mathbf{x}, \mathbf{C}(\mathbf{f}^*, \mathbf{x}))\mathbf{d}(\mathbf{C}(\mathbf{f}^*, \mathbf{x})) \tag{1a}$$

$$\mathbf{x}, \mathbf{f}^* \in E \tag{1b}$$

$$\mathbf{x}, \mathbf{f}^* \in T \tag{1c}$$

where:

- $\mathbf{x}$ is the vector of the design variables;
- w is the objective function;
- $\mathbf{f}^*$ is the vector of User Equilibrium link traffic flows;
- $\boldsymbol{\Delta}$ is the link-path incidence matrix;
- $\mathbf{P}$ is the path choice probability matrix;
- $\mathbf{C}$ is the vector of path costs;
- $\mathbf{d}$ is the vector of travel demand;
- Equation (1a) represents the consistency constraint among demand, flows and supply parameters (set of possible configurations of network flows)
- Equations (1b) and (1c) express the sets of supply parameters satisfying external (E) and technical (T) constraints.

In urban NDP the vector of decision variables can include traffic lights control parameters vector (g) and topological parameters vector (y) such as capacity of road links, lanes width and/or numbers, flow direction and so on.

3 The Optimization Model

In the problem presented in Eq. (1) we introduce also uncertain, incomplete information about design variables, external and technical constraints assuming the congested network case. For example possible constraints can be:

$$B \approx a \tag{2a}$$

$$ca_{l1} \gg ca_{l2} \tag{2b}$$

$$g_u/g_v \approx b \tag{2c}$$

where B is the available budget, ca_{l1} and ca_{l2} are the link l_1 and l_2 capacity, g_u and g_v are the effective green times of two phases traffic light regulation scheme, a and b are real numbers. These constraints represent the incomplete information and quantitatively formalize a linguistic/approximate expression.

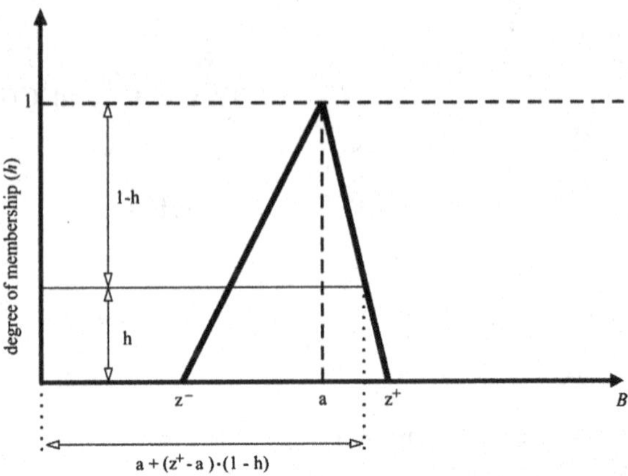

Fig. 1 Fuzzy set $B \approx a$

In this work we propose to specify these uncertain equations as fuzzy sets [2, 6]. In fuzzy logic a crisp number belongs to a set (fuzzy set) with a certain degree of membership, named also satisfaction *h*. The degree of membership is defined by a "Membership Function" (MF). For example, the constraint (2a) can describe the statement "the budget *B* is approximately equal to *a*". If no additional specific information is provided, a triangular MF *h(B)* can be assumed to represents the fuzzy constraint and so the previous statement is analytically defined by the fuzzy set depicted in Fig. 1. The z^-, a and z^+ values could be based on availability budget forecasts.

In fuzzy set theory, the closer to one the degree of membership is, the more the corresponding abscissa value belongs to the respective linguistic variable (fuzzy set). If the MFs are triangular then all the fuzzy constraints considered in problem (3) can be expressed as inequalities and depend on the satisfaction *h*.

In the example the inequalities representing the constraints are:

$$B \leq a + (z^+ - a) \cdot (1 - h); \quad B \geq a - (a - z^-) \cdot (1 - h)$$

The closer to one the satisfaction is, the more the constraints are fulfilled. Therefore, in order to find the optimal solution to problem (1) subject to certain and fuzzy constraints, it is necessary to maximize the satisfaction *h* of the fuzzy constraints and, at the same time, to optimize the value of the objective function. In fuzzy linear programming, the problem (1) is dual to the following problem (3), where the objective function to be maximized is the satisfaction *h*, and the optimization of the objective function becomes a further constraint to the problem [8]. This constraint depends on the type of optimization: if the problem (1) is a minimization then the constraint is the Eq. (3a) (i.e. MF in Fig. 2a); if the problem (1) is a maximization then the constraint is given by Eq. (3b) (i.e. MF in Fig. 2b):

$$Max\, h \tag{3}$$

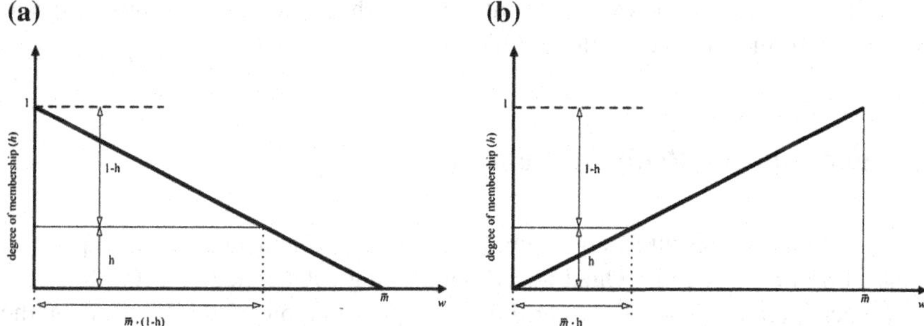

Fig. 2 Fuzzy set representing the expression "satisfactory w optimization"

subject to:

$$w(\mathbf{x}(h),\mathbf{f}^*) \leq \bar{m}\cdot(1-h) \tag{3a}$$

OR

$$w(\mathbf{x}(h),\mathbf{f}^*) \geq \bar{m}\cdot h \tag{3b}$$

$$\mathbf{f}^* = \boldsymbol{\Delta}(\mathbf{x}(h))\mathbf{P}(\mathbf{x},\mathbf{C}(\mathbf{f}^*,\mathbf{x}(h)))\mathbf{d}(\mathbf{C}(\mathbf{f}^*,\mathbf{x}(h))) \tag{3c}$$

$$x_e, \mathbf{f}^* \in E\,\forall e = 1,2,\ldots,p \tag{3d}$$

$$x_t, \mathbf{f}^* \in T\,\forall t = 1,2,\ldots,q \tag{3e}$$

$$x_i(h) \in F_i\,\forall i = 1,2,\ldots,n \tag{3f}$$

where:

- $p + q$ is the number of certain design variables;
- n is the number of uncertain design variables;
- $F_i = \{(w,h(w))|w \in W\}$ is the fuzzy set$i$;
- W is the set of all possible objective function values w.

In Fig. 2 the value of $\bar{m}$ represents the maximum acceptable value for the objective functions in problem (1).

The assumed fuzzy constraints will definitely depend on the same value of the satisfaction h. The closer to one the value of h (maximization of satisfaction) is, the more the objective function of the (1) is optimized (through the (3a) or (3b)) according to the consistency constraints (3c), with the certain relations (3d) and (3e) and with the fuzzy constraints (3f).

4 Algorithm for Problem Solution

In this section we present an algorithm to solve the problem specified in Eq. (3). It is based on the Sequential Quadratic Programming (SQP) method [16, 18].

In each iteration s of the algorithm an approximation of the Hessian of the Lagrangian function is used to generate a Quadratic Programming (QP) sub-problem whose solution is adopted to form a search direction for a line search procedure.

Within each optimization step of the generated QP sub-problem, the fixed point formulation in Eq. (3c) is solved to find the user equilibrium flows vector. In fact we assume the congested network case and thus a stochastic user equilibrium assignment model [18, 20]. In the constraint 3c, the assignment matrix $\mathbf{H}$ (i.e. $\mathbf{H} = \boldsymbol{\Delta} \cdot \mathbf{P}$) is assumed to be function of both the path costs vector $\mathbf{C}$ and design variables vector $\mathbf{x}$.

The solution algorithm of the assignment problem is based on the Method of Successive Averages—Flow Averaging (MSA—FA) [21] with variable step length in order to improve the convergence speed by the following steps:

Step 0 - initialize: $k = 1$ and $k_i = k$;

- choose the number of iterations after which the rise of the magnitude of step begins: N_i
- initialize the flows with a feasible solution corresponding to free flow costs: $\mathbf{f}^{k=1} = \mathbf{f}^{0}$;

Step 1 update iteration counter: $k = k + 1$;
Step 2 $\mathbf{c}^{k} = \mathbf{c}[\mathbf{f}^{k-1}, \mathbf{x}(h^{s})]$ (compute the link costs);
Step 3 $\mathbf{f}^{k}_{SNL} = \boldsymbol{\Delta}^{T}\ \mathbf{P}(\boldsymbol{\Delta}^{T}\ \mathbf{c}^{k}\ \mathbf{d}$ (compute the Stochastic Network Loading flows);
Step 4 calculate the average of flows: $\mathbf{f}^{k+1} = [(k\text{-}1)\mathbf{f}^{k} + \mathbf{f}^{k}_{SNL}]/k$;
Step 5 if $|f_i^{k+1} - f_{i\ SNL}^{k}|/f_i^{k+1} < \varepsilon_{MSA}$ then go to Step 8, else go to Step 6;
Step 6 if $k \le N_i$ go to Step 1, else go to Step 7;
Step 7 Change the iteration counter: $k_i = 2k$, $N_i = 2N_i$, $k = k_i$ and go to Step 2;
Step 8 Stop.

5 Numerical Application and Results

The following application aims to experimentally evaluate the performances of the proposed design model and algorithm. In this test we propose a network design optimization considering signal settings parameters as supply design variables.

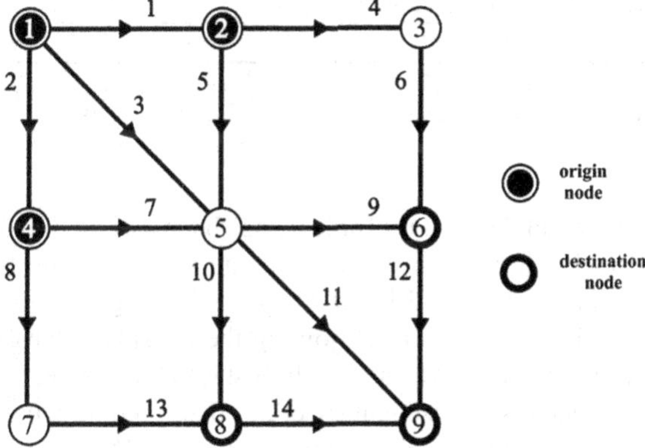

Fig. 3 Test network

The chosen approach to the problem is the global optimization of signal settings that consists in searching the optimal effective green time for all intersection signal settings (vector **g***); these values are obtained through the optimization of an objective function depending on signal settings and equilibrium flows as previously described in (1). The proposed model has been applied to the network used in [22].

The graph is made up of 9 nodes (3 origins and 3 destinations), and 14 links as depicted in Fig. 3. All the employed data are those proposed in [22] but the links length (i.e. L_l = 80 m) and free flow travel time. The signalized intersections are the node 5 with 3-phases regulation scheme and nodes 6 and 8 (2-phases). The effective cycle time is set to $C_t = 90$ s and the starting effective green time is equally divided for each phase (starting values—Case 0).

The link cost values c_l for the numerical tests are the sum of the free flow travel time and the waiting time due to the signalized intersections. The free flow travel time is calculated as ratio between link length and speed using an experimental speed-flow relationship [1]. The waiting time is estimated using the Doherty's delay function [1]:

$$
\begin{aligned}
t^l_{wa} &= 0.5 \cdot C_t(1-\mu)^2 + \frac{1980}{\mu \cdot s} \cdot \frac{f_l}{\mu \cdot s - f_l} \quad if\ f_l \leq 0.95 \cdot \mu \cdot s \\
t^l_{wa} &= 0.5 \cdot C_t(1-\mu)^2 - \frac{198.55}{\mu \cdots /3600} + \frac{220 \cdot f_l}{(\mu \cdot s)^2/3600} \quad if\, f_l > 0.95 \cdot \mu \cdot s
\end{aligned} \tag{4}
$$

where:

- t^l_{wa} is the waiting time at intersection on link l (s/veh);
- f_l is the traffic flow on link l (veh/h);
- s is the saturation flow (veh/h);
- g is the effective green time;
- μ is the effective green ratio (g/C_t).

Table 1 O-D Demand Flow vector d

O-D	1–6	1–8	1–9	2–6	2–8	2–9	4–6	4–8	4–9
i	1	2	3	4	5	6	7	8	9
d	120	150	100	130	200	90	80	180	110

The travel demand **d** (Table 1) has been assigned to the network using a Stochastic User Equilibrium traffic assignment model, with a Logit formulation for the path choice model (SUE-Logit Model) with Logit parameter value set as 1.5.

All the possible paths joining the considered origin and destination nodes have been considered. The aim of the test is to find the vector of effective green time for the considered signalized intersections that minimizes the equilibrium travel time on the considered links and the total users costs T^{uc} (i.e. the sum of the product between equilibrium link flows f_l* and equilibrium link costs c_l*) under uncertain constraints. We assume that the analyst aims at reducing, compared to the equally divided effective green time, the total travel time of links 9 and 10 end then along the relevant corridors. The linguistic assessment of the analyst in order to minimize these (for example) total travel time can be "$T_9^{uc} = c_9 * \cdot f_9 *$ $(and\ \ T_{10}^{uc} = c_{10} * \cdot f_{10} *)$ *are smaller than starting values* $T_9^{uc\_eq}(and\ T_{10}^{uc\_eq})$". If this problem is solved with a classical approach the optimization problem (1) becomes [1]:

$$\mathbf{g}^* = \arg\min_{\mathbf{g}} \sum_l f_l^* \cdot c_l^*(\mathbf{f}^*, \mathbf{g}) \tag{5}$$

subject to:

$$f_9^* \cdot c_9^* \leq T_9^{uc\_eq} \tag{5a}$$

$$f_{10}^* \cdot c_{10}^* \leq T_{10}^{uc\_eq} \tag{5b}$$

$$\mathbf{f}^* = \mathbf{\Delta}(\mathbf{g})\mathbf{P}(\mathbf{g}, \mathbf{C}(\mathbf{f}^*, \mathbf{g}))\mathbf{d}(\mathbf{C}(\mathbf{f}^*, \mathbf{g})) \tag{5c}$$

$$\sum_{in} g_l^{in} = C_t^{in} \quad \forall in \in \{5, 6, 8\} \tag{5d}$$

$$g_l^{in} \leq 85\ [\mathrm{s}] \quad \text{and} \quad g_l^{in} \geq 5\ [\mathrm{s}]\ \forall in \in \{5, 6, 8\} \tag{5e}$$

where:

- Equations (5a) and (5b) represent the crisp constraints set by analyst to the costs on link 9 and 10 respectively;
- Equation (5c) states the consistency among demand, flows and supply parameters;

- Equation (5d) ensures that the sum of the effective green time for each signalized intersection is equal to the prefixed effective cycle time;
- Equation (5e) ensures maximum and minimum values for the effective green time.

The vague (uncertain) expression for total user costs can be better formalized with fuzzy constraints through the proposed method. In this case, as previously discussed, the problem (5) can be formulated as follows:

$$\max h \tag{6}$$

subject to:

$$\sum_{l} f_l^* \cdot c_l^*(\mathbf{f}^*, \mathbf{g}(h)) \leq \bar{m} \cdot (1 - h) \tag{6a}$$

$$f_9^* \cdot c_9^* \leq T_9^{uc\_eq} \cdot (1 - h) \tag{6b}$$

$$f_{10}^* \cdot c_{10}^* \leq T_{10}^{uc\_eq} \cdot (1 - h) \tag{6c}$$

$$\mathbf{f}^* = \mathbf{\Delta}(\mathbf{g})\mathbf{P}(\mathbf{g}, \mathbf{C}(\mathbf{f}^*, \mathbf{g}))\mathbf{d}(\mathbf{C}(\mathbf{f}^*, \mathbf{g})) \tag{6d}$$

$$\sum_{in} g_l^{in} = C_t^{in} \; \forall in \in \{5, 6, 8\} \tag{6e}$$

$$g_l^{in} \leq 85 \text{ [s]} \quad \text{and} \quad g_l^{in} \geq 5 \text{ [s]} \;\; \forall in \in \{5, 6, 8\} \tag{6f}$$

where Eqs. (6a), (6b) and (6c) are fuzzy constraints with the triangular shaped MF of Fig. 2a.

The starting values of the proposed method and algorithm are:

- $\bar{m} = 600000$, $T_9{}^{uc\_eq} = 11000$, $T_{10}{}^{uc\_eq} = 3500$ (fixed from their starting values);
- algorithms starting values: $h^0 = 0.5$ and $N_i = 10$;
- algorithm stop test parameters: $\varepsilon_{SQP} = 10^{-6}$, $\varepsilon_{MSA} = 10^{-3}$.

In order to compare our method with the classical signal settings optimization we have carried out numerical applications using the classical model (Eq. 5) (Case 1) and numerical tests with the proposed method (Eq. 6) (Case 2).

For both the methods two sub-cases have been considered:

- sub-case (*a*): it considers the constraint just on link 9;
- sub-case (*b*): it considers the constraints on both links 9 and 10.

Table 2 Summary of the results

Case	h^*	g_3^5[s]	g_5^5[s]	g_7^5[s]	g_6^6[s]	g_9^6[s]	g_{10}^8[s]	g_{13}^8[s]	T_9^{uc}	T_{10}^{uc}	T^{uc}
0	–	30	30	30	45	45	45	45	56882	236757	3573251
1	–	5	46.66	38.34	55.14	34.86	38.59	51.41	237174	441717	2206192
a	–	5	49.56	35.44	40.12	49.88	37.03	52.97	56882	452228	2305237
1 *b*	–	5	47.75	37.25	41.12	48.88	47.39	42.61	56882	240000	2406767
2	0.45	5	47.69	37.31	52.51	37.49	38.07	51.93	197836	446913	2209434
a	0.41	5	50.96	34.04	36.77	53.23	36.64	53.36	35308	454948	2353882
2 *b*	0.35	5	47.04	37.96	38.63	51.37	53.85	36.15	39179	156715	2611920

Some results obtained in all cases for triangular MFs are summarized in Table 1. The optimization does not heavily affect the values of equilibrium link flows, but it affects the value of the costs on the network and the intersection green times.

The proposed method better satisfies the relative constraints compared to the method with rigid thresholds. Whereas in the classical method the algorithm stops when all the constraints are even slightly satisfied, the maximization of the satisfaction h leads to fulfill the fuzzy constraints as much consistently as possible with all the other constraints.

In Case 2 the total link user costs are generally lower than the relative values of Case 1, although the network total user costs are respectively the highest. Nevertheless the classical optimization is not able to reduce link costs as in the proposed method where the travel time reduction on the considered links is between 34.7–41.2 % with respect to the starting situation (Table 2).

These first results show that the fuzzy optimization is suitable when a high travel time reduction on specific links is required such as in emergency situations or transit systems.

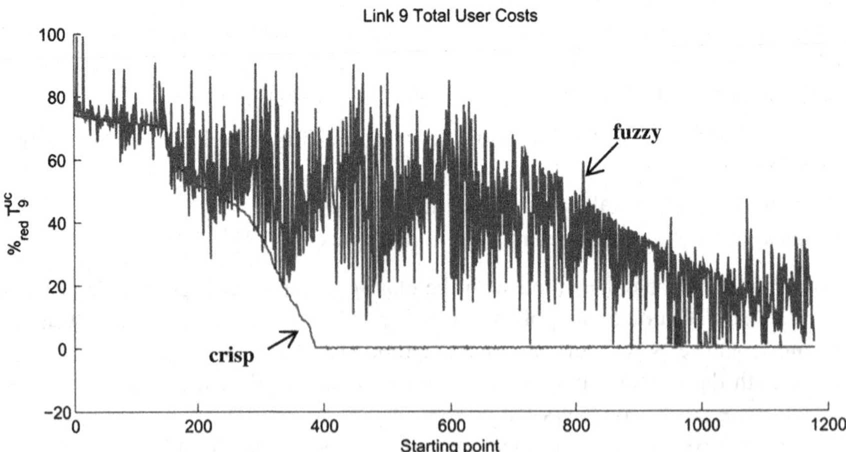

Fig. 4 Reduction of total user cost on link 9 (case *2b* versus case*1b*)

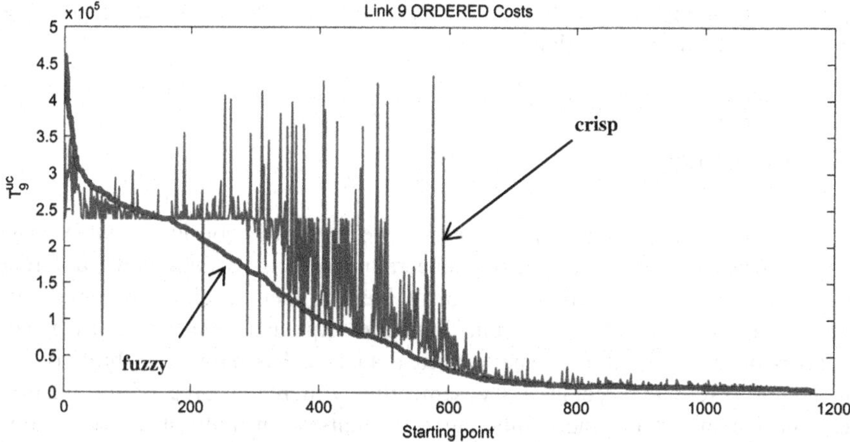

Fig. 5 Ordered values of travel time on link 9 in sub-case C (constraints on links 9 and 10)

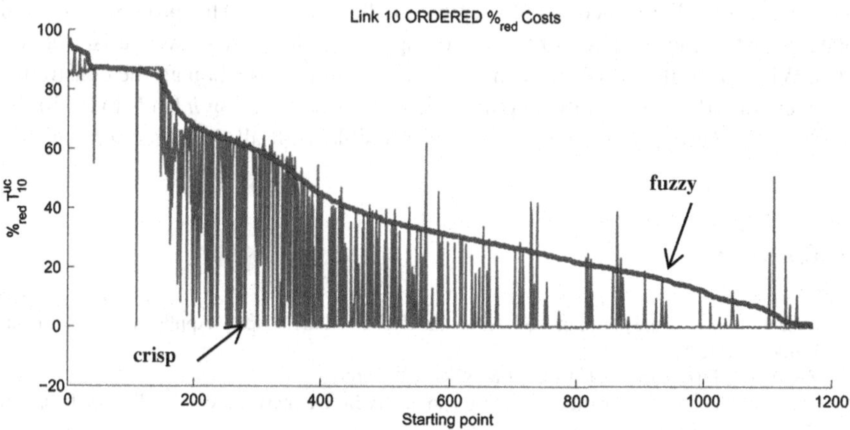

Fig. 6 Reduction of travel time on link 10 in the sub-case C (constraints on links 9 and 10)

A sensitivity analysis has been also carried out by varying the vector of the starting signal setting parameters (about 1200 starting points).

In Fig. 4 has been reported the plot of the total cost reduction (%) on link 9 considering the cases *b*. The continuous smoothed line represents the values of the proposed method.

It is observed that the better reduction of total travel time is performed by the fuzzy proposed approach.

In the Fig. 5 is represented the value of travel time on link 9 with respect to the starting situation in sub-case C, that is by assuming the constraints on both links 9 and 10. In this case the cost is lower than that considering crisp constraints.

In Fig. 6 is depicted the value of the travel time reduction (%) on link 10 with respect to the starting situation in sub-case C.

6 Conclusions

In this paper, jointly with crisp constraints, we suggest to consider flexible goals and constraints in NDP to include uncertain/imprecise values, with different uncertainty levels. The network design problem is then specified as fuzzy programming problem where uncertain data are explicitly represented by fuzzy numbers or fuzzy constraints. The numerical analysis has shown the ability of the method to take into account at the same time different source of information (certain and uncertain) and conflicting constraints as in multi-objective optimization. In fact, considering a set of fuzzy constraints the NDP is a kind of multiple criteria problem, where the single constraint is representative of a certain point of view and consequently the final solution is a compromise/optimal that satisfy (in different way) all the needs represented in the problem. The proposed method better satisfies the relative constraints compared to the method with crisp thresholds. Whereas in the classical methods the algorithm stops when all the constraints are even slightly satisfied, the maximization of the satisfaction h leads to fulfill the fuzzy constraints as much consistently as possible with all the other constraints.

References

1. Cascetta, E.: Transportation Systems Analysis: Models and Applications. Springer, Heidelberg (2009)
2. Zadeh, L.: Fuzzy sets. Inf. Control **8**, 338–353 (1965)
3. Meng, Q., Yang, H.: Benefit distribution and equity in road network design. Transp. Res. Part B **36**, 19–35 (2002)
4. Cantarella, G.E., Vitetta, A.: The multi-criteria road network design problem in an urban area. Transportation **33**, 567–588 (2006)
5. Paksoy, T., Özceylan, E., Weber, G.W.: A multi-objective mixed integer programming model for multi echelon supply chain network design and optimization. Syst. Res. Inf. Technol. **4**, 47–57 (2010)
6. Zimmermann, H.J.: Fuzzy Set Theory and Its Applications. Kluwer Academic Publishers, Dordrecht/London (1996)
7. Luo, X., Lee, J.H., Leung, H., Jennings, N.R.: Prioritised fuzzy constraint satisfaction problems: axioms, instantiation and validation. Int. J. Fuzzy Sets Syst. **136**(2), 155–188 (2003)
8. Teodorovic, D., Vukadinovic, K.: Traffic Control and Transport Planning: A Fuzzy Sets and Neural Networks Approach. Kluwer Academic Publishers, Boston (1998)
9. Das, S.K., Goswami, A., Alam, S.S.: Multiobjective transportation problem with interval cost, source and destination parameters. Eur. Jour. of Oper. Res. **117**, 110–112 (1999)
10. Mudchanatongsuk, S., Ordóñez, F., Liu, J.: Robust solutions for network design under transportation cost and demand uncertainty. J. Oper. Res. Soc. **59**, 652–662 (2008)

11. Selim, H., Ozkarahan, I.: A supply chain distribution network design model: an interactive fuzzy goal programming-based solution approach. Int. J. Adv. Manuf. Technol. **36**(3), 401–418 (2008)
12. Ghatee, M., Hashemi, S.M.: Application of fuzzy minimum cost flow problems to network design under uncertainty. Fuzzy sets syst. **160**, 3263–3289 (2009)
13. Kikuchi, S., Kronprasert, N.: Constructing transit origin-destination tables from fragmented data. Transp. Res. Rec. **2196**, 34–44 (2010)
14. Caggiani, L., Ottomanelli, M., Sassanelli, D.: A fixed point approach to origin-destination matrices estimation using uncertain data and fuzzy programming on congested networks. Transport. Res. Part C **28**, 130–141 (2013)
15. Caggiani, L., Ottomanelli, M.: Traffic equilibrium network design problem under uncertain constraints. Procedia: Soc. Behav. Sci. **20**, 372–380 (2011)
16. Marcianò, F.A., Musolino, G., Vitetta, A.: Signal setting design on a road network: application of a system of models in evacuation conditions. In: Brebbia C.A. (ed.) Proceedings of Risk Analysis VII & Brownfields V, pp. 443–454. WIT Press, Southampton (2010)
17. Schittkowski, K.: NLQPL: A FORTRAN-subroutine solving constrained nonlinear programming problems. Ann. Oper. Res. **5**, 485–500 (1985)
18. Bonnans, J.F., Gilbert, J.C., Lemarechal, C., Sagastizábal, C.A.: Numerical Optimization: Theoretical and Practical Aspects. Springer, Heidelberg (2006)
19. Chen, A., Kim, J., Lee, S., Choi, J.: Models and algorithm for stochastic network designs. Tsinghua Sci. Technol. **14**(3), 341–351 (2009)
20. Meng, Q., Lee, D.H., Yang, H., Huang, H.J.: Transportation network optimization problems with stochastic user equilibrium constraints. Transp. Res. Rec. **1882**, 113–119 (2004)
21. Cantarella, G.E.: A general fixed-point approach to multimode multi-user equilibrium assignment with elastic demand. Transp. Sci. **31**(2), 107–128 (1997)
22. Yang, H., Meng, Q., Bell, M.G.H.: Simultaneous estimation of the origin-destination matrices and travel-cost coefficient for congested networks in a stochastic user equilibrium. Transp. Sci. **35**, 107–123 (2001)

Application of Data Fusion for Route Choice Modelling by Route Choice Driving Simulator

Mauro Dell'Orco, Roberta Di Pace, Mario Marinelli and Francesco Galante

Abstract Modelling route choices is one of the most significant tasks in transportation models. Route choice models under Advanced Traveller Information Systems (ATIS) are often developed and calibrated by using, among other, Stated Preferences (SP) surveys. Different types of SP approaches can be adopted, alternatively based on Travel Simulators (TSs) or Driving Simulators (DSs). Here a pilot study is presented, aimed at setting up an SP-tool based on driving simulator developed at the Technical University of Bari. The obtained results are analysed in order to check the accordance with expectations in particular the results of application of data fusion technique are shown in order to explain how data collected by DSs, can be used to reduce the effect of choice of behaviour in unrealistic scenarios in TSs.

1 Introduction

The study of travellers' behaviour in Advanced Travellers Information Systems (ATIS) contexts is a crucial task in order to properly simulate phenomena like compliance with information, route choices in presence of information, etc.

M. Dell'Orco (✉) · M. Marinelli
D. I. C. A. T. E. Ch, Technical University of Bari, via Orabona 4, 70125 Bari, Italy
e-mail: dellorco@poliba.it

M. Marinelli
e-mail: m.marinelli@poliba.it

R. Di Pace
Department of Civil Engineering, University of Salerno, Salerno, Italy
e-mail: rdipace@unisa.it

F. Galante
Department of Transportation Engineering, University of Naples "Federico II", Naples, Italy
e-mail: francesco.galante@unina.it

V. Snášel et al. (eds.), *Soft Computing in Industrial Applications*,
Advances in Intelligent Systems and Computing 223, DOI: 10.1007/978-3-319-00930-8_27,

A correct simulation of these phenomena is crucial in appraising ATIS options, as evidenced for instances by [1, 25, 26]. In order to estimate models of the travellers' behaviour, observation of reactions is needed. The most adopted approach for collecting data is the Stated Preferences (SP) one. Two main types of tools for SP in ATIS contexts are the most popular: driving-simulators (DSs) and travel-simulators (TSs). Both methods are computer-based. DSs are characterised by a greater realism, provided that the respondents are asked to drive in order to implement their travel choices, as it happens in the real world. In TSs, travel choices are entered after having received a description of travel alternatives and of associated characteristics, without any driving. TSs compensate some lack of realism with a minor cost and with less burden for the respondents, thus allowing for many more trials by the same respondent. In most of the cases data have been collected by using TSs, as for instance in [2, 5, 11, 12, 18, 24] only a limited number of studies have been carried out by adopting DSs (e.g.: [7, 8, 21, 23, 28].) Provided the great effort required in setting up DS experiments and the burden to which both the analysts and the respondents are subject, it is crucial to carry out some pilot study in order to assess the ability of specific DSs environments in reproducing the effect of ATIS on travellers' choices. This is here done by comparing the results of a specific experiment with expectancies. Expectancies are assumed from authors experiences and common sense; however, it does not differ from empirical results obtained in other studies. For instance, [11] shows that the provision of accurate pre-trip information is beneficial for the ability of respondents to choose the best (shortest) route; it alleviates the detrimental effect caused by the natural random variation in network travel times. This comparison is here referred as assessment of the internal consistency of the simulation environments [13] by comparing the intuitions considering what kind of choices could be made in real life in context of choice that could be judged comparable. Moreover, an experiment has been carried out at the Technical University of Bari. Other than the assessment of the internal consistency of the experiment allows for validating the simulation environments, thus showing that more trials and experiments worth to be implemented.

The network reproduced in the virtual experiment refer to real one in Bari. Moreover, the respondents recruited for the experiments were travellers familiar with the networks. The network is proposed to respondents in the simulations in a double configuration, with and without ATIS. In turns, the configuration without ATIS was presented to respondents with some variants, reproducing different congestion levels and travel times, accordingly with their statistical distribution in the real world. This was aimed at avoiding the learning-phase (in the sense of warming-up), which is typical of SP-based ATIS experiments [11]. This allows for less trials for each respondent and so much more cheaper experiments oriented to the development of modelling frameworks. In fact, familiarisation is generally required in both TSs (for instance, 20 trials in [11] 10 trials in [3]) and in DSs (for instance 3 trials in [27] and up to 6 trials in [17]). It should be noted that familiarisation is a critical issue especially in DSs where each familiarisation trial induce a relevant extra-burden in the experimentation.

Moreover, dealing with known networks, allows for observing travellers' reactions to the introduction of ATIS in specific network contexts and thus for analyses with high and direct practical relevance.

This paper is organised as in following described: in Sect. 2 the simulator adopted for the experiment is briefly described and the experiment design is shown and discussed; in Sect. 3 preliminary analysis and the data fusion technique results are shown; conclusions and future work perspectives are discussed in Sect. 4.

2 Employed Simulation Tools and Design of the Experiments

In order to carry out the SP experiments a PC-based driving simulator of Technical University of Bari has been adopted (Fig. 1a).

The UC-win/Road driving simulator software was used. This software is developed by FORUM8, a Japanese company. UC-win/Road is plugin-based, allowing to extend software functionalities by using the UC-win/Road SDK Framework that allows for Delphi code. In our case, a plugin was created for data acquisition during driver's simulation, allowing to record for successive analyses (and in CSV format) data related to speed, position, steering, etc. In particular, we have employed data related to position in post-processing in order to observe route choices made by respondents.

The simulation system works on a single computer provided with NVidia Graphic Card (1 GB of graphic memory) and a Quad-Core CPU which guarantees very good real-time rendering and computation performances. The simulation is based on a steering wheel (*Logitech*^{TM} MOMO Racing Force Feedback Wheel), able to provide force feedback, as well as six programmable buttons (ignition, horn, turn signals, etc.), sequential stick shifters and paddle shifters. A 22" wide-screen monitor was used in order to have a good field of view, also showing internal car cockpit with tachometer and speedometer. Environmental sounds are reproduced to create a more realistic situation.

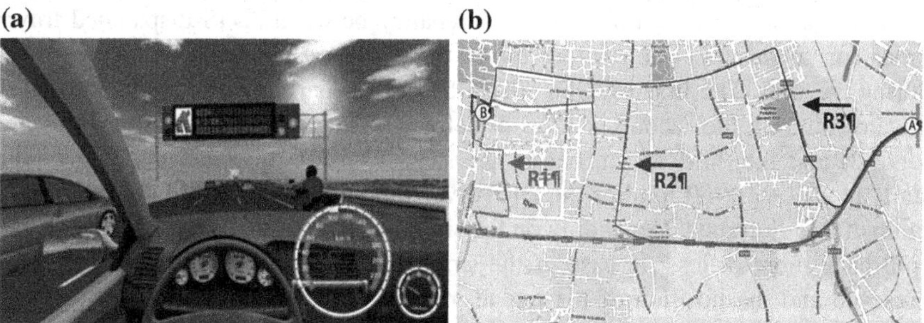

Fig. 1 **a** Snapshot of simulation; **b** Snapshot of network

During the experiment respondents have been asked for choosing a route among three alternatives. The context is configured in such a way that the choice can be assumed as a (possible) switching from a natural reference alternative. As already discussed, respondents were recruited for the experiment ensuring a familiarity with the experimental context. In fact, the simulated networks were part of a real network in Bari (Fig. 1b). The choice set can be viewed as composed by a main route (route 1) that connects the considered origin-destination pair. Depending on traffic conditions, the traffic could spill-back up to a later diversion node (detour toward route 2) or even up to an earlier diversion node (detour toward route 3). These three different conditions (straight route, later detour, earlier detour) are conventionally classified here as three different levels of congestion (free-flow/low congestion, intermediate congestion, high congestion). The experiment has been designed in order to have in most of the times (70 %) the system in the intermediate congestion pattern, even if extreme (low and high) congestion levels are less frequent but can be observed a not negligible number of times. Before starting the simulation, respondents can adapt themselves with the simulator by driving along each alternative route of the choice set, without ATIS and in free-flow traffic conditions. After this possible training, respondents are asked to make 6 successive trials, grouped in 3 driving sessions. At each session, respondents drive twice. The first time the VMSs representing the ATIS are not active, meaning that the system have not to dispatch information on particular incidents. The second trial is characterised by the activation of the ATIS as a consequence of an accident occurred, that perturbs the standard traffic pattern to an extent that depends on the accident severity. Trials when the Variable Message Signs (VMSs) are not activated are referred to in the following as trials without information; otherwise trials are said with information. Respondents are asked to first make their choices without information, with the main experimental aim to enforce their perception of the realism of the simulation, in terms of consistency with the real network he/she is used to. Then, at a second step, respondents are assisted by information. During the trials without information the respondents can encounter different congestion patterns. In case of information, the congestion level is assigned to the trial according to the following probabilities of being the shortest route: route 1, 13 %; route 2, 67 %; route 3, 20 %. Respondents are provided with information by VMSs (see Table 1), providing, in any case, fully accurate information. Full accuracy is here intended as a null discrepancy between ATIS-dispatched travel times (or suggestions) and actual travel times that respondents will experiment during the trial. The first VMS is 300 m before the early diversion (Mungivacca, toward route 3), the second is 1250 m before the late diversion (Carrassi, toward route 2) and the third 150 m before the same Carrassi diversion node. A queue starts in all cases 900 m after the later diversion node and 500 m before the exit-ramp of Poggiofranco. Depending on the simulated congestion level, the queue can spill back more or less. In case of the simulation of Bari the VMSs displays the presence and the position of the queue, not the length or the estimated queuing time.

Table 1 Scheme of VMSs, ramps and queue location

	From	To	Distance (m)
UC-win/Road	Entrance	1st VMS	400
	1st VMS	I Diversion, Exit 13A-Mungivacca	300
	I Diversion, Exit 13A-Mungivacca	2nd VMS	700
	2nd VMS	3rd VMS	1100
	3rd VMS	II Diversion, Exit 12-Carrassi	150
	II Diversione, Exit 12-Carrassi	Queue	900
	Queue	Exit 11-Poggiofranco	500

Recruitment of the sample was performed at the Polytechnic of Bari and 10 respondents were randomly selected. Such a small number of respondent is consistent with the pilot nature of the study.

3 Results

On the base of collected data the first analysis on respondents' reactions concerns the identification by the respondents of the familiar context of choice. Provided that respondents have no other sources of information and that 3 trials without information surely are too few to understand from the scratch how the simulated network performs, if respondents show a good ability in choosing the shortest route in absence of information, this should depend on their familiarity with the network and on the high level of realism of the simulation.

3.1 Preliminary Aggregate Analysis

First of all the research aim is verify the internal consistency of simulator in order to ensure the validation. With reference to others researches in case of travel simulators [13] and in case of driving simulator [11] internal validation can be made by observing the ability of drivers in choosing the shortest route and the effect of information when information can be considerate accurate.

In absence of information, with reference to the tendency to choose the shortest path the discrepancy between the actual share and the observed share has been evaluated (see Table 2); in case of scenarios with information the difference between the induced share by information system and observed share has been evaluated.

Moreover, with reference to [11] if the values of "Delta" in Table 2 are compared, the effect of accurate information can be evaluated in terms of drivers'

Table 2 Actual share versus observed share

Route	Without ATIS			With ATIS		
	Prob of being the shortest(%)	Observed Share(%)	Delta	Prob. of being the shortest(%)	Observed share delta(%)	Delta
1	73	87	+14.0	13.3	6.7	−6.6
2	20	10	−10.0	66.7	76.6	+10
3	7	3	−4.0	20.0	16.7	−3.3
Eulerian distance			312.0			154.45

ability in choosing the shortest route. In fact, Delta increases as discrepancy between observed choices and probability of being shortest with reference to each route, decreases. Differently than other study done by TSs [11], the realism of driving simulation environment, reduces the necessity of analysts to introduce more complex and complete information so this is more understandable.

3.2 Data Fusion to Model Route Choice Behavior

To incorporate information on system conditions in the choice process, we assume that drivers: (1) have some experience about the attributes of the transportation system; (2) use information to update his experience; (3) choose an alternative according to his updated experience.

Since the drivers' knowledge about the transportation system could be imprecise or approximate, it can be expressed in the same way we used for perceived information. So both drivers' knowledge and information can be expressed in terms of Possibility, like Fig. 2. To update knowledge of the system, drivers aggregate data coming both from their experience and from current information.

However, aggregation could be not always meaningful, since data coming from different sources can be far from each other, and thus not compatible. Therefore, a suitable aggregation function should include also a measure of compatibility.

In accordance with the literature several authors have been adopted alternative paradigms to random utility theory [4] in order to model travellers' behaviors in different contexts of choices [9, 15, 16]

In this work, we have used a route choice model based on uncertainty-based Information Theory as proposed by [14]. Acquired data has been fused using the method proposed by [29].

The information fusion model incorporates important aspects such as: (1) dynamic nature of information integration, since the perceived cost of an alternative is influenced by the user's previous experience and memory; (2) accuracy of the informative system, since the more accurate information is, the more important is the effect on the drivers' perception; (3) non-linear relationship between information and perception. Applications of this methodology can be useful in

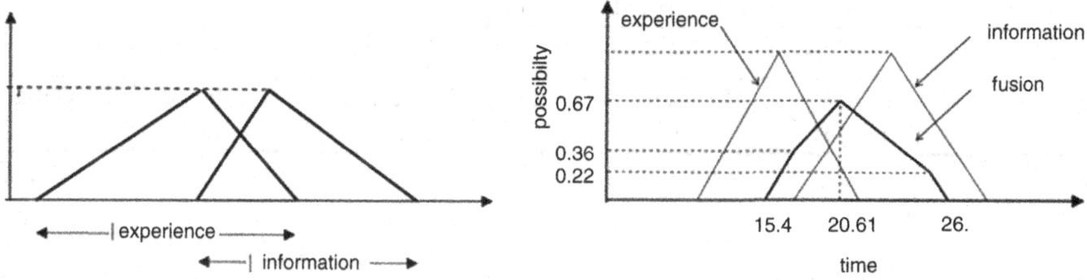

Fig. 2 Possibility distributions of experience and information-fusion of experience and perceived information

management of transportation networks: path choice models assume that users make choices comparing the costs of different alternatives. Thus, providing additional fortuitous costs through different VMS messages, en-route changes of pre-trip choices are possible. The result of information fusion is a subnormal fuzzy set because its height hf is less than 1 (e.g. 0.67) as reported in Fig. 2.

In order to interpret the information given by this fuzzy set, t_f must be redefined as follows: the values of t_f are increased by the same amount of $1 - h_f$. The interpretation of information contained in a subnormal fuzzy set is explained by [22]. Possibility is a useful concept in representing decision-maker's uncertainty about the attributes of individual alternatives, but cannot be used directly by analysts; for this reason, a conversion to Probability values on the basis of a justifiable principle is needed. To pass from Possibility to Probability we use the probabilistic normalization, $(\Sigma_i p_i = 1)$ along with the Principle of Uncertainty Invariance, systematized by [22]. This principle specifies that uncertainty in a given situation should be the same, whatever is the mathematical framework used to describe that situation. Under the requirement of normalization and uncertainty equivalence, we should use a transformation having two free coefficients. Thus, according to [22], we use the log-interval scale transformations. The model allows the quantitative calculation of users' compliance with information, and thus a realistic updating of expected travel time. For more details about this model, refer to [14]. We have applied fuzzy fusion to data acquired in Bari at the end of each simulation. Thus, we have modeled drivers' choice behavior according to uncertainty-based Information Theory. Results of route choice modeling are shown in Table 3, in which they are compared with share observed after experiments.

4 Conclusions and Future Work

This paper shows the results of a pilot study addressed to the internal validation of a driving simulator developed at Technical University of Bari (Italy). First of all the driving simulator has been validated by considering the ability of drivers in choosing the shortest route in case of scenario without information and with

Table 3 Observed share versus Predicted share by model

Route	Without ATIS			With ATIS		
	Observed share(%)	Predicted share(%)	Delta	Observed share(%)	Predicted share(%)	Delta
1	87	90.8	+3.8	13.3	34.5	+21.2
2	10	6.5%	−3.5	66.7%	57.7	−9.0
3	3	2.7%	−0.3	20.0	7.8	−12.2
Eulerian distance			16.39			125.44

information. Furthermore collected data have to be used in order to increase the effect of reduced realism of TSs. To this end the preliminary application of data fusion technique [14] has been made. In particular, expected travel times are updated according to results of data fusion and the influence of uncertainty on drivers' compliance with provided information is examined according to uncertainty-based Information Theory.

In future work authors would like to define a methodology of route choice modeling by mixed data set collect by TSs [17] and DSs. Furthermore the authors will introduce more test in order to validate the adopted modelling approach [19].

References

1. Avineri, E., Prashker, J.: The impact of travel time information on travellers' learning under uncertainty. Transportation **33**(4), 393–408 (2006)
2. Ben-Akiva, M.E., Morikawa, T., Shiroshi, F.: Analysis of the reliability of preference ranking data. J. Bus. Res. **23**, 253–268 (1991)
3. Ben-Elia, E., Di Pace, R., Bifulco, G.N., Shiftan, Y.: The impact of travel information's accuracy on route-choice. Transp. Res. Part C **26**, 146–159 (2013)
4. Bifulco, G.N., Cantarella, G.E., de Luca, S., Di Pace, R.: Analysis and modelling the effects of information accuracy on travellers' behavior, Intelligent Transportation Systems (ITSC), 2011 14th International IEEE Conference on 5–7 Oct. 2011, pp. 2098–2105, Washington, DC (2011)
5. Bifulco, G.N., Di Pace, R., Simonelli, F.: A simulation platform for the analysis of travel choices in ATIS context through Stated Preferences experiments EWGT Conference-Padua, Italy (2009)
6. Bifulco, G.N., Simonelli, F, Di Pace, R.: The role of the uncertainty in ATIS applications. In: 12th On-line Conference on Soft Computing in Industrial Applications—Sp. Sess. Traffic Systems, 16/10/2007 (2007)
7. Bonsall, P., Firmin, P., Anderson, M., Palmer, I., Balmforth, P.: Validating the results of route choice simulator. Transp. Res. Part C **5**(6), 371–387 (1997)
8. Bonsall, P., Parry, T.: Using an interactive route choice simulator to investigate drivers' compliance with route guidance advice. Transp. Res. Rec. **1306**, 59–68 (1991)
9. Cantarella, G.E., de Luca, S.: Multilayer feedforward networks for transportation mode choice analysis: an analysis and a comparison with random utility models. Transp. Res. Part C **13**(2), 121–155 (2005)

10. Cascetta, E.: Transportation Systems Engineering: Theory and Methods, Kluwer Academic publishers (2011)
11. Chang, H.L., Chen, P.C.: Impact of uncertain travel information on drivers' route choice behaviour in TRB 88th annual meeting compendium of Papers DVD, report n. 09 (2009)
12. Chorus, C.G., Arentze, T.A., Timmermans, H.J.P.: Traveler compliance with advice: a Bayesian utilitarian perspective. Transp. Res. Part E **45**(3), 486–500 (2009)
13. Chorus, C.G., Molin, E.J.E., Arentze, T.A., Hoogendoorn, S.P., Timmermans, H.J.P., Van Wee, G.P.: Validation of a multimodal travel simulator with travel information provision. Transp. Res. Part C: Emerg. Technol. **15**(3), 191–207 (2007)
14. Dell'Orco, M., Marinelli, M.: Fuzzy data fusion for updating information in modelling drivers' choice behavior. ICIC 2009, LNAI 5755: 1075–1084 (2009)
15. Di Pace R., Marinelli M., Bifulco G.N., Dell'Orco M.: Modeling risk perception in ATIS context through Fuzzy Logic. Procedia: Social & Behavioral Sciences, vol. 20, pp. 916–926. (2011) ISSN: 1877–0428. doi:10.1016/j.sbspro.2011.08.100
16. Di Pace R., Marinelli M., Bifulco G.N.: Dell'Orco M.: Modelling risk perception in ATIS context: a comparison of different Fuzzy Logic-based approaches In: Procedia: Social & Behavioral Sciences (2012)
17. Di Pace, R., Galante, F., Bifulco, G.N., Pernetti, M.: Collecting data in advanced traveller information system context: travel simulator platform versus choice driving simulator, Compendium of presented papers, 90th TRB Annual Meeting (2011)
18. Di Pace, R.: Analytical Tools for ATIS (Strumenti Analitici per Applicazioni ATIS). Ph.D. thesis, Università degli Studi di Napoli "Federico II"- Facoltà di Ingegneria (2008) http://www.fedoa.unina.it/view/people/Di_Pace,_Roberta.html
19. de Luca, S., Cantarella, G.E.: Validation and comparison of choice models. In: Saleh, W., Sammer, G. (eds.) Success and Failure of Travel Demand Management Measures, pp. 37–58. Ashgate publications, UK (2009)
20. Geer, J.F., Klir, G.J.: A mathematical analysis of information-preserving transformations between probabilistic and possibilistic formulations of Uncertainty. Int. J. Gen. Syst. **20**, 143–176 (1992)
21. Katsikopoulos, K.V., Duse-Anthony, Y., Fisher, D.L., Duffy, S.A.: Risk attitude reversals in drivers' route choice when range of travel time information is provided. Hum. Factors Ergon. Soc. **43**(3), 466–473 (2002)
22. Klir, G.J., Wang, Z.: Fuzzy Measure Theory. Plenum Press, New York (1992)
23. Koutsopoulos, H.N., Lotan, T., Yang, Q.: A driving simulator and its application for modelling route choice in the presence of information. Transp. Res. Part C: Emerg. Technol. **2**(2), 91–107 (1994)
24. Mahmassani, H.S., Jou, R.C.: Transferring insights into commuter behaviour dynamics from laboratory experiments to field surveys. Transp. Res. Part A: Policy Pract. **34**(4), 243–260 (1998)
25. Srinivasan, K.K., Mahmassani, H.S.: Dynamic decision and adjustment processes in commuter behavior under real-time information Technical Research Report SWUTC/02/167204-1, Center for Transportation Research, University of Texas at Austin, February 2002
26. Srinivasan, K.K., Mahmassani, H.S.: Role of congestion and information in tripmakers' dynamic decision processes: an experimental investigation. Transp. Res. Rec. **1676**, 43–52 (1999)
27. Tian, H., Fisher, D.L., Post, B.: Route choice behavior in a driving simulator with real-time information, 90th TRB annual meeting (2010)
28. Tian, H., Gao, S., Fisher, D.L., Post, B.: A mixed-logit latent-class model of strategic route choice behavior with real-time information transportation research board 91st annual meeting (2012)
29. Yager, R.R., Kelman, A.: Fusion of fuzzy information With consideration for compatibility, partial aggregation, and reinforcement. Int. J. Intell. Syst. **15**, 93–122 (1996)

Sustainability Evaluation of Transportation Policies: A Fuzzy-Based Method in a "What to" Analysis

Riccardo Rossi, Massimiliano Gastaldi and Gregorio Gecchele

Abstract The widely debated concepts of sustainability and sustainable development represent nowadays an essential aspect in transportation studies, in particular for the analyses of interactions between transportation and land-use systems. In this paper the three-dimensional concept of sustainability (social, economic and environmental sustainability) is formalized by a Fuzzy-Based Evaluation Method, which has already been applied for evaluating the sustainability of alternative transportation policies. The method is tested as a tool to interpret the preferences expressed by the decision makers, to identify the most important characteristics of alternative transportation policies and to support the design of hypothetical transportation services, following a "*What to*" analysis.

1 Introduction

The concepts of sustainable transportation and sustainable development have been widely debated in recent years in transportation studies. As a major result a multi-dimensional vision of sustainability [35, 36] has been identified and generally accepted as a common framework. From this perspective, a sustainable transportation system may be viewed as one which allows "the movement of people

R. Rossi (✉) · M. Gastaldi · G. Gecchele
Department of Civil, Architectural and Environmental Engineering,
University of Padova, Via Marzolo 9, 35131 Padova, Italy
e-mail: riccardo.rossi@unipd.it

M. Gastaldi
e-mail: massimiliano.gastaldi@dicea.unipd.itl
URL: http://www.dicea.unipd.it/

G. Gecchele
e-mail: gregorio.gecchele@dicea.unipd.it
URL: http://www.dicea.unipd.it/

V. Snášel et al. (eds.), *Soft Computing in Industrial Applications*,
Advances in Intelligent Systems and Computing 223, DOI: 10.1007/978-3-319-00930-8_28,

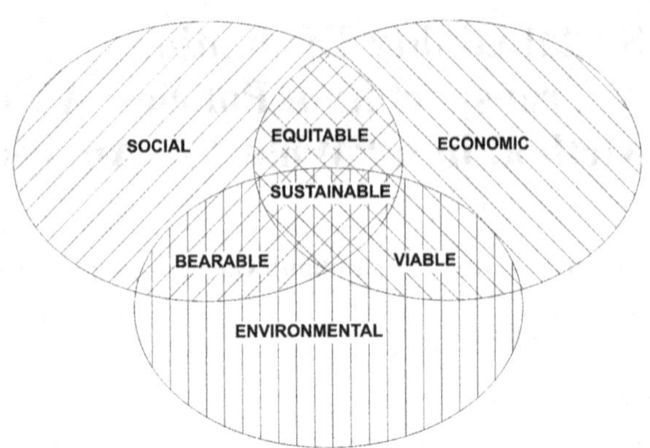

Fig. 1 The "three pillars" of sustainable development

and goods by modalities that are sustainable from an environmental, economic and social point of view" [22, 24]. Several visual representations have been proposed to better explain this concept [5, 18], including the so-called "three pillars of sustainability" or the "triple bottom line" (Fig. 1). If the development is bearable (socially and environmentally sustainable), equitable (socially and economically), and viable (environmentally and economically), it can be considered sustainable/ durable. Adopting such representation one can clearly observe that the concept of sustainability is the result of interactions among the three dimensions, which overlap and cannot be analysed separately from each other.

Fuzzy sets and systems theory can be an effective tool to deal with these conditions and assess the sustainability of a given action plan [1, 4, 7, 23], since it can formalise situations characterised by:

- non-homogeneous variables or quantities;
- uncertain and imprecise information on the system (present and future), in particular when judgements expressed by experts are included in the evaluation;
- interrelations among the dimensions of sustainability which tend to induce "overlaps" ("fuzzy" boundaries).

In this paper a Fuzzy-Based Evaluation Method (F-BEM) [26, 29], which formalises the three-dimensional concept of sustainability, is tested on a case study to evaluate its usefulness as a tool to interpret the preferences expressed by the decision makers, to identify the most important characteristics of alternative transportation policies and to support the design of hypothetical transportation services ("*What to*" analysis).

The paper is organized as follows. Section 2 briefly summarizes past studies concerning sustainability evaluation in the transportation field. In Sect. 3 a description of the architecture of the F-BEM method is provided, while Sect. 4 describes the case study. Concluding remarks are presented in Sect. 5.

2 Background

The evaluation of sustainability for transportation systems has been addressed by various approaches, which can be grouped in eight categories [2]:

1. Life-cycle analysis (LCA) [9], with limited applications in transportation systems;
2. Cost-Benefit Analysis (CBA) and Cost Effectiveness Analysis (CEA) [12, 17], which consider the monetary equivalent of positive and negative effects of project alternatives;
3. Environmental Impact Assessment (EIA), sometimes included in transportation evaluations [8, 37].
4. Optimisation models, applied in the context of sustainable transportation [41];
5. System Dynamics Models, which describe the relationships among the elements of the system by examining time-varying flows and feedback mechanisms [34];
6. Assessment indicator models, subdivided among composite index models (e.g. ecological footprint [3] or the green gross national product), multi-level index models, and multi-dimension matrix models [19];
7. The Data Analysis approach, which uses statistical techniques to evaluate sustainability.
8. Multi-Criteria Decision Analysis (MCDA) methods, that include well-known methodologies such as Multi-Attribute Utility Function Theory (MAUT) [13], Analytic Hierarchy Process (AHP) [32] and ELECTRE methods [30, 39].

MCDA methods are probably the most common approaches used for sustainability evaluation in the transportation field, however they appear inadequate to deal with complexity, uncertainties and impreciseness that characterise many analyses.

Based on this remarks, fuzzy sets and systems theory have been introduced in the framework of multicriteria analysis (fuzzy MCDA), with interesting applications in environmental management and social choice problems [20, 21]. With specific reference to the transportation field, Tangari et al. [33] adopted the NAIADE approach [20] for evaluating the transportation network that could better connect the Balkans to the European Union (EU): the fuzzy multicriteria methodology was used to determine the more efficient and the fairest solution among the alternatives considered. More recently Iannucci et al. [11] used a fuzzy logic-based method for identifying a ranking over transportation facilities and supporting decisions concerning future scenarios of facilities remodelling.

Coherently with these approaches, the authors have already adopted fuzzy sets and systems theory to overcome the above-mentioned limitations, developing the F-BEM [25–27, 29], based on similar applications in other research fields [1, 23].

In this paper the main interest is related to the application of the F-BEM as a tool to design hypothetical transportation services following a "*What to*" analysis, based on the identification of the most relevant characteristics expressed by the decision makers.

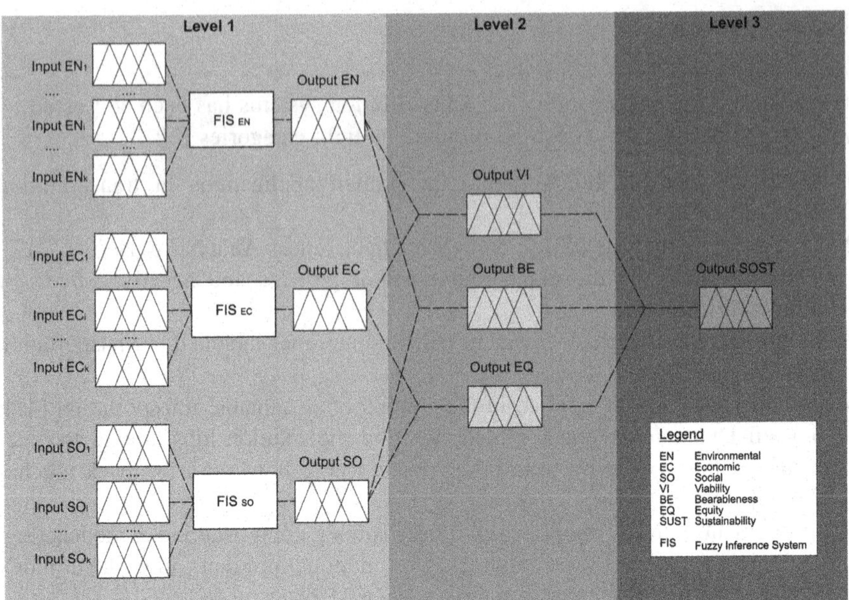

Fig. 2 Three-level structure of F-BEM

3 Methodology

The Fuzzy-Based Evaluation Method (F-BEM) [29] applied in the paper interprets the three-dimensional concept of sustainability working on three different levels (Fig. 2).

At the first level, three fuzzy inference systems (FIS_{EC}, FIS_{EN}, FIS_{SO}) process environmental (Input $EN_1, \ldots,$ Input EN_k), economic (Input $EC_1, \ldots,$ Input EC_k), and social (Input $SO_1, \ldots,$ Input SO_k) input variables (indicators), respectively. Each FIS produces a corresponding sustainability index (Output EN, Output EC, Output SO), defined on a two-level semantic scale: "unsustainable" and "sustainable" (Fig. 3).

The variable domain is subdivided into two parts: the "unsustainable zone" (values from 0.0 to 0.5), where unsustainability is higher than sustainability, and the "sustainable zone", (with values from 0.5 to 1.0) where sustainability is higher than unsustainability. The central value of the domain (0.5) represents the uncertainty of expressing the right judgement, since it has the same Grade of Membership (GoM) at both sustainable and unsustainable level (0.5). The rules employed in each FIS are defined by decision makers within a focus group [16] and are applied with Mamdani's sum-product inference in order to guarantee the monotonicity of the output [15].

At the second level, the fuzzy variables representing social, environmental and economic sustainability indices (Output EN, Output EC, Output SO) are examined

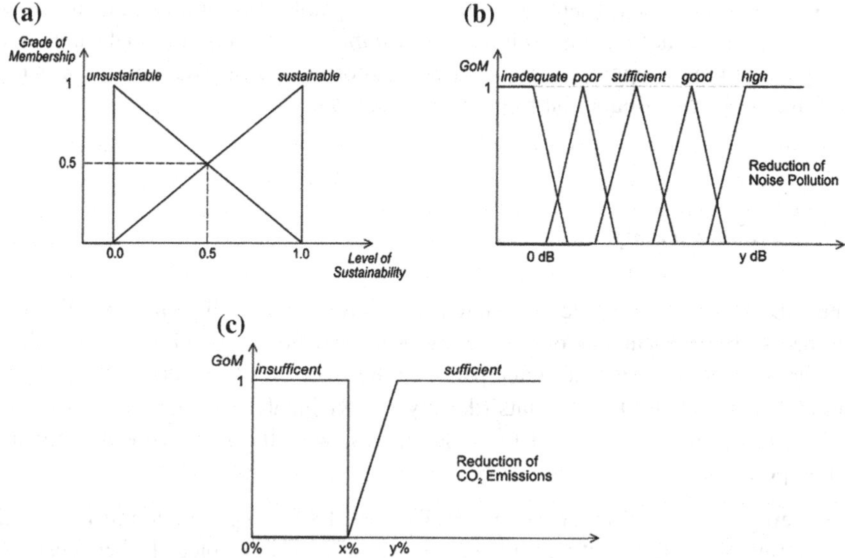

Fig. 3 Fuzzy representation of (**a**) index of sustainability and (**b, c**) various type of indicators

in pairs, in order to obtain the fuzzy indices of equity "Output EQ" (social-economic dimension), viability "Output VI" (economic-environmental dimension) and bearableness "Output BE" (social-environmental dimension). This is done by aggregating the fuzzy sets obtained at the first level using the weighted average [38]. This choice represents a way to model the importance of each input variable with its corresponding weight, although other kinds of aggregation can be adopted (more details can be found in Rossi et al.'s paper [29]).

At the third and final level, the aggregation is repeated for the second-level output fuzzy sets, producing the fuzzy index of "overall" sustainability "Output SUST" (social-environmental-economic dimension).

In the F-BEM, "indicators" are the input variables for the first-level fuzzy inference systems and measure the extent to which the stated objectives are achieved by each alternative with reference to the three dimensions of sustainability. Indicators are introduced in the F-BEM using a fuzzy representation, where the shape of Membership Functions (MFs) are built with the help of experts [14]. A large number of indicators have been proposed and applied in the past [10, 31, 40], therefore the choice of which indicators to adopt is complex and depends on the specific case study. Generally, the description is based on two types of structure (Fig. 3) [29]:

1. Semantic Scale with triangular MFs, with a variable number of membership functions, depending on the level of details needed (e.g. "Reduction of Noise Pollution" or "Operating Cost");

2. Scale for indicators where the minimum acceptable variation corresponds to a percentage reduction goal with respect to *status quo* conditions. Goal can be set by government agencies or by local authorities (e.g. "Reduction of CO_2 Emissions" or "Reduction in number of accidents").

4 Test Case Study

The F-BEM was tested on data adapted from a real case study, with the objective of analysing the capability of the method as a interpretation and a design tool.

The area of study was a municipality in the North East of Italy, with a population of about 40,000 inhabitants (density 393.81 inhabitants/km^2).

The local authorities tested two alternatives, with the main aim of reducing traffic pollution:

1. Alternative Urban Transit Service (UTS). A UTS linking the main districts and the railway station, with established bus-stops and timetables. Tickets cost 1.0 Euro per trip (free for people over 75 and those with train or bus passes).
2. Alternative Even-Odd Plate Number (EOPN). Excluding non-catalysed vehicles and an even-odd number plate rule in the mornings (8.0–10.0 a.m.) and in the evenings (4.0–7.0 p.m.) for catalytic vehicles, two days a week. An average reduction of 2.5 % was observed in traffic volumes in the week considered in the analysis.

However the results were considered unsatisfactory and local authorities asked for new alternatives, with the aim of enhancing existing public transportation services, reducing traffic pollution and improving equity among citizens.

4.1 Evaluation of Existing Alternatives

Each alternative was evaluated by indicators grouped by type (social, economic, environmental), paying particular attention to data availability. Table 1 lists the indicators considered, the type of membership functions adopted (see Fig. 3) and the performance obtained by the alternatives.

The indicators included the viewpoints of various stakeholders, such as the local authority ("Operating cost") the population ("Propensity towards service"), users ("User cost variation") and the community as a whole ("Community livability" and environmental indicators).

The performance was estimated by the F-BEM examining the data for one week in winter (weekdays only; Monday-Friday), considering the variations compared with the *status quo* for most indicators. Methods already established in the

Table 1 Performance of alternatives with reference to each indicator

Indicator	Unit of measurement	Memb. Func.	UTS	EOPN	MEAN	HYP
EN_1 Reduction of CO_2 Emissions[a]	Percent	c	+0.01	+2.27	+1.1	+2.27
EN_2 Reduction of CO Emissions[a]	Percent	c	+0.01	+8.58	+1.1	+8.58
EN_3 Reduction of fuel consumption[a]	Percent	c	+0.01	+2.28	+1.1	+2.28
EC_1 Operating cost	Euro per week	b	1260	860	1000	2000
EC_2 User cost variation[a]	Euro per trip	b	−4	0	−2	−2
SO_1 Propensity towards service[b]	Fuzzy intervals	b	2	1	1.5	2.5
SO_2 Community livability	Qualitative	b	4	2	3	4

(a) *referring to the status quo*
(b) *only the central value is given*

literature were followed, however further details are given concerning the calculation of indicator "Propensity towards service".

This indicator considers that the improvement in users' quality of life as a result of the introduction of a new transportation service, may be measured indirectly by users' propensity towards the new service. This propensity can be collected by a questionnaire, as part of an SP survey [28], in which interviewees express their "propensity" towards changing the current transportation service for a new one on a suitable semantic scale of responses. The semantic scale [28] is translated into a set of fuzzy intervals, and the average of the propensities, taken as fuzzy intervals, is calculated as an aggregate measurement of responses, maintaining the uncertainty associated to the expressions of propensity [6, 14].

Table 2 summarises the results of the evaluation, reporting values of the GoM for the extreme values of the dominion (GoM(0) and GoM(1)) for the sustainability indices calculated by the F-BEM. The results are included between the extreme cases of complete sustainability (GoM(1) = 1 and GoM(0) = 0), and complete unsustainability (GoM(1) = 0 and GoM(0) = 1).

4.2 Interpretation of Preferences

To evaluate the capability of F-BEM to interpret the preferences expressed by the decision makers, the method was applied considering an hypothetical alternative with intermediate characteristics between the original ones (MEAN alternative) adopting a traditional "*What if*" analysis (See Tables 1 and 2). Each indicator has been changed, one-by-one, taking sample values between the minimum and the maximum and evaluating the effects on fuzzy indices of sustainability, maintaining the original values for other indicators ("*ceteris paribus*" conditions).

The analysis highlights that F-BEM evaluates the alternative as expected by the decision makers, showing an increase in sustainability when better performance is

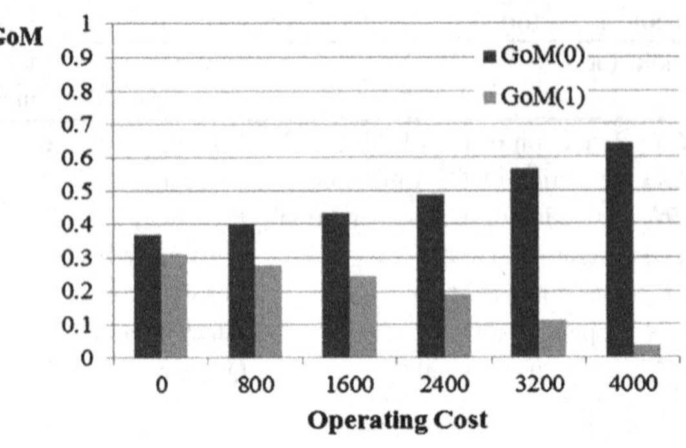

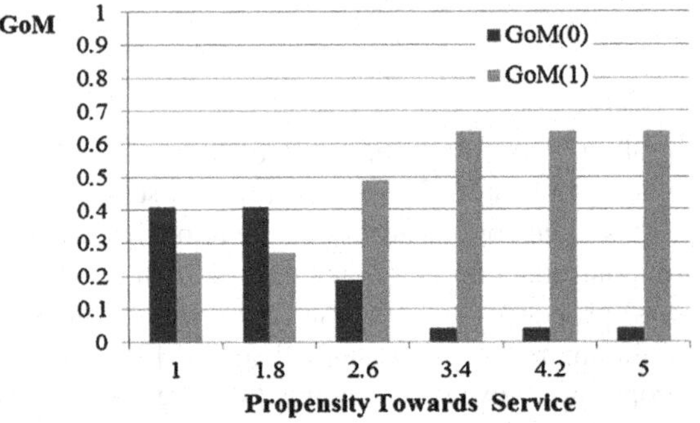

Fig. 4 Effects on the overall sustainability of changing indicators "Operating Costs" and "Propensity towards service"

assumed and, vice-versa, a decrease if there is a worse performance for the indicator under analysis. Furthermore, the influence of each indicator on the sustainability changes depending on the importance of the related dimension (that is the dimension weight): the larger the weight given to the dimension, the greater the change observed in the index of sustainability.

As an example Fig. 4 shows the effects of changing the "Operating Costs" and the "Propensity towards service" on the overall sustainability, expressed by the values of the GoM for the extremes of the dominion (GoM(0) and GoM(1) values).

As expected by the decision makers, if the Operating Cost increases, the overall sustainability decreases, since the cheaper is the service, the better it is considered. Similarly if the Propensity towards the proposed service increases, then the overall sustainability increases. The differing size of the effects observed is proportional to the importance given by the decision makers to the Economic and the Social dimensions, as expressed by the relative weights.

4.3 Identification of Hypothetical Alternatives

One of the main interest of local authorities was the design of new transportation alternatives, based on some clear requests. Therefore the F-BEM was applied following a "*What to*" analysis, in order to understand the characteristics of the

Table 2 Performance of alternatives. GoM(0) and GoM(1) of output fuzzy indices

Dimension	UTS		EOPN		MEAN		HYP	
	GoM(0)	GoM(1)	GoM(0)	GoM(1)	GoM(0)	GoM(1)	GoM(0)	GoM(1)
Environmental	0.00	2*E-8	0.00	0.45	0.00	0.10	0.00	0.45
Economic	0.06	0.94	0.21	0.79	0.15	0.85	0.30	0.70
Social	0.49	0.52	1.00	0.00	1.00	0.00	0.25	0.75
Bearableness	0.30	0.31	0.61	0.18	0.61	0.04	0.15	0.63
Viability	0.03	0.43	0.10	0.61	0.07	0.45	0.14	0.57
Equity	0.33	0.66	0.72	0.28	0.69	0.31	0.27	0.73
Overall sustainability	0.20	0.44	0.42	0.38	0.41	0.27	0.17	0.63

hypothetical alternatives. Given these characteristics, experts can identify a transportation service (or a series of measures) which guarantees the satisfaction of the requests.

In particular the local authorities would like to:

1. reduce the emission of pollutants (Environmental dimension) more than (or at least as done by) the EOPN alternative;
2. spend no more than 2000 Euro/week (Budget Constrains) for the new alternative;
3. reach the sustainability in each dimension at each level.

The F-BEM has been used to satisfy the requests expressed by local authorities (Table 2), defining the characteristics of the hypothetical service, from the overall sustainability to the first-level dimensions. The HYP alternative identified can be described based on the values of first-level indicators (Table 1):

- Economic Dimension. The budget constrain is satisfied and the "User Cost Variation" is set to −2.0, because the service needs to reduce the cost for the users in order to be chosen.
- Environmental Dimension. The reduction of pollutants is set to the values obtained by the EOPN alternative to satisfy the requests of the local authorities.
- Social Dimension. To change the users' choices the new alternative must obtain a quite high level of propensity (2.5) and increase the Community livability at least as the UTS alternative. The Social dimension is particularly important since it has a relevant impact on the overall sustainability.

The reduction of the pollutants is equivalent to the reduction of a certain amount of trips and, based on the length of trips, in vehicles travelling in the network. Lastly one can obtain the demand to be satisfied by the new alternative.

At this point the experts can consider various options to refine the results of the analysis and define a new measure or, more probably, a combination of measures such as:

1. Designing a survey to evaluate the mobility demand with reference to the types of trips that affects the municipalities, in particular internal trips, and trips with origin/destination in the municipality. A detailed survey, which should consider the most relevant origin/destination or exchange points (e.g. railway stations, bus stations, parking lots for commuters) can provide a better comprehension of the mobility needs of citizens.
2. Based on the demand survey, defining solutions which ease the modal shifts from private car to other "greener" modes. These solutions may include:
   - new flexible transportation services (such as Dial-a-Ride), which can substantially increase the social sustainability of transit system, giving particular attention to non-drivers, the low-income population, and elderly or disabled people;
   - integration of existing services (e.g. private car, bus, train) to guarantee high level of quality in transit systems in a global sense;
   - supporting non-vehicular mobility, in particular for internal trips.
3. Based on the demand survey, estimate the revenue that may be collected by fares, in order to come out with a correct definition of the solutions, also in terms of economic sustainability.

5 Conclusions

This paper deals with the application of a Fuzzy-Based Evaluation Method (F-BEM) as a tool to support the design of hypothetical transportation services, by the identification of the most important characteristics of alternative hypotheses. The F-BEM adopts fuzzy sets and systems theory to formalise the concept of the "three pillars of sustainability", managing the complexity and the uncertainties that characterise the sustainability assessment in the transportation field.

Some remarkable findings are that:

- F-BEM, applied in a "*What if*" analysis, is capable of representing the preferences expressed by the decision makers; the sustainability indices changes accordingly to the modifications of the values of the indicators respecting the relative importance given to the various dimensions of sustainability;
- F-BEM, applied in a "*What to*" analysis, can represent a simple but effective tool to identify new hypothetical alternatives which satisfy the decision makers' requests. Experts can benefit from the analysis by easily refining the results and defining a combination of new measures, as shown by the test case study presented in the paper.

References

1. Andriantiatsaholiniaina, L.A., Kouikoglou, V., Phillis, Y.: Evaluating strategies for sustainable development: fuzzy logic reasoning and sensitivity analysis. Ecol. Econ. **48**, 149–172 (2004)
2. Awasthi, A., Chauhan, S.S., Omrani, H.: Application of fuzzy TOPSIS in evaluating sustainable transportation systems. Expert Syst. Appl. **38**, 12270–12280 (2011)
3. Browne, D., O'Regan, B., Moles, R.: Use of ecological footprinting to explore alternative policy scenarios in an Irish cityregion. Transp. Res. Part D **13**, 315–322 (2008)
4. Cornelissen, A.M.G., van den Berg, J., Koops, W.J., Grossman, M., Udo, H.M.J.: Assessment of the contribution of sustainability indicators to sustainable development: a novel approach using fuzzy set theory. Agric. Ecosyst. Environ. **86**, 173–185 (2001)
5. Dalal-Clayton, B., Bass, S.: Sustainable Development Strategies, 1st edn. Earthscan Publications Ltd, London, p. 358 (2002)
6. Dubois, D., Prade, H.: Possibility theory. An approach to computerized processing of uncertainty. Plenum Ed, New York (1987)
7. Dunn, E.G., Keller, J.M., Marks, L.A., Ikerd, J.E., Gader, P.D., Gosey, L.D.: Extending the application of fuzzy sets to the problem of agricultural sustainability. In: Proceedings of 3rd International Symposium on Uncertainty Modelling and Analysis (ISUMA '95), pp. 497–502. IEEE Computer Society, Washington DC (1995)
8. ECMT: Assessment and decision making for sustainable transport. European Conference of Ministers of Transportation, Organization of Economic Coordination and Development (2004) http://www.oecd.org
9. Guine, J.B.: Handbook on life cycle assessment. An operational guide to the ISO standard. Kluwer, London, p. 704 (2002)
10. Haghshenas, H., Vaziri, M.: Urban sustainable transportation indicators for global comparison. Ecol. Indic. **15**, 115–121 (2012)
11. Iannucci, G., Ottomanelli, M., Sassanelli, D.: A fuzzy logic-based methodology for ranking transport infrastructures. AISC **96**, 369–377 (2011)
12. INFRAS, CE Delft, ISI, University of Gdansk: Handbook on estimation of external costs in the transport sector. Report for the European Commission, Produced within the Study Internalisation Measures and Policies for All External Costs of Transport (IMPACT)(2007)
13. Keeney, R.L., Raiffa, H.: Decisions with Multiple Objectives. Cambridge University Press, Cambridge (1993)
14. Klir, G.J., Yuan, B.: Fuzzy Sets and Fuzzy Logic. Theory and Applications. Prentice-Hall PTR, Upper Saddle River (1995)
15. Kouikoglou, V.S., Phillis, Y.A.: On the monotonicity of hierarchical sum-product fuzzy systems. Fuzzy Set. Syst. **160**(24), 3530–3538 (2009)
16. Krueger, R.A., Casey, M.A.: Focus Groups: A Practical Guide for Applied Research, 3rd edn. Sage Publications Inc., Thousand Oaks (2008)
17. Kunreuther, H., Grossi, P., Seeber, N., Smith, A.: A Framework for Evaluating the Cost-Effectiveness of Mitigation Measures. Columbia University, USA (2003)
18. Lozano, R.: Envisioning sustainability three-dimensionally. J. Clean Prod. **16**, 1838–1846 (2008)
19. Mori, K., Christodoulou, A.: Review of sustainability indices and indicators: towards a new City Sustainability Index (CSI). Environ. Impact. Assess. Rev. **32**, 94–106 (2012)
20. Munda, G.: Multicriteria Evaluation in a Fuzzy Environment. Theory and Applications in Ecological Economics. Physica-Verlag, Heidelberg (1995)
21. Munda, G.: Social multi-criteria evaluation: methodological foundations and operational consequences. Eur. J. Oper. Res. **158**(3), 662–677 (2004)
22. Organization of Economic Cooperation and Development (OECD): Towards sustainable transportation. OECD Proceedings of the Vancouver Conference, OECD (1996)

23. Phillis, Y., Andriantiatsaholiniaina, L.A.: Sustainability: an ill-defined concept and its assessment using fuzzy logic. Ecol. Econ. **37**, 435–456 (2001)
24. Rassafi, A.A., Vaziri, M.: Sustainable transport indicators: definition and integration. Int. J. Environ. Sci. Technol. **21**, 83–96 (2005)
25. Rossi, R., Gastaldi, M., Vescovi, R.: A methodological approach to evaluating the sustainability level of a transportation service. Sustain. Dev. Plan. **4**(2), 411–424 (2009) WITPress, ISBN: 978-1-84564-181-8
26. Rossi, R., Gastaldi, M., Gecchele, G., Vescovi, R.: An improvement of a fuzzy three-level model to evaluating transport systems sustainability considering decision maker's attitude. Proceedings of the XIII EWGT Meeting, Padova, Italy, 23–25 September 2009, Padova University Press, ISBN: 978-88-903541-4-4 (2009)
27. Rossi, R., Gastaldi, M., Gecchele, G., Vescovi, R.: Using a fuzzy approach for evaluating sustainability of transportation system pollution-reducing policies: a case study. Proceedings of TRB 89th Annual Meeting, Washington D.C., 10–14 January 2010. CD-ROM (2010)
28. Rossi, R., Gastaldi, M., Gecchele, G.: Fuzzy systems approach versus possibility theory approach for representing customers' stated preferences on freight transport services. In: Mussone, L., Crisalli, U. (eds.) Transport Management and Land-Use Effects in Presence of Unusual Demand. Selected Papers, vol. 1797.38, pp. 275–296. Franco Angeli, Milan, ISBN: 978-88-568-4174-9 (2011)
29. Rossi, R., Gastaldi, M., Gecchele, G.: Comparison of fuzzy-based and AHP methods in sustainability evaluation: a case of traffic pollution-reducing policies. Eur. Transp. Res. Rev. (2012) In Press doi:10.1007/s12544-012-0086-5
30. Roy, B., Hugonnard, D.: Ranking of suburban line extension projects on the Paris Metro System by a multi-criteria method. Transp. Res. Rec. **16A**(4), 301–312 (1982)
31. Russo, F., Comi, A.: Measures for sustainable freight transportation at urban scale: expected goals and tested results. Eur. J. Urban Plan. Dev. **137**(2), 142–152 (2011)
32. Saaty, T.L.: The Analytic Hierarchy Process. McGraw-Hill, New York (1980)
33. Tangari, L., Ottomanelli, M., Sassanelli, D.: Multicriteria fuzzy methodology for feasibility study of transport projects case study of southeastern trans-european transport axes. Transp. Res. Rec. **2048**, 26–34 (2008)
34. Tao, C.-C., Hung, C.-C.: A comparative approach of the quantitative models for sustainable transportation. J. East. Asia Soc. Transp. Stud. 5, 3329–3344, http://www.easts.info/2003journal/papers/3329.pdf(2003)
35. Transportation Research Board TRB: Toward a sustainable future; addressing the long-term effects of motor vehicle transportation on climate and ecology. TRB Special Report 251, National Academy Press, Washington, DC (1997)
36. United Nations World Commission on Environment and Development: Our Common Future. Oxford University Press, Oxford (1987)
37. Wood, C.: Environmental Impact Assessment: A Comparative Review, vol. 405, 2nd edn. Prentice-Hall, UK (2002)
38. Yager, R.R.: Fuzzy decision making including unequal objectives. Fuzzy Set. Syst. **1**, 87–95 (1978)
39. Yu, W.: ELECTRE Tri: Aspects methodologiques et manuels d'utilisation. Document de LAMSADE, 74, Universit Paris-Dauphine (1992)
40. Zito, P., Salvo, G.: Toward an urban transport sustainability index: a European comparison. Eur. Transp. Res. Rev. **3**, 179–195 (2011)
41. Zuidgeest, M.H.P.: Sustainable urban transport development: a dynamic optimization approach, Ph.D. thesis, University of Twente, Enschede. (2005) http://doc.utwente.nl/57439

Artificial Bee Colony-Based Algorithm for Optimising Traffic Signal Timings

Mauro Dell'Orco, Özgür Başkan and Mario Marinelli

Abstract This study proposed Artificial Bee Colony (ABC) algorithm for finding optimal setting of traffic signals in coordinated signalized networks for given fixed set of link flows. For optimizing traffic signal timings in coordinated signalized networks, ABC with TRANSYT-7F (ABCTRANS) model is developed. The ABC algorithm is a new population-based metaheuristic approach, and it is inspired by the foraging behavior of honeybee swarm. TRANSYT-7F traffic model is used to estimate total network performance index (PI). The ABCTRANS is tested on medium sized signalized road network. Results showed that the proposed model is slightly better in signal timing optimization in terms of final values of PI when it is compared with TRANSYT-7F in which Genetic Algorithm (GA) and Hill-climbing (HC) methods are exist. Results also showed that the ABCTRANS model improves the medium sized network's PI by 2.4 and 2.7 % when it is compared with GA and HC methods.

1 Introduction

In urban networks, traffic signals are used to control vehicle movements so as to reduce congestion, improve safety, and enable specific strategies such as minimizing delays, improving environmental pollution, etc. [1]. Signal systems that

M. Dell'Orco (✉) · M. Marinelli
Technical University of Bari, D.I.C.A.T.E.Ch., Bari, Italy
e-mail: dellorco@poliba.it

M. Marinelli
e-mail: m.marinelli@poliba.it

Ö. Başkan
Pamukkale University, Department of Civil Engineering, Denizli, Turkey
e-mail: obaskan@pau.edu.tr

V. Snášel et al. (eds.), *Soft Computing in Industrial Applications*,
Advances in Intelligent Systems and Computing 223, DOI: 10.1007/978-3-319-00930-8_29,

control road junctions are operated according to the type of junction. Although the optimization of signal timings for an isolated junction is relatively easy, the optimization of signal timings in coordinated road networks requires further research due to the "*offset*" term. Early methods such as that of Webster [2] only considered an isolated signalized junction. Later, fixed time strategies were developed that optimizing a group of signalized junctions using historical flow data [3]. For the Area Traffic Control (ATC), TRANSYT-7F is one of the most useful network study software tools for optimizing signal timing and also the most widely used program of its type. It simulates traffic in a network of signalized intersections to produce a cyclic flow profile of arrivals at each intersection that is used to compute a Performance Index (*PI*) for a given signal timing and staging plan. Optimization in TRANSYT-7F consists of a series of trial simulation runs, using the TRANSYT-7F simulation engine. Each simulation run is assigned a unique signal timing plan by the optimization processor. The optimizer applies the Hill-Climbing (HC) or Genetic Algorithm (GA) searching strategies. Although the GA is mathematically better suited for determining the absolute or global optimal solution, relative to HC optimization, it generally requires longer program running times, relative to HC optimization (TRANSYT-7F Release 11.3 Users Guide, 2008).

Wong in [4] proposed group-based optimization of signal timings for area traffic control. Heydecker in [5] decomposed the optimization of traffic signal timings into two levels; first, optimizing the signal timings at the individual junction level using the group-based approach, and second, combining the results from individual junction level with network level decision variables such as offset and common cycle time. Wong et al. in [6] developed a time-dependent TRANSYT traffic model for the evaluation of *PI*. Wong et al. in [7] developed a time-dependent TRANSYT traffic model which is a weighted combination of the estimated delay and number of stops. Girianna and Benekohal in [8] presented two different GA techniques which are applied on signal coordination for oversaturated networks. Similarly, Ceylan in [9] developed a GA with TRANSYT-HC optimization routine, and proposed a method for decreasing the search space to solve the ATC problem. Chen and Xu in [10] investigated the application of Particle Swarm Optimization algorithm to solve signal timing optimization problem. Similarly, Chiou in [11] presented a computation algorithm based on the projected Quasi-Newton method to effectively solve the ATC. Dan and Xiaohong in [12] developed a real-coded improved GA with microscopic traffic simulation model to find optimal signal plans for the ATC. Li in [13] presented an arterial signal optimization model that consider queue blockage among intersection lane groups under oversaturated conditions. Although there are many studies in literature with different heuristic methods to optimize traffic signal timings, there is no application of ABC optimization method to this area. Thus, in this study, **A**rtificial **B**ee **C**olony Optimization **TRANS**YT-7F (ABCTRANS) model, in which ABC and TRANSYT-7F are combined for solving the ATC, was developed.

2 Artificial Bee Colony Algorithm

The foraging behaviors of honeybees have recently been one of the most interesting research areas in swarm intelligence. Some approaches have been proposed to model the specific intelligent behaviours of honeybee swarms and they have been applied for solving optimization problems. Lucic and Teodorovic in [14] and Teodorovic in [15] suggested to use bee swarm intelligence aimed at solving complex problems in traffic and transportation. Teodorovic and Dell'Orco in [16] proposed the bee colony optimization to solve combinatorial problems characterized by uncertainty, as well as deterministic combinatorial problems. Teodorovic and Dell'Orco in [17] presented an application of the bee colony optimization, efficient in solving the ride-matching problem. Their results showed that proposed metaheuristic appears very promising, and indicated that the development of new models based on swarm intelligence principles could significantly contribute to the solution of a wide range of complex engineering and management problems. The Artificial Bee Colony (ABC) algorithm is a new population-based metaheuristic approach proposed by Karaboga in [18] and further developed by Karaboga and Basturk [19–21]. It is inspired by the foraging behavior of honeybee swarm. The foraging bees are classified into three categories employed, onlookers and scouts. Half of the bee colony consists of employed bees, and another half consists of onlookers. In ABC algorithm, the position of a food source represents a possible solution to the optimization problem and the nectar amount of a food source corresponds to the quality (fitness) of the associated solution [22]. The number of employed bees or the onlooker bees is equal to the number of solutions in the population. Employed bees are responsible for searching available food sources and gathering required information. They also pass their food information to onlooker bees. The onlookers select good food sources from those found by the employed bees to further search the foods. When the quality of the food source is not improved through a predetermined number of cycles, the food source is abandoned by its employed bee. In this case, the employed bee becomes a scout and starts to search for a new food source in the vicinity of the hive.

In ABC algorithm, each cycle of the search consists of three steps. At the initialization step, the ABC algorithm generates a randomly distributed initial population as number of *SN*, where *SN* denotes the number of employed bees or onlooker bees. Each initial solution $x_i (i = 1, 2, \ldots, SN)$ is a *D*-dimensional vector which *D* is the number of decision variables of a given optimization problem. At the first step of the cycle, employed bees come into the hive and share with the bees waiting on the dance area information about nectar sources. A bee waiting on the dance area is called onlooker, and is responsible for making decision about the choice of a food source. At the second step, an onlooker chooses a food source area depending on the nectar information distributed by the employed bees on the dance area. As the nectar amount of a food source increases, the probability of choice of that food source increases as well. The determination of a new food source is carried out by the bees based on a visual comparison process of positions of food sources. At the third step of the cycle, when a food source is abandoned by the bees, a new food source is randomly determined by a scout bee and replaces the abandoned one. These three steps are repeated until a predetermined number of cycles, called Maximum Cycle Number (*MCN*), is reached. An onlooker bee chooses a food source depending on the probability value, p_i, as follows:

$$p_i = \frac{fit_i}{\sum_{n=1}^{SN} fit_n} \tag{1}$$

where fit_i is the fitness value of solution i. In this way, the employed bees exchange their information with the onlookers. In order to share the information of nectar amount of the food sources, the employed bees use a proportional selection method known as "roulette wheel selection".

In order to produce a candidate food location from the old one in population, the following equation is used.

$$v_{ij} = x_{ij} + \phi_{ij}(x_{ij} - x_{kj}) \tag{2}$$

where v_{ij} is the candidate food position which can replace the old one in the memory and ϕ_{ij} is a random number generated in the interval $[-1, 1]$. The values $k = 1, 2, \ldots, SN$ and $j = 1, 2, \ldots, D$ are randomly chosen indexes. Of course, k must be different from i, to avoid that old and new location coincide, in order to find food sources having more nectar amount than the old one. The parameter ϕ_{ij} controls the production of neighbour food sources around x_{ij}, and represents the visual comparison of two food positions carried out by a bee. If a parameter value determined using Eq. (2) exceeds the constraints of the decision variables, the parameter is set to its upper and lower boundary, depending on which constraint has been exceeded.

As mentioned above, the food source abandoned by the bees is replaced with a new food source by the scouts at the third step of the cycle. In ABC algorithm, this is simulated by generating a random location and replacing the abandoned one with it. If a location cannot be further improved in a predetermined number of cycles, then that food source is assumed to be abandoned. The value of predetermined number of cycles, called "limit", is an important control parameter of the algorithm. Karaboga and Akay proposed to determine this value as *SN*D* [22]. This operation can be done using Eq. (3).

$$x_i^j = x_{\min}^j + \text{rand}\,[0, 1](x_{\max}^j - x_{\min}^j) \tag{3}$$

After each candidate source location v_{ij} is generated, its performance is compared with that of the old one. If the new food source has equal or better nectar than the old source, it replaces the old one in the memory. Otherwise, the old one is retained in the memory. In other words, a greedy selection mechanism is employed as selection between the old and the candidate location. The steps of the ABC algorithm are given as:

Step 0: Initialize the population of solutions $x_i, i = 1, 2, \ldots, SN$, and evaluate them

Step 1: Generate new solutions v_i for the employed bees by Eq. (2) and determine the quality of the solutions

Step 2: Apply the greedy selection process for the generated new solutions in Step 1

Step 3: Calculate the probability values p_i for the solutions x_i by Eq. (1)

Step 4: Generate the new solutions for the onlookers from the solutions x_i due to the probabilities p_i using roulette wheel selection

Step 5: Apply the greedy selection process for the onlookers

Step 6: Determine the solution for the scout bee and replace it with produced solution x_i by Eq. (3)

Step 7: If the number of cycle is reached to *MCN*, the algorithm is terminated. Else go to Step 1.

3 Model Formulation

The proposed ABCTRANS model consists of two main parts namely ABC algorithm and TRANSYT-7F traffic model. ABC algorithm optimizes traffic signal timings under fixed set of link flows. TRANSYT-7F traffic model is used to compute *PI*, which is called objective function, for a given signal timing and staging plan in network. The network Disutility Index (*DI*), one of the TRANSYT-7F's *PI*, is used as objective function. The *DI* is a measure of disadvantageous operation; that is stops, delay, fuel consumption, etc. The standard TRANSYT-7F's *DI* is linear combination of delay and stops. The objective function and corresponding constraints are given in Eq. (4).

$$PI = \underset{\psi,\, q = fixed}{Min\ DI} = \sum_{a \in L} \left[w_{d_a} \cdot d_a(\psi) + K \cdot w_{s_a} \cdot S_a(\psi) \right] \tag{4}$$

subject to

$$\psi(c,\theta,\phi) \in \Omega_0; \left\{ \begin{array}{ll} c_{\min} \le c \le c_{\max} & \text{cycle time constraints} \\ 0 \le \theta \le c & \text{values of offset constraints} \\ \phi_{\min} \le \phi \le c & \text{green time constraints} \\ \sum_{i=1}^{z} (\phi + I)_i = c & \end{array} \right\}$$

where d_a is delay on link a (L set of links), w_{d_a} is link-specific weighting factor for delay d, K is stop penalty factor to express the importance of stops relative to delay, S_a is stop on link a, w_{s_a} is link-specific weighting factor for stops S on link a, q is fixed set of link flows, ψ is signal setting parameters, c is common cycle time (sec), θ is offset time (sec), ϕ is green time (sec), Ω_0 is feasible region for signal timings, I is intergreen time (sec), and z is number of stages at each signalized intersection in a given road network.

The green timings can be distributed to all signal stages in a road network according to Eq. (5) in order to provide the cycle time constraint [23].

$$\phi_i = \phi_{\min,i} + \frac{p_i}{\sum_{k=1}^{z} p_i}\left(c - \sum_{k=1}^{z} I_k - \sum_{k=1}^{z} \phi_{\min,k}\right) \qquad i = 1, 2, \ldots, z \tag{5}$$

where ϕ_i is the green time (sec) for stage i, $\phi_{\min,i}$ is minimum green time (sec) for stage i, p_i is generated randomly green timings (sec) for stage i, z is the number of stages and I is intergreen time (sec) between signal stages and c is the common cycle time of the network (sec).

In the ABCTRANS, optimization steps can be given in the following way:

Step 0: Initialization. Define the user specified parameters; the number of decision variables (n) (this number is sum of the number of green times as stage numbers at each intersection, the number of offset times as intersection numbers and common cycle time), the constraints for each decision variable, the size of bee colony (*SN*), maximum cycle number (*MCN*).

Step 1: Set $t = 1$.

Step 2: Generate the random initial signal timings within the constraints of decision variables.

Step 3: Distribute to the green timings to the stages according to distribution rule as mentioned above. At this step, randomly generated green timings at Step 2 are distributed to the stages according to generated cycle time at the same step, minimum green and intergreen time.

Step 4: Get the network data and fixed set of link flows for TRANSYT-7F traffic model.

Step 5: Run TRANSYT-7F.

Step 6: Get the network *PI*. At this step, the *PI* is determined using TRANSYT-7F traffic model.

Step 7: If $t = MCN$then terminate the algorithm; otherwise, $t = t{+}1$ and go to Step 2.

The flowchart of the ABCTRANS can be seen in Fig. 1.

4 Application

In order to test the ABCTRANS model's performance, it is also applied to medium sized road network. The network is taken from [24], and given in Fig. 2. This network includes 23 links and 21 signal setting variables at six signal-controlled

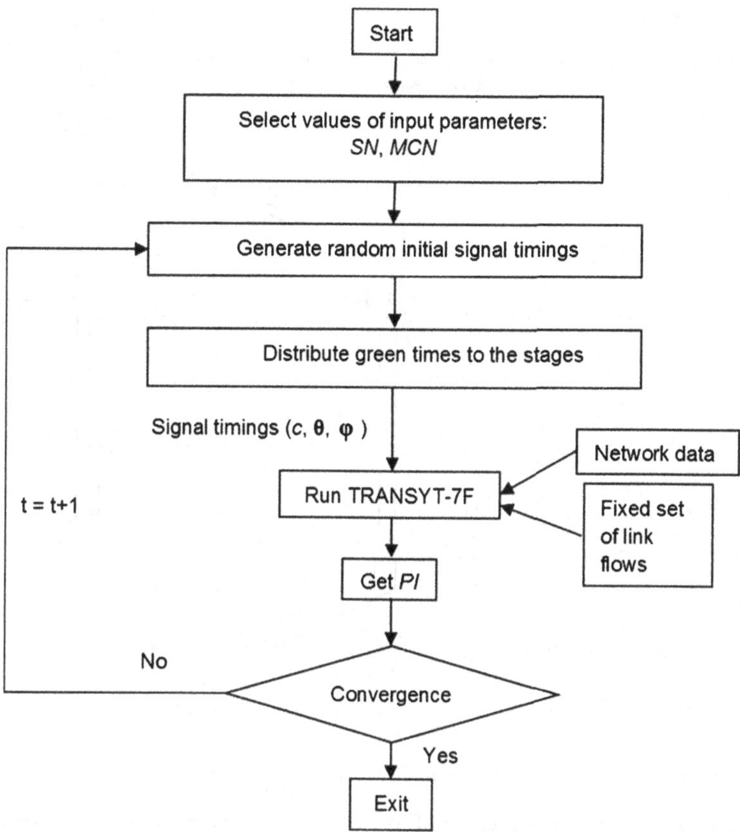

Fig. 1 The flowcart of the ABCTRANS

junctions. The fixed set of link flows, which are taken from [25], is given in Table 1. The constraints on signal timings are set as follows:

$36 \leq c \leq 140$	cycle time constraint
$0 \leq \theta \leq c$	offsets
$7 \leq \phi \leq c$	green split
$I_{1-2} = I_{2-1} = 5$	intergreen time

The ABCTRANS starts the solution process according to random generated signal timings and it was found that the value of *PI* is about 580. The significant improvement on the objective function takes place in the first few cycle. After that, small improvements to the objective function takes place since the ABC creates new solution vectors on the different search directions. Finally, the minimum

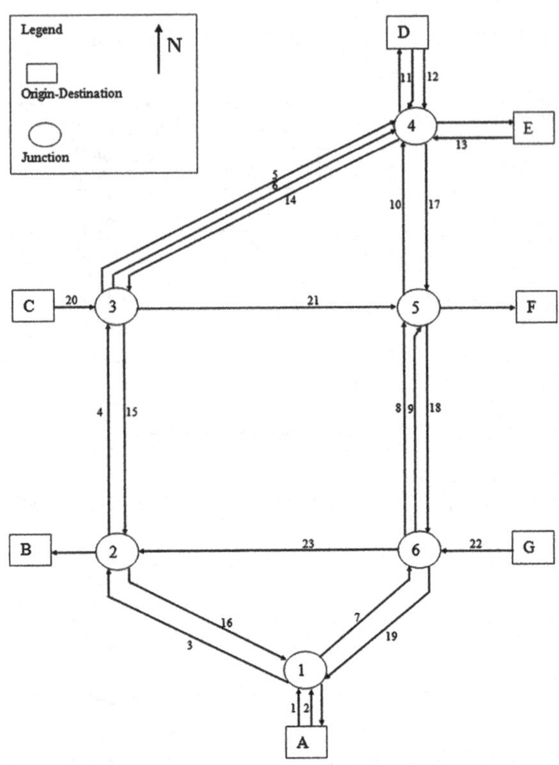

Fig. 2 Example test network

number of *PI* reached to the value of 398.0 after 200 cycles. In order to overcome non-convexity, the ABC starts with a large base of solutions, each of which provided that the solution converges to the optimum. The ABCTRANS is able to achieve global optimum or near global optimum to optimise signal timings.

The common network cycle time obtained from the ABCTRANS is 102 s. Moreover, medium sized network is optimized using TRANSYT-7F, which are GA and HC optimization tools. For studied network, the ABCTRANS and TRANSYT-7F optimizers' results are given in Table 2.

The best *PI* is found as 407.9 in TRANSYT-7F with GA while its value is obtained as 409.2 in TRANSYT-7F with HC. The common network cycle time is 114 and 108 s in TRANSYT-7F with HC and GA, respectively. The ABCTRANS improves network's *PI* about 2.4 and 2.7 % when it is compared with TRANSYT-7F with GA and HC. These results showed that the ABCTRANS model illustrates good performance for finding near optimal solutions of traffic signal timings in coordinated networks with fixed set of link flows.

Table 1 Fixed set of link flows on example network

Link number	Link flow (veh/h)	Saturation flow (veh/h)	Free-flow travel time (sec)
1	716	2000	1
2	463	1600	1
3	716	3200	10
4	569	3200	15
5	636	1800	20
6	173	1850	20
7	462	1800	10
8	478	1850	15
9	120	1700	15
10	479	2200	10
11	499	2000	1
12	250	1800	1
13	450	2200	1
14	789	3200	20
15	790	2600	15
16	663	2900	10
17	409	1700	10
18	350	1700	15
19	625	1500	10
20	1290	2800	1
21	1057	3200	15
22	1250	3600	1
23	837	3200	15

Table 2 The results for example network

	PI	Cycle Time c (s)	Junction number i	Duration of stages (s)			Offsets (s) θ_i
				Stage 1 $\phi_{i,1}$	Stage 2 $\phi_{i,2}$	Stage 3 $\phi_{i,3}$	
			1	48	54	-	0
			2	60	42	-	92
			3	59	43	-	70
ABCTRANS	398.0	102	4	36	33	33	35
			5	16	32	54	36
			6	33	69	-	71
			1	46	68	-	0
			2	64	50	-	82
			3	69	45	-	110
TRANSYT-7F with HC	409.2	114	4	38	34	42	22
			5	15	33	66	24
			6	34	80	-	60
			1	54	54	-	0
			2	70	38	-	96
			3	62	46	-	10
TRANSYT-7F with GA	407.9	108	4	38	35	35	36
			5	18	33	57	38
			6	34	74	-	74

5 Conclusion and Comments

This work deals with the area traffic control problem using the ABC algorithm. TRANSYT-7F is used to compute PI for a given set of signal timing and staging plan in network. The ABCTRANS is tested on example road network, which contains six junctions, in order to show its effectiveness. Results showed that the ABCTRANS improves network's PI by 2.4 and 2.7 % according to TRANSYT-7F with GA and HC. The ABCTRANS provides an alternative to the GA and HC optimization tools in TRANSYT-7F. As a result, the proposed model may be used to optimize traffic signal timings at coordinated signalized network.

References

1. Teklu, F., Sumalee, A., Watling, D.: A genetic algorithm approach for optimizing traffic control signals considering routing. Comput. Aided Civil Infrastruct. Eng. **22**, 31–43 (2007)
2. Webster, F.V.: Traffic signal settings road research technical paper. No. 39, HMSO, London (1958)
3. Robertson, D.: TRANSYT' method for area traffic control. Traffic Eng. Control **10**, 276–281 (1969)
4. Wong, S.C.: Derivatives of the performance index for the traffic model from TRANSYT. Trans. Res. Part B **29**(5), 303–327 (1995)
5. Heydecker, B.G.: A decomposed approach for signal optimization in road networks. Trans. Res. Part B **30**(2), 99–114 (1996)
6. Wong, S.C., Wong, W.T., Xu, J., Tong, C.O.: A Time-dependent TRANSYT traffic model for area traffic control. In: Proceedings of the Second International Conference on Transportation and Traffic Studies. ICTTS, pp. 578–585 (2000)
7. Wong, S.C., Wong, W.T., Leung, C.M., Tong, C.O.: Group-based optimization of a time-dependent TRANSYT traffic model for area traffic control. Trans. Res. Part B **36**, 291–312 (2002)
8. Girianna, M., Benekohal, R.F.: Application of genetic algorithms to generate optimum signal coordination for congested networks. In: Proceedings of the Seventh International Conference on Applications of Advanced Technologies in Transportation, pp. 762–769 (2002)
9. Ceylan, H.: Developing combined genetic algorithm–hill-climbing optimization method for area traffic control. J. Trans. Eng. **132**(8), 663–671 (2006)
10. Chen, J, Xu, L.: Road-junction traffic signal timing optimization by an adaptive Particle swarm algorithm. In: 9th International conference on control automation robotics and vision, Vol. **1–5**, pp. 1103–1109 (2006)
11. Chiou, S.-W.: A hybrid optimization algorithm for area traffic control problem. J. Oper. Res. Soc. **58**, 816–823 (2007)
12. Dan, C., Xiaohong, G.: Study on intelligent control of traffic signal of Urban Area and microscopic simulation. In: Proceedings of the Eighth International Conference of Chinese Logistics and Transportation Professionals, Logistics. The Emerging Frontiers of Transportation and Development in China, pp. 4597–4604 (2008)
13. Li, Z.: Modeling arterial signal optimization with enhanced cell transmission formulations. J. Trans. Eng. **137**(7), 445–454 (2011)
14. Lucic, P., Teodorovic, D.: Transportation modeling: an artificial life approach. In: ICTAI, pp. 216–223 (2002)

15. Teodorovic, D.: Transport modeling by multi-agent systems: a swarm intelligence approach. Trans. Planning Technol. **26**(4), 289–312 (2003)
16. Teodorovic, D., Dell'Orco, M.: Bee colony optimization-a cooperative learning approach to complex transportation problems. In: 10th EWGT Meeting, Poznan, 3–16 Sep 2005
17. Teodorovic, D., Dell'Orco, M.: Mitigating traffic congestion: solving the ride-matching problem by bee colony optimization. Trans. Planning Technol. **31**(2), 135–152 (2008)
18. Karaboga, D.: An idea based on honeybee swarm for numerical optimization. Technical Report TR06, Erciyes University, Engineering Faculty, Computer Engineering Department, Turkey (2005)
19. Karaboga, D., Basturk, B.: A powerful and efficient algorithm for numerical function optimization: artificial bee colony (ABC) algorithm. J. Global Optim. **39**, 459–471 (2007a)
20. Karaboga, D., Basturk, B.: Artificial bee colony (ABC) optimization algorithm for solving constrained optimization problems. Lecture Notes in Artificial Intelligence, vol. 4529, pp. 789–798, Springer-Verlag, Berlin (2007b)
21. Karaboga, D., Basturk, B.: On the performance of artificial bee colony (ABC) algorithm. Appl. Soft Comput. **8**, 687–697 (2008)
22. Karaboga, D., Akay, B.: A comparative study of artificial bee colony algorithm. Appl. Math. Comput. **214**, 108–132 (2009)
23. Ceylan, H., Bell, M.G.H.: Traffic signal timing optimisation based on genetic algorithm approach, including drivers' routing. Trans. Res. Part B **38**(4), 329–342 (2004)
24. Allsop, R.E., Charlesworth, J.A.: Traffic in a signal-controlled road network: an example of different signal timings including different routings. Traffic Eng. Control **18**(5), 262–264 (1977)
25. Ceylan H (2002) A genetic algorithm approach to the equilibrium network design problem. Ph.D. Thesis, University of Newcastle upon Tyne, UK

Use of Fuzzy Logic Traffic Signal Control Approach as Dual Lane Ramp Metering Model for Freeways

Yetis Sazi Murat, Ziya Cakici and Gokce Yaslan

Abstract Metering of merging traffic flows from on-ramp section of freeways is an important research issue for traffic engineers. Although metering signal is one of the recent applications for the subject, assignment of signal timing is problematic. The problem is based on dynamic structure of traffic flows and uncertainties coming up from driver behaviors. Because of variations in car following behavior and perception-reaction times of drivers, uncertainties are occurred. To handle these uncertainties, fuzzy logic approach is preferred in this research. A **Fu**zzy **L**ogicControl based Dual Lane **R**amp **Me**tering (FuLCRMe) Model is proposed. The model considers following parameters as inputs; arrival headways of mainline, queue length at ramp and red time of ramp. Decision about red signal timing is made using these parameters. Based on this decision the final red time is assigned. The FuLCRMe model is tested by a simulation developed in Microsoft Excel program considering different cases. Results of the comparisons show that the FuLCRMe model provides significant decrease in delays, queue length, cycle time, CO_2 emission, fuel consumption, travel time and total cost.

Y. S. Murat (✉) · Z. Cakici
Pamukkale University, Faculty of Engineering-Civil Engineering Department,
Denizli, Turkey
e-mail: ysmurat@pau.edu.tr

Z. Cakici
e-mail: zcakici@pau.edu.tr

G. Yaslan
Yaslan Engineering and Construction Consultant Agency, Mugla, Turkey
e-mail: gokce4848@hotmail.com

V. Snášel et al. (eds.), *Soft Computing in Industrial Applications*,
Advances in Intelligent Systems and Computing 223, DOI: 10.1007/978-3-319-00930-8_30,

1 Introduction

Ramp meters are used for controlling traffic at entrances to freeways by traffic signals. The main objective is to control the number of vehicles that are allowed to enter the freeway in ramp metering. On the other hand, reducing freeway demand is also aimed. The purpose of these objectives is to ensure that the total traffic entering a freeway section remains below the operational or bottleneck capacity of that section. A secondary objective of ramp metering is to introduce controlled delay (cost) to vehicles wishing to enter the freeway, and as a result, reduce the incentive to use the freeway for short trips during peak hour. If the ramp metering is applied properly, expected benefits can be achieved, such as increased speeds, decreased fuel consumption and emissions, safer operation and decreased travel times etc [1–3]. Otherwise, the results expected can not be satisfied. In conventional ramp metering approaches, determination of signal timing have uncertainties and it cannot meet fluctuations in traffic flow pattern. Signal timings are pre-determined considering limited time of observations. These observations can include traffic flows that are trying to merge freeway only within corresponding period. But it is not the same all the day and fluctuate in times of a day. Traffic signal timings are assigned discarding these variations. Therefore traffic flows can not be controlled efficiently. On the other hand, delays of vehicles that are arriving to ramp can be excessive, because of unbalanced assignment of traffic signal timings. This unbalanced and rigid control scheme yields problems related to capacity (i.e. overcapacity or under capacity cases). To remove these deficiencies, fuzzy logic (a flexible or soft) approach is preferred in this study (Fig. 1).

2 Fuzzy Logic Control Model for Ramp Metering

To handle the problem of excessive delay and unbalanced signal timing assignment for ramp flows, fuzzy logic approach is used and the FuLCRMe Model is developed. In FuLCRMe model, fuzzifications of the parameters are made using the membership functions that are determined with respect to previous field studies and experience [4–7]. Mamdani's inference mechanism and centroid method are used as inference and defuzzification procedure of the FuLCRMe model. The Fuzzy Logic Toolbox of MATLAB program is used in developed FuLCRMe model. In this model, red time of ramp is determined by the parameters and rule base. The parameters used in the FuLCRMe model are defined in the following:

- Arrival Headway of traffic flows at mainline (ARHE)
- Queue length at ramp (QULE)
- Rate of remaining red time for ramp (REMRED)
- Decision of red signal time for ramp (SIGDEC)

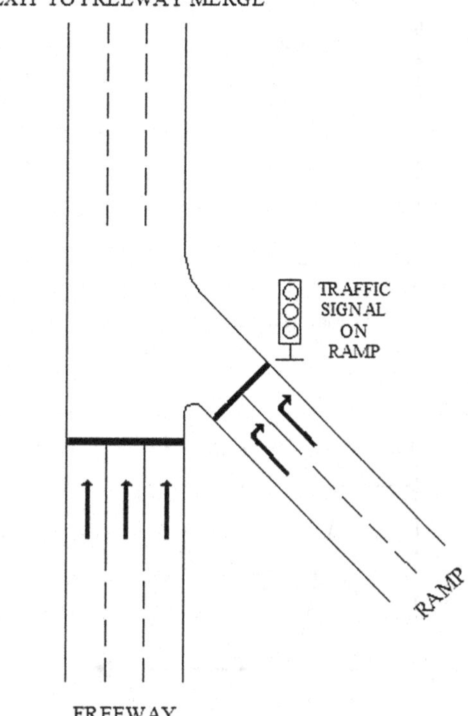

Fig. 1 Illustration of ramp metering signal

2.1 Arrival Headway of Traffic Flows at Mainline (ARHE)

Headways of traffic flows provide useful information about traffic conditions. Arrival Headway of traffic flows at mainline is considered as one of the input parameters in the FuLCRMe Model. It is used in signal timing decision (red time) of ramp. The boundaries of membership functions are determined considering results of previous researches and field studies [8, 9]. (Fig. 2)

2.2 Queue Length at Ramp (QULE)

Queue length at ramp is one of the key parameters in decision making. Traffic flows can be managed properly if signal timing is assigned considering queue length at ramp. Otherwise, excessive delays can be occurred for the vehicles on ramp. The red time of ramp is determined regarding queue length at ramp and arrival headways of mainline traffic flows. Membership functions of the parameter are defined in the Fig. 2.

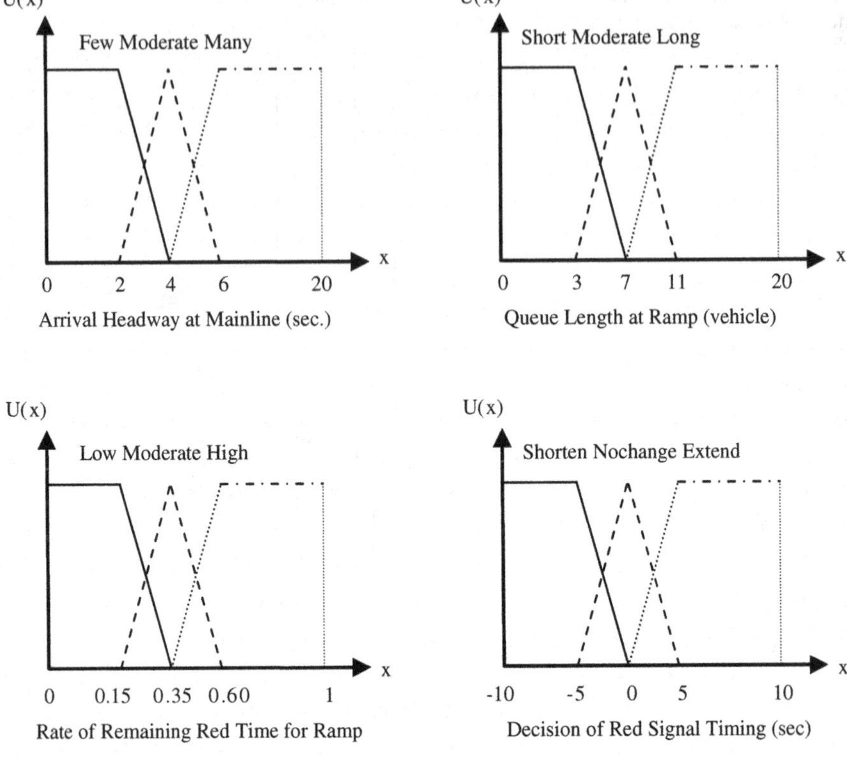

Fig. 2 Membership functions of the FuLCRMe model input and output parameters

2.3 Rate of Remaining Red Time for Ramp (REMRED)

Rate of remaining red time for ramp is a control parameter used in signal decision. Membership functions of this parameter are shown in Fig. 2. Rate of **Rem**aining **Red** Time parameter is obtained by following formula:

RemRed = (Remaining red time to switch into the green signal) / (Total red time in the cycle) (1)

2.4 Decision of Red Signal Time for Ramp (SIGDEC)

Decision of red signal time for ramp is critical in ramp metering. It can be determined based on trial-error approach or conventional signal timing methods.

Table 1 A few examples of FuLCRMe model rule base

Sample no.	IF	Arrival headway at mainline (sec.)	Queue length at ramp (veh.)	Rate of Remred for ramp	Signal decision
2		Many	Long	Moderate	Shorten
8		Many	Short	Moderate	Shorten
16		Moderate	Short	Low	Extend
23		Few	Moderate	Moderate	No change
25		Few	Short	Low	Extend

Based on this time, vehicles on ramp are controlled. Therefore it is considered as output parameter of the FuLCRMe Model. Decision about red signal timing for ramp is made considering the input parameters and the rule base. The FuLCRMe Model parameters and membership functions are given in the Fig. 2. In addition to this, a few examples of FuLCRMe Model Rule Base are given in the Table 1.

3 Analysis

3.1 Design of Experiments

The FuLCRMe model is tested regarding different cases. A total of 75 exercises were conducted. 15 sample cases which are provides best results are taken into consideration in the analysis. These sample cases are selected after analysis of numerous combinations. For each case, signal timing is computed both by fuzzy approach and by the conventional approach. Average delay, operational cost, 95 % back of queue, travel time, cycle time, degree of saturation value and capacity value are considered as parameters of performance index. In the analysis, two merging lanes for traffic flows on ramp and three through lanes for mainline flows are taken into account as the geometry for all cases. Illustration of geometry is shown in Fig. 1. In the analysis, it is assumed that traffic volumes given in the Table 2 are shared equally for lanes of each approach.

Table 2 Traffic volumes samples

Case no.	Traffic volumes (vph)		Case no.	Traffic volumes (vph)		Case no.	Traffic volumes (vph)	
	Mainline	Ramp		Mainline	Ramp		Mainline	Ramp
1	2400	1500	6	3000	700	11	3500	300
2	2550	1300	7	3000	1100	12	3500	600
3	2700	900	8	3100	700	13	3700	600
4	2850	700	9	3300	300	14	3800	300
5	2850	900	10	3300	600	15	3800	600

Table 3 Performance results comparisons for intersection

Cases no.	Traffic volumes	Performance results (FuLCRMe model/conventional model)-intersection						
		Average delay (s)	Operating cost (usd/h)	Travel time (s)	Travel speed (km/h))	95 (%) back of queue-vehicle	95 (%) back of queue-distance (m)	Cycle time (s)
1	2400–1500	8.9/14.3	1249.0/1426.0	34.1/39.5	44.3/38.2	24.7/37.2	173.0/206.6	96/114
2	2550–1300	7.7/11.6	1158.1/1285.0	32.0/35.9	45.6/40.7	18.4/26.9	128.9/188.0	79/94
3	2700–900	5.1/7.1	944.2/1005.1	27.9/29.9	49.1/45.8	8.6/12.3	60.2/86.3	49/60
4	2850–700	3.9/5.7	861.8/912.2	25.8/27.5	50.8/47.7	5.9/8.9	41.4/62.3	41/53
5	2850–900	5.3/7.6	982.8/1056.6	27.9/30.3	48.6/44.9	9.7/14.2	68.2/99.4	55/68
6	3000–700	3.5/5.8	880.6/949.1	25.3/27.6	51.6/47.3	5.7/9.9	39.6/69.2	42/60
7	3000–1100	9.8/13.1	1244.1/1365.2	33.0/36.3	42.1/38.3	23.6/30.3	165.5/211.8	102/114
8	3100–700	4.5/6.5	927.4/992.7	26.1/28.1	49.8/46.1	7.6/11.3	53.3/79.1	51/65
9	3300–300	2.0/2.6	735.3/753.0	21.9/22.5	54.4/52.9	2.7/3.7	18.8/26.0	37/48
10	3300–600	4.0/5.8	912.7/970.4	25.1/26.9	50.5/47.1	6.9/10.2	48.1/71.4	53/69
11	3500–300	2.2/2.9	778.9/799.9	22.0/22.7	53.9/52.3	3.2/4.4	22.7/30.9	43/55
12	3500–600	5.5/6.8	1003.8/1047.7	26.5/27.7	47.5/45.3	10.5/12.8	73.4/89.7	74/84
13	3700–600	6.6/8.0	1084.4/1134.6	27.5/28.8	45.5/43.4	13.7/16.3	96.0/114.4	96/107
14	3800–300	2.8/3.3	855.2/869.0	22.5/23.0	52.4/51.4	4.8/5.6	33.9/39.3	64/72
15	3800–600	7.1/9.2	1124.4/1203.6	27.9/30.0	44.7/41.5	15.3/19.3	107.1/135.3	107/123

Table 4 Comparisons of degree of saturation values and effective intersection capacity values

Cases no.	Traffic volumes (mainline–ramp) Samples	Comparisons of degree of saturation values (FuLCRMe Model/ Conventional approach) Intersection	Comparisons of effective intersection capacity values (vph) (FuLCRMe Model/ conventional approach) Intersection
1	2400–1500	0.694/0.824	5619/4732
2	2550–1300	0.664/0.790	5798/4873
3	2700–900	0.585/0.716	6159/5030
4	2850–700	0.543/ 0.703	6532/5053
5	2850–900	0.596/0.737	6287/5085
6	3000–700	0.547/0.696	6764/5317
7	3000–1100	0.744/0.831	5514/4933
8	3100–700	0.591/0.754	6424/5041
9	3300–300	0.602/0.636	5983/5658
10	3300–600	0.602/0.732	6481/5330
11	3500–300	0.638/0.729	5954/5213
12	3500–600	0.692/0.786	5922/5217
13	3700–600	0.727/0.810	5914/5306
14	3800–300	0.693/0.716	5917/5728
15	3800–600	0.740/0.851	5946/5173

3.2 System Architecture and Calculation Process

In the analysis, SIDRA Intersection program is used as a test environment for comparisons of the conventional and fuzzy logic based models. The FuLCRMe model is developed using Matlab program and it is worked interactively with the simulation program developed in Microsoft Excel environment. In first stage of calculation procedure, using the traffic volumes given in Table 2, cycle time, red time and green time are calculated by SIDRA for conventional ramp metering approach. In the second stage, the calculated cycle times for each case is used as starting cycle time in simulation of the FuLCRMe (fuzzy) model. In FuLCRMe model, each case is simulated using MS Excel regarding ARHE, QULE and REMRED parameters and decision about red signal time for ramp is made by cycle basis. Each case is simulated 15 min time periods and the results are reported.

3.3 SIDRA Intersection Program

The SIDRA Intersection is a mesoscopic simulation program that is used for both intersection design and research aid. Ramp metering is one of the useful tools of the SIDRA. In ramp metering tool, the cycle time can be used either calculating by

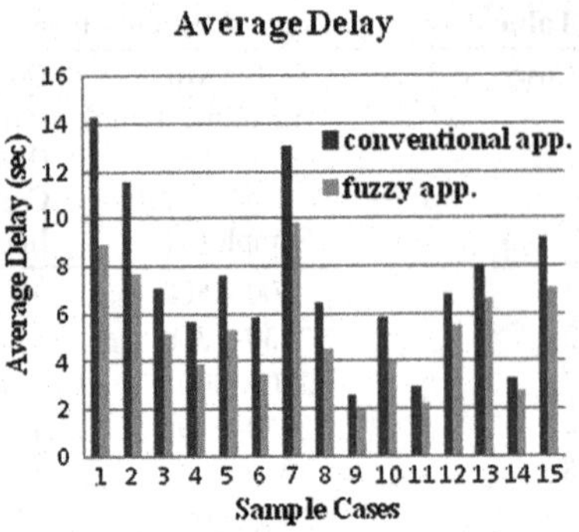

Fig. 3 Comparisons of average delay

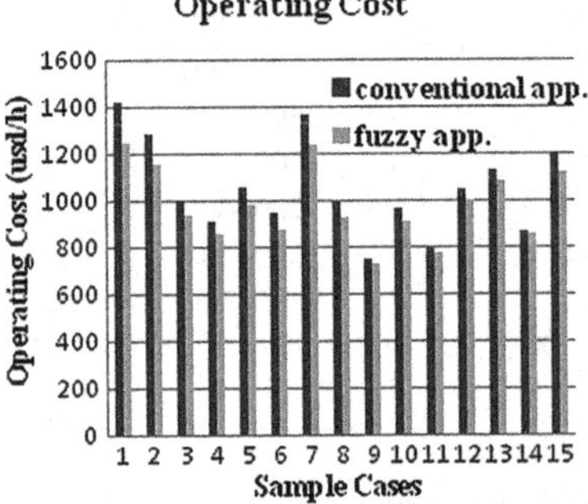

Fig. 4 Comparisons of operating cost

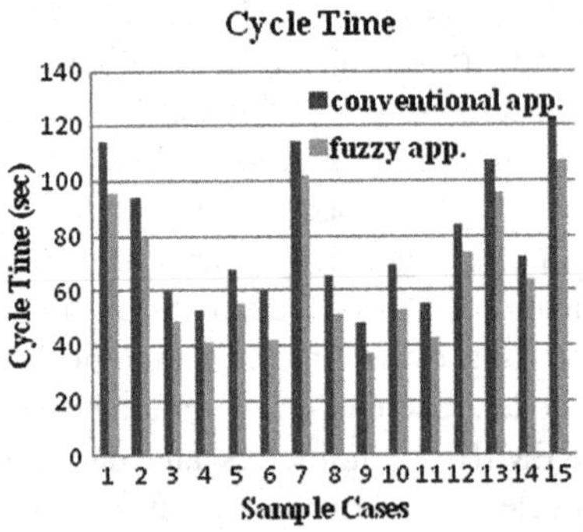

Fig. 5 Comparisons of cycle time length

the program or entering by user. Cycle time, red time and green time are calculated by the program for given traffic volumes. Cycle time calculation is based on conventional signal timing methods by Akcelik. Average delay, fuel consumption, queue length, operational cost, travel time, travel speed etc. are performance index parameters of the program that is developed by Akcelik.

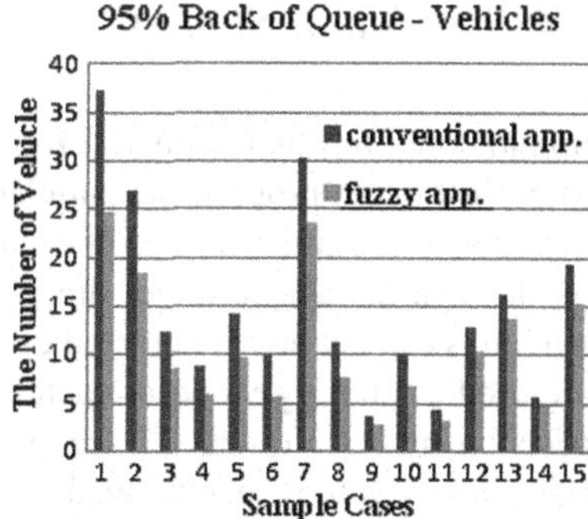

Fig. 6 Comparisons of 95 % back of queue—vehicles

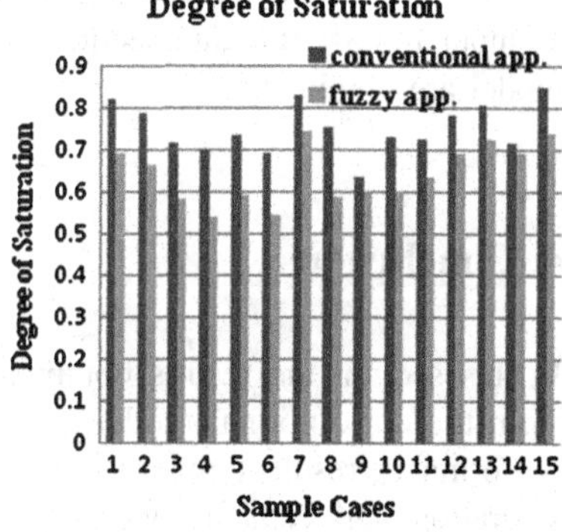

Fig. 7 Comparisons of degree of saturation value

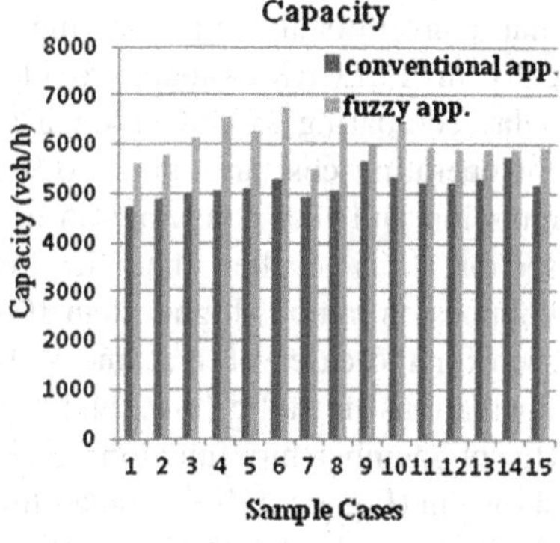

Fig. 8 Comparisons of capacity

Table 5 Paired-t test's results

Parameters	t stat	Critical t-value
Average delay	1.9226	1.7011
Operating cost	1.0431	1.7011
Cycle time	1.4508	1.7011
95% Back of queue	1.3347	1.7011
Degree of saturation	4.7021	1.7011
Capacity	7.8844	1.7011

3.4 Simulation Studies

Simulation study is conducted for testing the FuLCRMe model performance by Microsoft Excel program. In simulation, firstly random numbers are generated and then converted to headways of vehicles considering given traffic volumes. Based on these headways, vehicles are generated. Using these generated vehicles, simulation is carried out. Calculated cycle times of conventional ramp metering model are used as starting value of each case in simulation. FuLCRMe model produced decisions in three times of each cycle; half time of cycle, 75 % time of cycle and last 5 s of cycle respectively. Using these decisions, cycle time is changed and timing is assigned for each case. For each case, at the end of simulation period, average cycle time and red time is calculated and these balanced timings are used as input for SIDRA intersection program and performance values of FuLCRMe model are obtained.

4 Conclusions

In this section the results obtained from FuLCRMe model with these obtained from the conventional model are compared. Table 3 presents intersection based performance results comparisons while Table 4 presents comparisons of degree of saturation and capacity values. The results achieved from these analyses are remarkable. As seen on tables and figures, the FuLCRMe model performs better than conventional ramp metering approach for all performance criteria. Specifically, the FuLCRMe model provides about 30 % improvements in average vehicle delay comparing to conventional approach (Fig. 3). The improvements obtained for operating cost are about 10 % (Fig. 4) and improvements obtained for CO_2 emission and fuel consumption are about only around 5 % for the whole intersection. It is resulted that, the FuLCRMe model performs better if the traffic volumes on ramp is higher than 1000 vehicles per hour. This finding can easily be seen for the cases of 1, 2 and 7. For these cases, the FuLCRMe model met the fluctuations in traffic flows and regulates uncertainties by elastic control scheme (fuzzy membership functions and rule base). Comparisons of cycle times are shown in (Fig. 5). It is seen that, the cycle times obtained by FuLCRMe model are shorter than that of the conventional model yields for all cases. The FuLCRMe model provides about average 18 % decrease in cycle times with respect to all sample cases. Besides to improvements mentioned above, as seen on (Fig. 6) that FuLCRMe model decreases 95 % back of queue-vehicles values about 30 % (average), regarding conventional approach. To measure efficiency of the proposed model, in addition to the parameters given above, degree of saturation and capacity values are taken into account. Average benefit rate for degree of saturation is about 14 % (Fig. 7) and it is 15 % for capacity (Fig. 8). These values are meaningful for traffic engineering point of view and it is showed that, use of fuzzy logic approach for ramp metering removed deficiencies and balanced signal timing.

Paired-t test has been applied for the results provided from the Figs. (3–8). Different 6 parameters are investigated with the Paired-t test. These parameters are Average Delay, Operating Cost, Cycle Time, 95 % Back of Queue, Degree of Saturation and Capacity. Paired-t test's results are given in the Table 5.

In this study, α value is taken into account as 0.1 for the Paired-t test. As seen on the table, Average Delay, Degree of Saturation, Capacity values are statistically significant.

Acknowledgments Authors of this paper would like to thank Dr. Rahmi Akcelik for providing SIDRA Intersection program. On the other hand, support of Pamukkale University Scientific Research Project Coordination Department by the project number 2010FBE061 is appreciated.

References

1. Papageorgiou, M., Blosseville, J.M., Hadj-Salem, H.: Modelling and real-time control of traffic flow on the southern part of Boulevard Peripherique in Paris-Part I modeling. Transp. Res. A **24**, 345–359 (1990)
2. Papageorgiou, M., Hadj-Salem, H., Middelham, F.: ALINEA local ramp metering summary of field results. Transportation Research Record Journal of the Transportation Research Board No 1603, TRB of National Academies, Washington D.C., 90–98, (1997)
3. Taylor, C., Meldrum, D.: Freeway traffic data prediction via artificial neural networks for use in a fuzzy logic ramp metering algorithm. In: Proceedings of the Intelligent Vehicles '94 Symposium, pp. 308–313, 24–26 Oct 1994.
4. Murat, Y.S., Gedizlioğlu, E.: A fuzzy logic multi-phased signal control model for isolated junctions. Transp. Res. Part C Emerg. Technol. **13/1**:19–36, Pergamon Press (2005).
5. Murat, Y.S., Gedizlioğlu, E.: Investigation of vehicle time headways in Turkey. In: Proceedings Of the Institution Of Civil Engineers-Transport, Vol. 160(2), pp. 73–78 (2007)
6. Murat, Y.S., Uludag, N.: Route choice modelling in urban transportation networks using fuzzy logic and logistic regression methods. J. Sci. Ind. Res. **67**(1), 19–27 (2008)
7. Murat, Y.S., Kulak, O.: Use of information axiom for route choice in transportation networks. Pamukkale Univ. J. Eng. Sci. (PAJES) **11**(3), 425–435 (2005)
8. Taylor, C., Meldrum, D., Jacobson, L.: Fuzzy ramp metering design overview and simulation results transportation research record. Journal of Transportation Research Board No 1634, TRB of National Academies, Washington D.C., 10–19 (1998)
9. Chen, X., Tian, A.: Research on fuzzy on-ramp metering and simulation in urban expressway. In: International Conference on Optoelectronics and Image Processing (ICOIP), 2010, Vol. 2, pp. 221–224, (2010) (Publication Year)

The Variable Neighborhood Search Heuristic for the Containers Drayage Problem with Time Windows

D. Popović, M. Vidović and M. Nikolić

Abstract The containers drayage problem studied here arise in International Standards Organization (ISO) container distribution and collecting processes, in regions which are oriented to container sea ports or inland terminals. Containers of different sizes, but mostly 20 ft, and 40 ft empty and loaded should be delivered to, or collected from the customers. Therefore, the problem studied here is closely related to the vehicle routing problem with the time windows where an optimal set of routes is obtained. Both delivery and pickup demands can be satisfied in a single route. The specificity of the containers drayage problem analyzed here lies in the fact that a truck may simultaneously carry one 40 ft, or two 20 ft containers, using an appropriate trailer type. This means that in one route there can be one, two, three or four nodes, which is equivalent to the problem of matching nodes in single routes. This paper presents the Variable Neighborhood Search (VNS) heuristic for solving the Containers Drayage Problem with Time Windows (CDPTW). The results from the VNS heuristic are compared with the two optimal MIP mathematical formulations that were introduced in our previous research papers.

D. Popović (✉) · M. Vidović
Faculty of Transport and Traffic Engineering, Department of Logistics,
University of Belgrade, Belgrade, Serbia
e-mail: d.popovic@sf.bg.ac.rs

M. Vidović
e-mail: m.vidovic@sf.bg.ac.rs

M. Nikolić
Faculty of Transport and Traffic Engineering, Department of Operations Research in Traffic,
University of Belgrade, Belgrade, Serbia
e-mail: m.nikolic@sf.bg.ac.rs

V. Snášel et al. (eds.), *Soft Computing in Industrial Applications*,
Advances in Intelligent Systems and Computing 223, DOI: 10.1007/978-3-319-00930-8_31,

1 Introduction

The routing problem studied here is a typical for the intermodal transportation systems where containers are delivered by trucks to customers oriented to a container sea port or an inland container terminal. In the intermodal transportation the major part of the cargo's journey is performed by rail, inland waterway or sea, while the initial and/or final legs, distribution and collection of containers, are typically carried out by road (for more details about intermodal transportation systems see [1]). Truck should deliver a loaded container that has arrived at a terminal (import i.e. inbound container) to a consignee, and to pickup and haul back the loaded container from consignor to the terminal (export i.e. outbound container). In the case when a part of a container terminal serves as an empty containers' depot, in addition to pickup/delivery operations with loaded containers, the empty containers also needs to be delivered to a shipper for loading and hauled back empty to the terminal after unloading goods at the consignee site. Container transportation within a local region oriented to an intermodal terminal with a few possible types of container moves is also known as drayage operations [2]. Drayage operations are driven by the need to fulfill customer demands while satisfying various constraints imposed by the technology and customers' requirements. Drayage includes regional movements of loaded and empty equipment (trailers and containers) by tractors between terminals, shippers, consignees, and equipment yards. In general, drayage operations involve not only the provision of containers but also empty trailers [2], while in this research only the problem of containers pickup and delivery is considered.

Most intermodal containers are sized according to International Standards Organization (ISO). Based on ISO, containers are classified in several groups (10, 20, 30, 40, 45 ft and since recently 48 and 53), where 20 and 40 ft containers are the most frequently used all over the world. In Europe (except Finland and Sweden) and Asia, road vehicles are restricted to transport only 20 and 40 ft containers, only few countries allow 45 ft containers, while larger containers are in use only in the USA and Canada [3]. In most countries transport of two loaded 20 ft containers by a standard vehicle is not permitted, except in the case when the weight limitation of 26 tons is not exceeded. In the USA, Australia, Canada, Finland and Sweden the vehicles in use are the ones that offer the possibility of transporting two fully loaded 20 ft containers, while the EU has set up regulations which permit certain types of vehicles called "modular concept vehicles", offering the possibility of transporting two fully loaded 20 ft containers using special combined chassis. In the containers drayage operations realized by combined chassis vehicles, up to four nodes can be visited in a single route starting and ending in container terminal or depot which is assumed here to be part of the terminal. In turn, the use of those technical solutions provides different opportunities for improving the efficiency of container transportation by merging different pickup and delivery operations in a single route.

Drayage operations and especially container truck transportation account for a significant portion of the total transportation cost. Therefore, it is very important to improve the efficiency of container transportation through the optimization of such transportation processes, which leads to necessity of solving the truck scheduling problem in containers drayage operation [4].

Optimization of containers operation has received increased attention over the past decade due to its importance in intermodal freight transportation. Jula et al. [5] formulated the problem of container movement with time windows at origins and destinations as asymmetric multiple traveling salesman problem and proposed three solving approaches. Coslovich et al. [6] investigated containers drayage operation with the present and future operating costs minimized. Imai et al. [7] formulated containers drayage problem as a pickup and delivery and proposed Lagrangian relaxation to solve the problem. Chung et al. [8] built several mathematical models of container truck transportation. They formulate the basic problem where every vehicle can transport exactly one container at a time, and the multi-commodity problem with a combined chassis used in transporting two 20 ft containers or one 40 ft container. To solve the problem a solution algorithm based on the Insertion Heuristic was proposed. Namboothiri and Erera [9] studied the management of a fleet of trucks providing container pickup and delivery service (drayage) to a port with an appointment-based access control system. Zhang et al. [4], considered a truck scheduling problem for container transportation in a local area with multiple depots and multiple terminals. They proposed an approach based on an integer programming heuristic determines pickup and delivery sequences for daily drayage operations with minimum transportation cost. Savelsbergh and Sol [10] show that container transportation problems belong to pickup and delivery problems, and because of the nature of the problem, drayage operations also corresponds to multi-stop Vehicle Routing Problems with Backhauls (VRPB). A more detailed insight in VRPB, as well as in Vehicle Routing Problems with Pickup and Delivery (VRPPD), can be found in recent comprehensive overview given by [11] and [12].

The purpose of this paper is to propose the heuristic approach for finding the optimal trucks' routing in containers drayage in the case when pickup and delivery nodes may be visited only during a certain predefined time intervals (Containers Drayage Problem with Time Windows—CDPTW). In this way, our previous research [13–15] is extended by introducing a Variable Neighborhood Search heuristic for solving the CDPTW. The main reason for developing the heuristic approach is solving a large problem instances that cannot be solved optimally in reasonable CPU time. A VNS heuristic approach was originally developed by [16] and it mostly depends on the quality of the local search and shaking procedure. For an insight in methods and application of VNS we recommend the paper [17]. For the local search procedure we applied Variable Neighborhood Descend (VND) with two neighborhoods: reallocating and interchange of tasks between routes. As for the shaking procedure, we use a method of the "destruction" of randomly chosen route/routes with multiple tasks and reconstruction of a single task routes from those tasks.

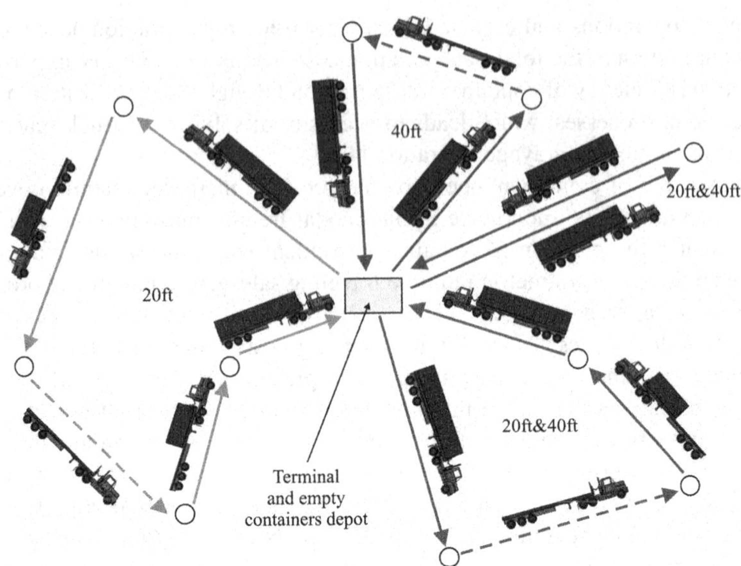

Fig. 1 Some of routing possibilities when drayage is realized by combined chassis

The remainder of this paper is organized as follows. Section 2 presents a brief problem description. Section 3 presents the mathematical formulation for the CDPTW. The VNS heuristic is presented in Sect. 4. Computational results are presented in Sect. 5, and Sect. 6 gives some concluding remarks.

2 Problem Description

The problem of distributing/collecting ISO containers (20, and 40 ft) may be described as a variant of VRPB in which a truck visits up to four nodes until return to terminal. In this paper we observe the CDPTW where empty and loaded containers from terminal should be distributed to costumers, and empty and loaded containers should be picked up at customers' sites and hauled back to the terminal. Vehicle fleet is of unlimited size and homogenous, with capacity of 2 twenty-foot equivalent units (TEUs). Therefore, when truck tow combined chassis is used, matching possibilities include all feasible combinations of 20 ft, and 40 ft containers (there are fifteen possible matchings of task nodes into merged routes, and four direct pickup or delivery routes) that should be transported from/to terminal and customers. A few possible matchings of task nodes are presented in Fig. 1 Obviously, it is worthwhile to choose those resulting with minimal length. Additionally, pickup and delivery nodes may be visited only during a certain predefined time windows.

3 Mathematical Formulation for the CDPTW

Vidović et al. [15] presented two mathematical formulations for the CDPTW: multiple assignment formulation, and general mixed integer programming (MIP) formulation. While the first, multiple assignment formulation, has been proposed in [14], the second, general MIP formulation presented here is adopted from [18]. The results from these two optimal MIP mathematical formulations are used as benchmarks for the evaluation of the VNS heuristic proposed in this paper. For the purpose of mathematical presentation of the CDPTW we use the general MIP formulation because of its clarity and easiness of understanding.

Let G(N,E) be a graph, where N is the set of nodes $i \in N$ with containers move requests, and $E = \{(i,j)|i \neq j; i,j \in N\}$ is the edge set. All container move requests are known at the beginning of the planning horizon (usually one day), and all vehicles start from the terminal. Node with index "0" represents the terminal node. It is assumed that any node may simultaneously have both, containers demand and supply move requests. This graph is transformed into another graph, in which each customer node is replaced with task nodes. In this way all task nodes of the transformed graph have single move request, either pickup or delivery. It is assumed that the transport costs between task nodes which belong to the same customer node may be neglected. In the model we used the following notation:

K—total number of used vehicles (routes)
Q—vehicle capacity (expressed in number of TEUs)
c_{ij}—transport costs between customer nodes i and j (Euclidian distance)
x_{ijk}—binary decision variable (1 if vehicle k travel from the task node i to node j)
y_{ij}—integer variable (number of pickup TEUs transported in arc $i,js \in$E)
z_{ij}—integer variable (number of delivery TEUs transported in arc $i,j \in$E)
s_{ik}—time of beginning of service at task node i by the vehicle k
a_i—start time of a time window for task node i
b_i—end time of a time window for task node i
t_i—service time in the task node i
t_{ij}—travel time between task nodes i and j
p_j—pickup demand of task node j
d_j—delivery demand of task node j
M—a sufficiently big number

Objective function:

$$MIN \rightarrow \sum_{i=0}^{N} \sum_{j=0}^{N} \sum_{k=1}^{K} c_{ij} x_{ijk} \tag{1}$$

st.

$$\sum_{i=0}^{N} \sum_{k=1}^{K} x_{ijk} = 1 \;\; \forall j = \overline{1,N} \tag{2}$$

$$\sum_{i=0}^{N} x_{ijk} - \sum_{i=0}^{N} x_{jik} = 0 \ \forall j = \overline{1,N}; \forall k = \overline{1,K} \tag{3}$$

$$\sum_{j=1}^{N} x_{0jk} \leq 1 \ \forall k = \overline{1,K} \tag{4}$$

$$\sum_{i=0}^{N} y_{ji} - \sum_{i=0}^{N} y_{ij} = p_j \ \forall j = \overline{1,N} \tag{5}$$

$$\sum_{i=0}^{N} z_{ij} - \sum_{i=0}^{N} z_{ji} = d_j \ \forall j = \overline{1,N} \tag{6}$$

$$y_{ij} + z_{ij} \leq Q \sum_{k=1}^{K} x_{ijk} \ \forall i = \overline{0,N}; \forall j = \overline{0,N}; i \neq j; \forall k = \overline{1,K} \tag{7}$$

$$s_{ik} + t_i + t_{ij} - M(1 - x_{ijk}) \leq s_{jk} \ \forall i = \overline{0,N}; \forall j = \overline{1,N}; i \neq j; \forall k = \overline{1,K} \tag{8}$$

$$a_i \leq s_{ik} \leq b_i \ \forall k = \overline{1,K}; \forall i = \overline{1,N} \tag{9}$$

$$x_{ijk} \in \{0,1\}; y_{ij} \geq 0; z_{ij} \geq 0; i = \overline{0,N}; j = \overline{0,N}; i \neq j; k = \overline{1,K} \tag{10}$$

The objective function (1) seeks to minimize total transport costs. Constraints (2) ensure that each task node is visited exactly once from exactly one task node, while constraints (3) guarantee that the same vehicle arrives and departs from each task node it serves. Constraints (4) prevent multiple departure of the vehicle from the terminal, in the same route. Constraints (5) and (6) are flow equations for pickup and delivery demands, respectively. Constraints (7) prevent vehicle overloading. Constraints (8) and (9) are time windows constraints. Constraints (10) define variables domains.

4 VNS Heuristic for the CDPTW

Mladenovic and Hansen [16] developed the VNS algorithm as a new metaheuristic concept with the basic idea of a systematic change of a neighborhood within a local search algorithm. After the initial solution construction, the VNS heuristic uses local search and shaking procedure to iteratively improve the current best solution until a stopping criterion is met. The VNS concept is based on the use of several neighborhoods in a search for a solution improvement, where each succeeding neighborhood covers a larger search space. Shaking procedure is responsible for obtaining a new starting point for a local search with the main purpose of overcoming the local optima "trap". For the purpose of solving large scale problem instances of CDPTW we introduce the VNS heuristic. The initial solution is obtained by the sweep method [19], where routes are constructed starting from the task that has the biggest "polar gap" to its nearest preceding task (effective for the multi-compartment vehicle routing [20]).

For the local search procedure we applied Variable Neighborhood Descend (VND) with two neighborhoods: reallocating and interchange of tasks between routes. The local search procedure examines all possible changes in a neighborhood of a current solution (*Current_sol*) for improvement of its objective function (*Current_obj_val*). If a change in observed neighborhood incurs improvement of a *Current_obj_val*, then both *Current_sol* and*Current_obj_val* are updated and the local search is restarted within observed neighborhood. The reallocation of tasks between routes observes all possible transfers of a task from one route to another. The interchange of tasks between routes observes all possible exchanges of a task from one route with a task from another route. The local search VND algorithm is presented in Fig. 2.

```
improvement ← True;
LOCAL_SEARCH_PROCEDURE ( Current_sol, Current_obj_val ) :
while improvement:
    improvement ← False;
    REALLOCATE _PROCEDURE:
    reallocate_improvement ← True;
    while reallocate_improvement:
        reallocate_improvement ← False;
        for route_1 in all_routes:
            for task_1 in route_1:
                for route_2 in all_routes/route_1:
                    if task_1 reallocation to route_2 is feasible:
                        if the reallocation incurs lower Current_obj_val:
                            improvement ← True;
                            reallocate_improvement ← True;
                            update Current_sol, Current_obj_val;
                            break to REALLOCATE_PROCEDURE;
    INTERCHANGE_PROCEDURE:
    interchange_improvement ← True;
    while interchange_improvement:
        interchange_improvement ← False;
        for route_1 in all_routes:
            for task_1 in route_1:
                for route_2 in all_routes/route_1:
                    for task_2 in route_2:
                        if the interchange of the task_1 and task_2 is feasible:
                            if the interchange incurs lower Current_obj_val:
                                improvement ← True;
                                interchange_improvement ← True;
                                update Current_sol, Current_obj_val;
                                break to INTERCHANGE PROCEDURE;
```

Fig. 2 The algorithm of the local search procedure (VND)

```
SHAKING_PROCEDURE ( Current_sol, Sh_size_max ) :
counter ← 0;
while counter < Sh_size_max:
    counter ← counter + 1;
    if route with multiple stops exists:
        rnd_route ← chose a random multiple stops route;
        for stop in rnd_route: /*stop represents a P/D location*/
            add direct route terminal-stop to Current_sol;
        delete rnd_route from Current_sol;
    else break;
```

Fig. 3 The shaking procedure

As for the shaking procedure, we use a method of the "destruction" of randomly chosen route/routes with multiple stops and reconstruction of single stop (direct) routes to each of those stops. For example, if we select two routes servicing five different locations in total, after shaking procedure we would have five direct routes (each servicing one location). The number of routes to be "destructed" is denoted as *Sh_size*, and the number of repeated passes for given *Sh_size* is denoted by *Sh_pass*. By changing the maximum values for these two parameters (*Sh_size_max*, and *Sh_pass_max*) we define the total number of shaking neighborhoods in the VNS heuristic. The shaking procedure used in proposed VNS heuristic is presented in Fig. 3.

For both local search and shaking procedure, only feasible changes to current solution are allowed. The feasibility is defined by time windows and possible matchings of task nodes into routes. For example, it is not feasible for a vehicle to load 40 ft container at the terminal, to load additional 20 ft container at a pickup node, then to unload 40 ft container to a delivery node before returning to the terminal (constraint of a vehicle capacity is violated). The flowchart of the proposed VNS heuristic is presented in Fig. 4. The stopping criterion of the VNS heuristic becomes active when the shaking and local search procedures for the last neighborhood (*Sh_size*=*Sh_size_max*, and *Sh_pass*=*Sh_pass_max*) does not incur improvement in current best solution value.

5 Computational Results

Testing the quality of the VNS heuristic to solving the containers drayage problem when combined chassis is used has been carried out on the several problem instances. Our idea was to test the VNS heuristic on three group of instances presented in [15] for which optimal solutions exists. Additionally, we generated fourth group of large scale problem instances that cannot be solved optimally in reasonable CPU time and that are solved only by VNS heuristics. All instances are generated according to the Solomon VRPTW benchmark problems (http://web.cba.neu.edu/~msolomon/problems.htm). There are six different sets of

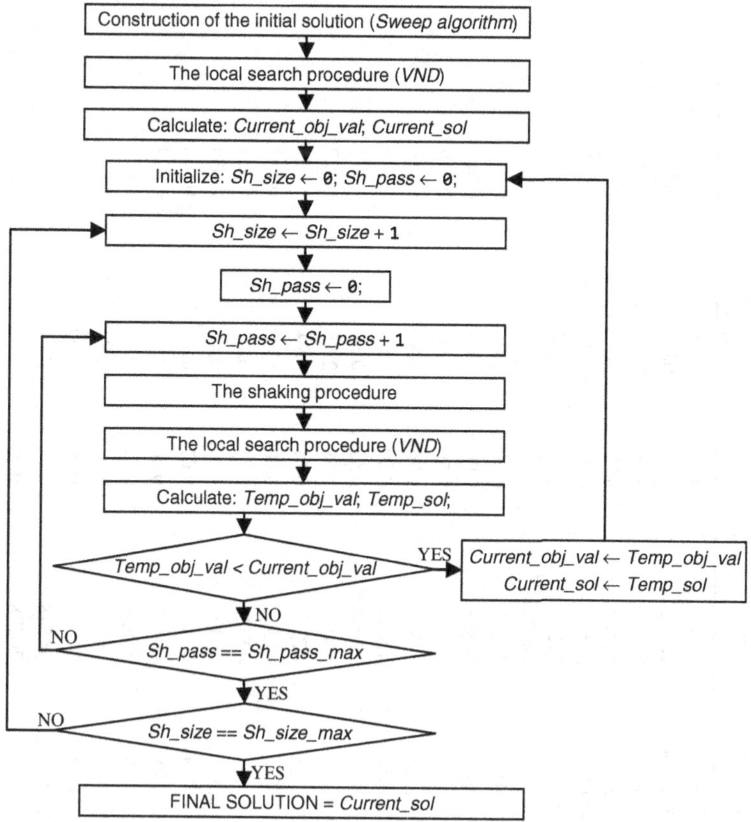

Fig. 4 The flowchart of the VNS heuristic

problem instances. Task nodes are randomly generated in problem sets R1 and R2, clustered in problem sets C1 and C2, and both, randomized and clustered in problem sets RC1 and RC2. Also, problem sets R1, C1 and RC1 have a short scheduling horizon (few possible costumers per route) while problem sets R2, C2 and RC2 have a long scheduling horizon. Demand at each task node in original Solomon instances can have up to 50 units. In the CDPTW we observe only several containers that need to be picked from or delivered to each task node. Therefore, we have transformed the original Solomon instances to have both pickup and delivery task, to have fewer tasks per each node and to have four types of tasks. Total number of tasks in each node is obtained by dividing the original Solomon demand by 20 and rounding that number to greater integer value. Then, we randomly allocate derived tasks to 20 or 40 ft containers, as pickup or delivery demand. Total number of pickup/delivery locations (nodes) is denoted with N', and total number of containers (tasks) is denoted with N.

Table 1 Results for small scale problem instances (results from two mathematical formulations are taken from [15]

N'	Instance	N	Multiple assignment formulation		General MIP formulation		VNS heuristic			
			Solution	CPU time [sec]	Solution	CPU time [sec]	Average solution	StDev [%]	Average Err [%]	Avg. CPU time [sec]
10	C1 01	11	16964	0.010	16964	0.320	16964	0.00	0.00	0.010
	C1 02	11	22427	0.080	*22427	1800.000	22438	0.10	0.05	0.013
	C2 01	11	34312	0.020	34312	0.630	34312	0.00	0.00	0.012
	C2 02	11	32869	0.001	32869	4.020	32869	0.00	0.00	0.010
	R1 01	11	33572	0.020	33572	0.110	33572	0.00	0.00	0.011
	R1 02	11	31214	0.010	31214	72.750	31235	0.48	0.07	0.012
	R2 01	11	29931	0.030	29931	4.910	29931	0.00	0.00	0.011
	R2 02	11	30128	0.020	30128	2.060	30128	0.00	0.00	0.011
	RC1 01	13	40883	0.030	40883	773.810	40883	0.00	0.00	0.018
	RC1 02	13	46712	0.020	46712	1519.540	47499	1.55	1.66	0.019
	RC2 01	13	50344	0.020	50344	256.760	51301	1.76	1.87	0.017
	RC2 02	13	48646	0.030	*48646	1800.000	48646	0.00	0.00	0.015
Average			*34833.5*	*0.024*	*34833.5*	*519.578*	*34981.5*	*0.32*	*0.30*	*0.013*
15	R1 01	17	51797	0.050	51797	1.530	53043	1.68	2.35	0.027
	R1 02	17	46853	0.130	46853	743.730	47296	0.61	0.94	0.033
	R2 01	17	42562	0.030	42562	17.660	42562	0.00	0.00	0.031
	R2 02	17	42856	0.050	42856	671.060	42929	0.71	0.17	0.034
	C1 01	18	40310	0.050	40310	1347.190	40493	1.37	0.45	0.039
	C1 02	18	43257	0.170	*43257	1800.000	43305	0.07	0.11	0.030
	C2 01	18	69170	0.020	69170	60.760	69170	0.00	0.00	0.024
	C2 02	18	55820	0.020	55820	722.270	55850	0.37	0.05	0.028
	RC1 01	19	63943	0.030	63943	26.780	64488	0.68	0.84	0.029
	RC1 02	19	71521	0.050	*71652	1800.000	71521	0.00	0.00	0.027
	RC2 01	19	74242	0.050	*74242	1800.000	74548	0.32	0.41	0.039
	RC2 02	19	70742	0.200	*73245	1800.000	70778	0.05	0.05	0.043
Average			*56089.4*	*0.071*	*56308.9*	*899.258*	*56332.0*	*0.49*	*0.45*	*0.032*

Solution obtained after 1800 sec of CPU time (CPLEX parameter *timelimit* is set to 1800)

Table 2 Results for medium scale problem instances; general MIP formulation cannot solve medium scale problem instances in reasonable CPU time (optimal results taken from [15])

N'	Instance	N	Multiple assignment formulation		VNS heuristic			
			Solution	CPU time [sec]	Average solution	StDev[%]	Average Error [%]	Avg. CPU time [sec]
50	C1 01	59	141884	2.510	142234	0.36	0.25	0.415
	C1 02	59	140580	87.230	141604	0.32	0.72	0.584
	C2 01	59	184631	1.660	184824	0.27	0.10	0.574
	C2 02	59	167359	23.010	169335	0.51	1.17	0.461
	R1 01	60	204776	0.360	204894	0.07	0.06	0.506
	R1 02	60	190796	25.460	193627	1.00	1.46	0.456
	R2 01	60	174357	11.630	176667	0.70	1.31	0.630
	R2 02	60	163288	123.510	166787	0.53	2.10	0.657
	RC1 01	63	256907	1.420	262256	0.81	2.04	0.654
	RC1 02	63	264194	4.840	264680	0.07	0.18	0.347
	RC2 01	63	244955	10.680	250231	1.15	2.11	0.679
	RC2 02	63	244982	78.580	245471	0.45	0.20	0.823
Average			*198225.8*	*30.908*	*200217.5*	*0.52*	*0.97*	*0.566*

We observe following four cases: small scale problems with N' = 10 and N' = 15 nodes, medium scale problems with N' = 50 nodes (only first 10, 15 or 50 nodes are observed in each instance), and large scale problems with N' = 100 nodes (to be solved only by the VNS heuristic). In each case we observe only the first two instances from the problem set (to provide clarity of the results presentation and to save paper space). For all four cases, each instance is solved in 50 iterations by the VNS heuristic with the following values of the neighborhood parameters: *Sh_size_max*=5, and*Sh_pass_max*=10. The results for the small scale problem are presented in Table 1, for the medium scale problem in Table 2, and for the large scale problem in Table 3.

Large scale problem instances with N' = 100 could not be optimally solved in reasonable CPU time and therefore these instances are solved only sub-optimally by the VNS heuristic. The CPU time needed to obtain the solution for the CDPTW is very important for real life applications. The reason lies in the fact that a routing plan for all pickup/delivery tasks needs to be obtained on day to day basis. Mathematical models were implemented by the CPLEX 12.2. on the Intel(R) Core(TM) i3 CPU M380 2.53 Ghz with 6 GB RAM, while the VNS model was implemented by C++ and Microsoft Visual Studio 2010 (64-bit).

Table 3 Results for large scale problem instances (large scale problem instances cannot be optimally solved in reasonable CPU time)

N'	Instance	N	VNS heuristic		
			Avg. solution	StDev [%]	Avg. CPU time [sec]
100	R1 01	123	371052	0.55	2.901
	R1 02	123	338771	0.61	2.465
	R2 01	123	337175	0.56	2.869
	R2 02	123	297029	0.61	3.882
	C1 01	124	379768	0.88	3.364
	C1 02	124	348699	0.55	2.291
	C2 01	124	393768	0.47	2.103
	C2 02	124	360699	0.42	3.544
	RC1 01	125	441104	0.54	2.363
	RC1 02	125	410643	0.56	2.555
	RC2 01	125	416730	0.68	2.868
	RC2 02	125	416326	0.23	2.351
Average			*375980.4*	*0.55*	*2.796*

6 Concluding Remarks

Because the CDPTW is NP-hard problem, for obtaining a real-life solutions we propose the VNS heuristic approach. For small and medium scale problem instances we compare the heuristic results with the results of two mathematical formulations presented in [15]. Tables 1 and 2 shows that the VNS heuristic has relatively small average error compared to optimal solutions (0.30, 0.45, and 0.97 % for cases with 10, 15, and 50 nodes respectively). At the same time solution CPU time of the VNS heuristic is considerably lower than that from the mathematical models (e.g. for the medium scale problem instances general MIP formulation is not capable of finding the optimal solution in reasonable CPU time, multiple assignment formulation takes in average 30.91 sec per instance, and the VNS heuristic takes in average 0.57 sec per instance with average error of only 0.97 %). Additionally, standard deviation of heuristic solutions is in average less than 0.60 % for all four groups of instances which shows high level of solution convergence.

Wang and Regan [21] stated that typical sub-fleet of trucks consists of less than 20 trucks and is able to handle at most 75 containers a day. Table 3 shows that the VNS heuristic is capable of finding solutions for real life scale problems in a few seconds.

Although we observe 20 and 40 ft containers in our model (because they are the most frequently used ones), the proposed VNS heuristic can be applied to the CDPTW with other container dimensions where a modular concept vehicle can carry up to two containers of any dimension.

Further research should focus on more extensive testing, introducing additional constraints such as heterogeneous vehicles and multiple use of vehicles, and additional improvement of the VNS heuristic (neighborhoods for local search procedure, as well as techniques for shaking procedure).

Acknowledgments This research was partially supported by the Ministry of Education, Science and Technological Development, Government of the Republic of Serbia, through the project TR 36006, for the period 2011-2014.

References

1. Crainic, T.G., Kim, K.H.: Intermodal transportation. In: Barnhart, C., Laporte, G. (eds.) Handbook in OR & MS, vol. 14, pp. 467–537. Elsevier B.V (2007)
2. Macharis, C., Bontekoning, Y.M.: Opportunities for OR in intermodal freight transport research: a review. EJOR **153**, 400–416 (2004)
3. Nagl. P.: Longer combination vehicles (LCV) for Asia and Pacific region: some economic implications, UNESCAP Working Papers, No.2, UN (2007)
4. Zhang, R., Yun, W.Y., Kopfer, H.: Heuristic-based truck scheduling for inland container transportation. OR Spectrum **32**, 787–808 (2010)
5. Jula, H., Dessouky, M., Ioannou, P., Chassiakos, A.: Container movement by trucks in metropolitan networks: modeling and optimization. Transp. Res. E: Logistics Transp. Rev. **41**, 235–259 (2005)
6. Coslovich, L., Pesenti, R., Ukovich, W.: Minimizing fleet operating costs for a container transportation company. EJOR **171**, 776–786 (2006)
7. Imai, A., Nishimura, E., Current, J.: A Lagrangian relaxation-based heuristic for the vehicle routing with full container load. EJOR **176**, 87–105 (2007)
8. Chung, K.H., Ko, C.S., Shin, J.Y., Hwang, H., Kim, K.H.: Development of mathematical models for the container road transportation in Korean trucking industries. Comput. Ind. Eng. **53**, 252–262 (2007)
9. Namboothiri, R., Erera, A.L.: Planning local container drayage operations given a port access appointment system. Transp. Res. E: Logistics Transp. Rev. **44**, 185–202 (2008)
10. Savelsbergh, M.W.P., Sol, M.: The general pickup and delivery problem. Transp. Sci. **29**, 17–29 (1995)
11. Parragh, S., Doerner, K., Hartl, R.: A survey on pickup and delivery problems Part I: Transportation between customers and depot. J. für Betriebswirtschaft **58**, 21–51 (2008a). doi:10.1007/s11301-008-0033-7
12. Parragh, S., Doerner, K., Hartl, R.: A survey on pickup and delivery problems Part II: Transportation between pickup and delivery locations. J. für Betriebswirtschaft **58**, 81–117 (2008b). doi:10.1007/s11301-008-0036-4
13. Vidović, M., Radivojević, G., Ratković, B.: Vehicle routing in containers pickup up and delivery processes. Procedia Social Behav. Sci. **20**, 335–343 (2011)
14. Vidović, M., Radivojević, G., Ratković, B., Bjelić, N., Popović, D.: (2012a) Containers drayage problem with time windows. In: CD Book of Proceedings of the 15th International Conference on Transport Science ICTS 2012, Portoroz, Slovenia
15. Vidović, M., Nikolić, M., Popović, D.: (2012b) Two mathematical formulations for the containers drayage problem with time windows. In: 2nd International Conference on Supply Chains ICSC 2012, Katerini, Greece
16. Mladenovic, N., Hansen, P.: Variable neighborhood search. Comput. Oper. Res. **24**(11), 1097–1100 (1997)

17. Hansen, P., Mladenovic, N., Perez, J.A.M.: Variable neighbourhood search: methods and applications. Ann. Oper. Res. **175**, 367–407 (2010)
18. Lai, M., Cao, E.: An improved differential evolution algorithm for vehicle routing problem with simultaneous pickups and deliveries and time windows. Eng. Appl. Artif. Intell. **23**, 188–195 (2010)
19. Gillett, B.E., Miller, L.R.: A heuristic algorithm for the vehicle-dispatch problem. Oper. Res. **22**(2), 340–349 (1974)
20. Derigs, U., Gottlieb, J., Kalkoff, J., Piesche, M., Rothlauf, F., Vogel, U.: Vehicle routing with compartments: applications, modelling and heuristics. OR Spectrum **334**, 885–914 (2010)
21. Wang, X., Regan, C.A.: Local truckload pickup and delivery with hard time window constraints. Transp. Res. B: methodological **36**, 97–112 (2002)

Solving the Team Orienteering Problem: Developing a Solution Tool Using a Genetic Algorithm Approach

João Ferreira, Artur Quintas, José A. Oliveira, Guilherme A. B. Pereira and Luis Dias

Abstract Nowadays, the collection of separated solid waste for recycling is still an expensive process, specially when performed in large-scale. One main problem resides in fleet-management, since the currently applied strategies usually have low efficiency. The waste collection process can be modelled as a vehicle routing problem, in particular as a Team Orienteering Problem (TOP). In the TOP, a vehicle fleet is assigned to visit a set of customers, while executing optimized routes that maximize total profit and minimize resources needed. The objective of this work is to optimize the waste collection process while addressing the specific issues around fleet-management. This should be achieved by developing a software tool that implements a genetic algorithm to solve the TOP. We were able to accomplish the proposed task, as our computational tests have produced some challenging results in comparison to previous work around this subject of study. Specifically, our results attained 60% of the best known scores in a selection of 24 TOP benchmark instances, with an average error of 18.7 in the remaining instances. The usage of a genetic algorithm to solve the TOP proved to be an efficient method by outputting good results in an acceptable time.

J. Ferreira (✉) · J. A. Oliveira · G. A. B. Pereira · L. Dias
Centre Algoritmi, Universidade do Minho, Braga, Portugal
e-mail: joao.aoferreira@gmail.com

J. A. Oliveira
e-mail: zan@dps.uminho.pt

G. A. B. Pereira
e-mail: gui@dps.uminho.pt

L. Dias
e-mail: lsd@dps.uminho.pt

A. Quintas
Graduation in Informatics Engineering, Universidade do Minho, Braga, Portugal
e-mail: a51836@alunos.uminho.pt

V. Snášel et al. (eds.), *Soft Computing in Industrial Applications*,
Advances in Intelligent Systems and Computing 223, DOI: 10.1007/978-3-319-00930-8_32,

1 Introduction

In nature, there is no other living creature that endangers more the environment than us humans. Our way of living and eating habits represent the major source of solid waste production, along with the level of technology consumption. In order to manage and control this continuous waste generation, some decisions were taken. Giving a better and useful utilization to waste materials (recycling) is one of them, along with the corresponding reduction of the total amount of waste produced. In this case, waste separation is critical in order to generate a special stream of solid waste aside from the common waste. Therefore, a new collection system was developed and different collection points were made available to the population. There are companies specially dedicated to the separated waste collection for recycling, and that task is usually performed using a vehicle fleet, with fixed routes and schedules.

The real problem lies, not on designing an easily accessible network of collection points, but yet on the development of efficient methods for performing waste collection, where constant resource management is vital. One issue that often arises is how to find a way to perform waste collection with a limited fleet of vehicles while obtaining the highest possible profit. An approximation to this issue is the well-known Vehicle Routing Problem (VRP), described in the literature as the problem of designing the least-costly routes from a depot to a set of customers of known demand, where each customer is visited exactly once, without violating capacity constraints and aiming to minimize the number of vehicles required and the total distance travelled. However, the majority of real-world applications require more flexible systems that may lead to the selection of customers. The Team Orienteering Problem (TOP) models can be applied to solve that issue. In the TOP, each customer has an associated profit, and the routes have maximum durations or distances. The choice of customers is made by balancing their profits and their contributions for the route duration or distance. The objective is to maximize the total reward collected in all routes while satisfying the time or distance limit.

The TOP is a fairly recent concept, first suggested by [1] under the name of Multiple Tour Maximum Collection Problem. Later, [2] formally introduced the problem and designed one of the most frequently used sets of benchmark instances. In [3], achieved many of the currently best-known solutions for the TOP instances by presenting two versions of Tabu Search, along with two metaheuristics based on Variable Neighbourhood Search (VNS). Other competitive approaches were carried out by [4], with two Ant Colony Optimization (ACO) variations; [5], with a VNS-based heuristic; and more recently, [6] designed two variants of Greedy Randomized Adaptive Search Procedure with Path Relinking.

The work presented in this paper is part of a series of experiments integrated in the R&D project named Genetic Algorithm for Team Orienteering Problem (GATOP), which was approved by the Portuguese Foundation for Science and Technology (FCT). It involves five combined tasks to accomplish the desired goal which is the development of a more complete and efficient solution for several real-life multi-level Vehicle Routing Problems, with emphasis on the waste collection management. Within the GATOP project, the main task is to solve the TOP, and the development of heuristic solutions based on a genetic algorithm (GA) is suggested. The simplicity of a GA in modelling more complex problems and its easy integration with other optimization methods were the factors considered for its choice. We believe it can be applied to solve the TOP, since it was also used for the Orienteering Problem by [7].

In this work we propose to solve medium-to-large-scale TOP instances considering a time constraint. We intend to verify whether it is possible to develop a method, based on a GA, that optimizes the TOP by achieving equal or better results as presented in previous studies.

2 Problem Formulation

The aim of the present study is to solve the TOP, which means to develop a method that determines *P* paths which start in the same location and have the same destination, in order to maximize the total profit made in each path, while respecting a time constraint. Then, the generated paths are assigned to a limited vehicle fleet, usually one path to each vehicle available.

Following the mathematical formulation suggested by [4], the objective function of the TOP is presented in Eq. 1, where n is the total number of customers, m is the number of vehicles available, the value y shows if customer i is visited or not by a vehicle k, and finally, r is the reward associated to a certain customer i. The objective function consists of finding m feasible routes that maximize the total reward or profit.

$$max \sum_{i=2}^{n-1} \sum_{k=1}^{m} r_i \cdot y_{ik} \tag{1}$$

3 Developed Tool

3.1 The Genetic Algorithm

The Genetic Algorithm (GA) is a search heuristic that imitates the natural process of evolution as it is believed to happen to all the species of living beings. This method uses nature-inspired techniques such as mutation, crossover, inheritance and selection, to generate solutions for optimization problems. The success of a GA depends on the type of problem to which the algorithm is applied and its complexity.

In a GA, the chromosomes or individuals are represented as strings which encode candidate solutions for an optimization problem, that later evolve towards better solutions. Designing a GA requires a genetic representation of the solution domain, as well as a fitness function to evaluate the solutions produced. The GA evolutionary process starts off by initializing a population of solutions (usually randomly), which will evolve and improve during three main steps:

- Selection: a portion of each successive generation is selected, based on their fitness, in order to breed the new, and probably better fit, generation.
- Reproduction: the selected solutions produce the next generation through mutation and/or crossover, propagating the most crucial changes to the future generations by inheritance.
- Termination: once a stopping condition is met, the evolutionary process ends.

3.2 Algorithm Description

The developed genetic algorithm consists on three components. The most elementary one is the chromosome, which represents a set of vehicles and their respective routes. The next component is the evolutionary process, responsible for executing crossovers and mutations within the population of chromosomes. The third and last component of the algorithm controls the evolutionary process, ensures the validation of chromosomes according to the limitations imposed by the TOP and its instances, and also carries out the evaluation of the chromosomes based on a fitness function.

In the GA presented here, a possible solution for the TOP is represented in the form of a chromosome. At each new generation, a renewed population of chromosomes is obtained, each one being a valid solution. A valid solution must contain one route for each vehicle available, and each route includes a sequence of customers to be visited under a given time limit. An example of valid solution is showed in Fig. 1. There are six customers denoted as numbered vertices and two vehicles are available. There is also a starting point (S) and an ending point (E). In this case, vertex 6 is not included in any route because otherwise it would turn the solution invalid by violating the time constraint.

While addressing the team orienteering problem, the fitness function of the algorithm was set to correspond to the sum of all the collected rewards in each customer visited during the identified routes.

In order to produce solutions for the TOP, the algorithm starts off by generating an initial population of valid chromosomes. Then, some genetic operators are applied to the population in order to promote their evolution towards better fitness levels. This evolution process is repeated until a stopping criteria is met.

As explained before, the chromosomes in the GA contain routes to be assigned to the available vehicles in a certain TOP instance. Consequently, the creation of a new random chromosome is in fact the creation of a group of random valid routes. All these routes include the required starting and ending points. In order to

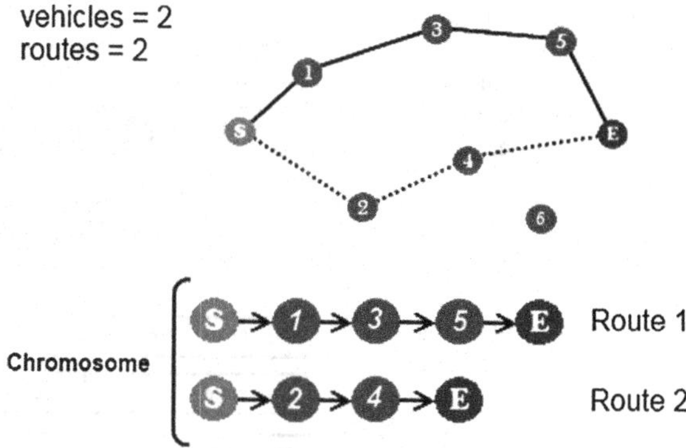

Fig. 1 Representation of an example valid solution for an instance with six vertices and two vehicles available

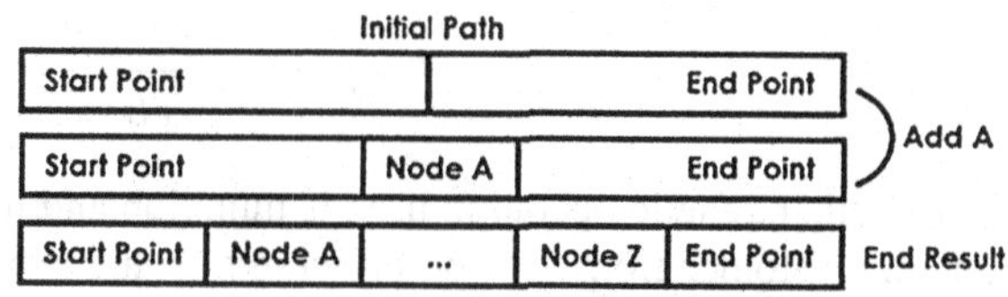

Fig. 2 Creation of a random route

assemble a route, customers are added in from an availability list that is common to all routes. The attempt to add a customer to a route is only successful if that addition keeps that route feasible while not exceeding the given time budget. Once added to a route, a customer is then removed from the list and marked as checked. Each and every customer in the list is tested for an insertion in the current route. A chromosome must contain as much routes as the number of vehicles available in a chosen TOP instance. Figure 2 presents a simple scheme of how a random route is created. The algorithm uses parallel processing in order to assemble all the routes in a chromosome.

In respect to the evolutionary process, it includes two genetic operations: crossover and mutation. The crossover procedure is done by exchanging routes between two chromosomes, resulting in the creation of two new chromosomes (see Fig. 3). The routes to be exchanged are randomly selected, yet entire blocks of consecutive routes are copied to the new chromosomes, as it can be observed in Fig. 3.

The chromosomes used for crossover are chosen based on a roulette-wheel selection, also known as fitness proportionate selection, which assesses the probability of a chromosome being used in combinatorial methods. Therefore, a chromosome with high fitness level is more likely to be selected as a parent for the next generation. The number of selected chromosomes must be an even number since these chromosomes are distributed in two lists of parents: Parents *A* and Parents *B*. In Eq. 2, the probability *p* of a chromosome *i* to be selected is denoted as p_i, the fitness of that chromosome is f_i and *N* is the population size.

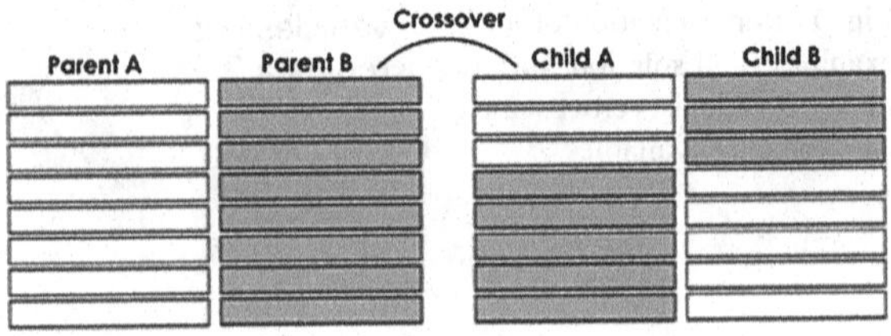

Fig. 3 Crossover procedure between two parents, resulting in two children chromosomes (each bar represents the route of a vehicle)

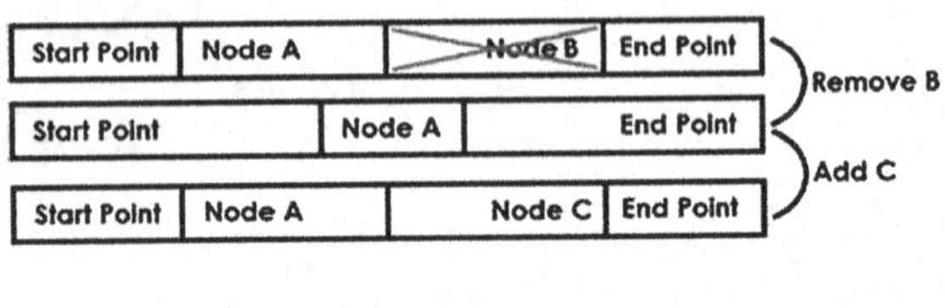

Fig. 4 Scheme of a single-switch mutation

$$p_i = \frac{f_i}{\sum_{j=1}^{N} f_j} \tag{2}$$

The other genetic operation used is mutation and it consists on the removal of a random customer from a randomly chosen route within a chromosome. Then, an attempt is made in order to insert one or more customers from the availability list of the chromosome. The customers in this list are checked one by one in a random order, and when the current customer represents a valid option, it is added to the route. During this process, an attempt is made in order to add as much new customers to the route as possible. In Fig. 4, a simple mutation is presented, where only one customer is removed from a route. There is also the possibility to perform more complex mutations by removing more than one customer from a route, and the process is executed in a similar way as in the simple mutation.

There are two special classes of chromosomes within the population, the elite and the sub-elite groups, to which different rules are applied. The elite class is the group of the most fit chromosomes within the population in a certain generation, and these chromosomes are immune to mutation until they are replaced by new chromosomes in further generations. As for the sub-elite class, it includes the fittest chromosomes immediately after the elite ones. This group is kept intact during the crossover phase but suffers mutation right after. During the evolutionary process, the resultant chromosomes from both crossover and mutation processes are kept, even if they have less fitness than the chromosomes that originated them.

3.3 Software Developed

The presented algorithm was implemented using the JAVA Swing Framework, and the resulting software tool incorporates a Graphical User Interface that allows adjustment of various parameters. A representation of the software functional process is given in Fig. 5.

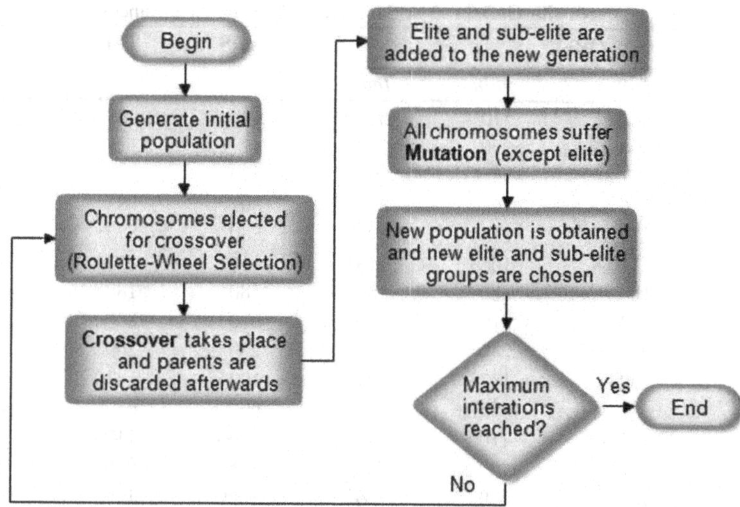

Fig. 5 Simplified representation of the software functional process

4 Computational Test Results and Discussion

In order to assess the performance of the algorithm, a series of computational experiments around the TOP were performed using the developed software tool. These experiments were conducted on 24 of the 320 benchmark TOP instances published by [2]. The instances were chosen in a semi-random way within four different sets, aiming to introduce diversity and different degrees of difficulty while testing the algorithm.

The results achieved during the tests were matched against the ones obtained by the [2], hereafter referred to as CGW. Comparisons were also carried out with the results presented by [8], hereafter referred to as TMH, and also the results produced by the algorithms presented by [3], hereafter referred to as AHS. The tests were run on a desktop computer with an Intel Pentium Core2Quad Q6600 2.40 GHz processor and 4 GB of RAM.

During the tests, the maximum number of generations to be produced for each instance was set to be the stopping criterion, with its value limited to 10.000. The number of vertices (customers) in each set of instances includes the starting and ending points. There are 100 vertices in the first set, 66 in the second, 64 in the third, and 102 in the fourth set. An instance is characterized by a number of vehicles, varying between 2 and 4, and by a time limit (Tmax).

In order to execute the tests, it was considered a total population of 220 chromosomes, with 10 being in the Elite and another 10 in the Sub-Elite. Also, per each generational cycle, 100 Crossovers and 210 Mutations occurred.

Table 1 Results achieved with GATOP-2 in the selected benchmark instances

Instance	GATOP-2			AHS	TMH	CGW
	Avg	*fmin*	*fmax*			
p4.2.e	598.5	570	618	618	593	580
p4.2.n	1063.5	990	1115	1171	1150	1112
p4.3.f	571.8	555	579	579	579	552
p4.3.i	786.6	768	806	807	785	798
p4.4.I	841.8	815	879	880	875	847
p4.4.q	1110.8	1063	1131	1161	1124	1084
p.5.2.e	180	180	180	180	180	175
p5.2.m	852.5	810	860	860	860	855
p.5.3.h	260	260	260	260	260	255
p.5.3.v	1393	1375	1415	1425	1410	1400
p.5.4.o	682.5	675	690	690	680	675
p.5.4.t	1129	1100	1160	1160	1100	1160
p.6.2.j	936.6	900	948	948	936	942
p.6.2.n	1228.8	1200	1242	1260	1260	1242
p.6.3.i	642	642	642	642	612	642
p6.3.I	985.2	966	1002	1002	990	972
p6.4.k	528	528	528	528	522	546
p6.4.n	1068	1068	1068	1068	1068	1068
p7.2.e	290	290	290	290	290	275
p7.2.r	1035.7	974	1077	1094	1067	1082
p7.3.g	344	344	344	344	344	338
p7.3.s	1033	1004	1064	1081	1061	1064
p7.4.t	1041.2	1014	1058	1077	1067	1066

We ran our algorithm ten times on each selected benchmark instance. The results achieved with our algorithm, hereafter referred to as GATOP-2,[1] in the tested instances, are presented in Table 1. The values *fmin* and *fmax* are respectively denoted as the minimum and maximum fitness values obtained for each instance. The value *fmin* can be considered a guaranteed value, reflecting the overall performance of the algorithm, along with the average fitness value. The value *fmax* represents the ability of the algorithm to reach good solutions. It is obtained by running the algorithm more than once and taking benefit of the randomness at the expense of a larger computational time.

In Table 1, the best scores found for each instance are displayed in bold print, and were mostly set by AHS. As it can be observed, on 14 of the 24 instances, our algorithm equalled the best scores. In particular, our scores were equal to AHS 14 times, and scored the same as TMH and CGW in 7 and 5 instances respectively. In addition, GATOP-2 outperformed TMH for 14 times and CGW for 16 times.

[1] GATOP-2 is preceded by GATOP-1—a previous work in the GATOP project yet to be presented in ICORES 2013 under the title "Developing tools for the team orienteering problem—A simple genetic algorithm".

Table 2 Statistics on the results achieved with GATOP-2 in the tests

Equal to best Score	14
Equal to AHS	14
Equal to TMH	7
Equal to CGW	5
Better than TMH	14
Better than CGW	16
Max.Error in respect to Best Score	56
Average Error in respect to Best Score	18.6
Max.gap(*fmax-fmin*)	125
Average gap(*fmax-fmin*)	36.17

In respect to the best scores, GATOP-2 fell behind on 10 instances, with a maximum error of 56 and an average error of 18.7, considering solely the results on those same ten instances. These error values are inferior to the ones produced by TMH and CGW, while comparing them to the best scores.

An overview of the results produced with GATOP-2 may be sufficient to infer about its performance, particularly in terms of consistency. We confirmed the occurrence of a gap between the *fmax* and *fmin* values in 17 instances, with a maximum value of 125. The average value for the gaps is 36.17. Comparing these values, for example, to the ones outputted by the overall best algorithm presented by AHS, the SLOW VNS_FEASIBLE, GATOP-2 produced much higher gaps between the maximum and minimum scores in the tested instances, since the SLOW VNS_FEASIBLE presents a highest gap value of 18 and an average gap of 6. Therefore, GATOP-2 is less consistent.

During the tests, the maximum computational time of a run with the adopted settings did not exceed 21 min, but since the best results were achieved in less than 5.000 iterations, in a practical sense, it can be assume GATOP-2 would achieve the same results in half the time spent on the tests.

We believe the crossover procedure to be the main reason for our algorithm to fail in achieving better results, since the chromosomes that are designated to be parents are discarded immediately after all the crossovers are performed, even if they have a higher fitness than their children. This may lead to a considerable loss of potentially good solutions that would benefit from a later mutation and probably evolve to better solutions.

In respect to the mutation procedure, we opted to decrease the complexity and computational time during the tests, and so we kept the single-switch mutation. By choosing this method, the mutations were accomplished much faster but with lower optimization when compared to a multi-switch mutation, where multiple customers from one or more routes are replaced. Only further tests would confirm if GATOP-2 would produce better results while using more complex mutations (Table 2).

Another aspect that have an impact on the results is the population size and also the number of elite and sub-elite chromosomes. A series of preparatory tests confirmed that assumption. During those tests, the values of the considered parameters were increased from test to test, and GATOP-2 outputted better scores accordingly. Although those values could be raised until the best scores were achieved, that would require a large amount of computational time, turning it into a non-practical method to solve the TOP in a real-world situation. This additional computational time is due to the large amount of crossovers and mutations that occur during each generation. One way to tackle this problem is to determine a better balance between the population size and the quantity of crossovers and mutations. That change would speed up the reproduction process in the algorithm.

5 Conclusions and Future Work

In this paper we presented a genetic algorithm as a solution method for the Team Orienteering Problem (TOP). The achieved results demonstrate that our algorithm is fairly efficient, attaining the best-known solutions in more than half of the tested instances, and scoring near the best in the remaining instances. There are still some enhancements that can be done to improve the results, like modifying the crossover and mutations procedures. Improvements can also be achieved by finding a better balance between parameters such as the total population size and the number of elite and sub-elite chromosomes. A possible way of doing this is to use dynamic parameters to set the behaviour of the evolution process within the genetic algorithm. This could be achieved by implementing a Machine Learning algorithm that would tune the parameters of the genetic algorithm by evaluating its performance during the tests and would apply the best parameter configuration to overcome adversities while aiming for better results.

The assessment of the software tool developed for the presented study was important in order to identify its functionalities, advantages and limitations. Future experimentations will focus on the usage of the C++ programming language, which might perform faster than JAVA.

With these experiments, we were able to improve our knowledge in the TOP, which is certainly a good contribute for the GATOP project. In future work, it is our plan to continue on the development of new strategies to apply in genetic algorithms in order achieve better solutions for the TOP and its variants with more constraints.

Acknowledgments This study was partially supported by the project GATOP - Genetic Algorithms for Team Orienteering Problem (Ref PTDC/EME-GIN/ 120761/2010), financed by national funds by FCT / MCTES, and co-funded by the European Social Development Fund (FEDER) through the COMPETE Programa Operacional Fatores de Competitividade (POFC) Ref FCOMP-01-0124-FEDER-020609.

References

1. Butt, S.E., Cavalier, T.M.: A heuristic for the multiple tour maximum collection problem. Comput. Oper. Res. **21**, 101–111 (1994)
2. Chao, I.M., Golden, B., Wasil, E.A.: Theory and methodology—the team orienteering problem. Eur. J. Oper. Res. **88**, 464–474 (1996)
3. Archetti, C., Hertz, A., Speranza, M.G.: Metaheuristics for the team orienteering problem. J. Heuristics **13**, 49–76 (2006)
4. Ke, L., Archetti, C., Feng, Z.: Ants can solve the team orienteering problem. Comput. Ind. Eng. **54**, 648–665 (2008)
5. Vansteewegen, P., Souffriau, W., Van Oudheusden, D.: A guided local search metaheuristic for the team orienteering problem. Eur. J. Oper. Res. **196**(1), 118–127 (2009)
6. Souffriau, W., Vansteenwegen, P., Vanden Berghe, G., Van Oudheusden, D.: A path relinking approach for the team orienteering problem. Comput. Oper. Res. (Metaheuristics for Logistics and Vehicle Routing) **37**, 1853–1859 (2010)
7. Tasgetiren, M.F.: A genetic algorithm with an adaptive penalty function for the orienteering problem. J. Econ. Soc. Res. **4**(2), 1–26 (2002)
8. Tang, H., Miller-Hooks, E.: A tabu search heuristic for the team orienteering problem. Comput. Oper. Res. **32**, 1379–1407 (2005)

Use of Fuzzy Optimization and Linear Goal Programming Approaches in Urban Bus Lines Organization

Yetis Sazi Murat, Sabit Kutluhan and Nurcan Uludag

Abstract Determination of bus stop locations and bus stop frequencies are important issues in public transportation planning. This study analyzes the relationships among demand, travel time, bus stop locations, frequency, fleet size and passenger capacity parameters and develops models for bus stop locations and bus service frequency using fuzzy linear programming and linear goal programming approaches. The models are microscopic and applied to determine the bus stop locations and bus service frequency in the city of Izmir, Turkey, where 26 bus routes pass through two stops in the center city. The fuzzy optimization model minimizes the passenger access time and in-vehicle travel time. The reduction of the values of the bus service frequency and time parameters derived by the two proposed models are validated by a cost function. Encouraging results are obtained.

1 Introduction

The locations of bus stops and the number of stops and scheduling directly affects transit system's performance and operation efficiency. For operations the most essential criteria is optimal bus frequency and correct bus schedules. Correct bus schedules affect waiting times in bus stops, passenger demand and comfort of travel especially in peak periods. The higher (or lower) bus frequencies yield higher operation costs and less demand.

Y. S. Murat (✉) · S. Kutluhan
Faculty of Engineering, Pamukkale University, Denizli, Turkey
e-mail: ysmurat@pau.edu.tr

N. Uludag
Yorum Building and Construction Inc., İstanbul, Turkey
e-mail: nurcanuludag@hotmail.com

V. Snášel et al. (eds.), *Soft Computing in Industrial Applications*,
Advances in Intelligent Systems and Computing 223, DOI: 10.1007/978-3-319-00930-8_33,

The question is how to select the best combination of the values of the parameters. Each transit system is unique in the demand pattern and the route characteristics, and operating practices. Thus the problem cannot be completely sanitized and solved in an abstract form. Rather, it is desirable to solve the problem considering unique circumstances, constraints, and analyst's judgement in the process.

Bus stop locations and frequency have been considered by many researchers in literature. Chien and Qina [1], studied on a mathematical model in order to improve bus service accessibility. Furth and Rahbee [2], developed an approach to examine a bus route with alternation of the bus stop locations; on a simple geographic model and the demand distribution is carried out. Saka [3], developed an optimal bus spacing model, in contemplation of minimizing access time of passengers and operation costs at the same time. Dell'Olio et al. [4], developed a bus stop location model based on optimizing a bus transit operation cost function. LeBlanc [5], developed a model for determining frequencies, using a modal-split assignment programming model.Wirasinghe [6], studied on the validity of frequency determination method by Newell. Alp [7], modeled the frequency of bus transit network in Istanbul with linear goal programming, using passenger data, travel times, vehicle capacities. Guihaire and Hao [8] reviewed transit network design and scheduling studies. Chien et al. [9] proposed Genetic Algorithm approach for transit route planning and design. Caggiani and Ottomanelli [10] used fuzzy-non linear programming approach for equilibrium network design problem in urban areas.

In this paper an approach that incorporates the ambiguity of the analyst as to the selection of specific parameter values and vague boundaries of constraints in the optimization problem is proposed. The approach accommodates the analyst ambivalence in the selection of the exact value. Within the sets of values or range of values of different parameters, a best set is chosen such that the satisfaction requirements of the parameters are met.

This study is employed in order to understand if the bus service frequency is appropriate for the passenger demand regarding optimized bus stop locations. The bus frequency model is developed using Linear Goal Programming. In the models, first, optimum frequencies are determined for the lines passing through two stops by using passenger demand, bus stop numbers and capacities, and the lines' travel time data. Then, the bus stop locations on these same lines are examined and optimal bus stop locations that minimizing access time to bus stops and in-vehicle travel time is found by Fuzzy Linear Programming. The existing case and the modeling case for frequency and stop locations regarding the access times and in vehicle travel times are compared by a cost function.

2 Bus Stops: Locations and Spacing

The bus stop location problem brings out bus stop spacing problem which directly affects access time to bus stops and in-vehicle travel time of the bus transit system. Closer bus stop design decreases the access time and distance to stops for the passengers, but increases in-vehicle travel time. Bus stop location problem should be handled as a balancing the expectations of passengers and operators. While the passengers' expectation is minimizing the access time [11]; for the operational conditions, operation costs, safe service and customer satisfaction is essential [12]. The walking distances should be appropriate for the passengers' access, in-vehicle travel time and operational conditions.

3 Bus Service Frequency

The bus service frequency should be determined with an evaluation of service standards, experience, and passenger demand together [13]. A service standard, generally called as congestion level, is generally denoted as permitted standees passenger number, upper and lower limits of headways. A prudent and logical public transit system should balance the increasing service frequency and its investment requirement. This paper is about microscopic analysis of determination of bus stop location and frequency at a limited city center where many bus lines converge. These are many specific constraints site specific and general constraints related to stop location, and bus frequency of individual lines. The process becomes adjustment problem of each parameter such that the overall the best situation can be derived.

For this types of problems a range for each parameter are introduced as constraints but the desirability within the range is also specified.

4 Bus Systems Analysis Model

4.1 Modeling Bus Service Frequency by Linear Goal Programming

In bus service frequency model, the data of 26 routes passing through Lozan and Montrö stops for morning peak period (08:00–11:00) were used. ESHOT, (Municipality of Izmir, Electrical, Water, Air Gas, Bus and Trolley Operations' Authority) data were used in model. The frequencies of the 26 routes' data were examined in meaning of bus fleet size, route travel times, passenger demand and a linear goal programming model for the optimum frequency was developed.

4.1.1 Model Development

The optimum bus service frequency model with Linear Goal Programming for the network was developed by determining decision variables, system and goal constraints, success functions and objective functions and solved by WINQSB program. The system constraints in the model were determined considering the concept of satisfying the passenger demand in terms of bus service frequency and bus passenger capacity. In other words, system constraint of the model is formed with the conception that the bus fleet's serving all the passengers on the route with the determined service frequency value.

The system constraint can be denoted as Si*Ci $\geq$ Ki

where,

S1, S2,...Si : Bus service frequencies for the routes,

C1, C2,...Ci : The capacity of the buses used

K1, K2,...Ki : The passenger demand on the routes

The goal constraints of the model were determined as *Time Goal Constraint* and *Capacity Goal Constraint.* The time goal constraints were determined due to the equivalence between the number of the buses in the fleet and the duration of each trip and its frequency in the period of 08:00–11:00 a.m. Similar approach is used in some previous studies [7].

The equations used for the *Time Goal Constraint* and the *Capacity Goal Constraint* are given in the follow.

Time goal constraint	*Capacity goal constraint*
$\sum_{i=1}^{26} t_i X_i + d_1^+ + d_1^- = T * B \quad (1)$	$\sum_{i=1}^{26} C_i X_i + d_2^+ + d_2^- = \sum_{i=1}^{26} K_i \quad (2)$
where;	where;
ti : Bus travel time (two way) (min)	C_i : The capacity of the buses used in route i during T
B : Total number of buses used	
Xi : Number of trips required for route i	Xi : Number of trips required for route i
T : Total time period (180 min)	d_2 : Capacity constraint deviation
d_1 : Time constraint deviation	K_i : the passenger demand on route i

In goal programming, it is tried to minimize the deviations from the objective function. The object of the model is, minimizing the objective function (S) which has a meaning of the sum of the deviations.

For the objective function in the model the time and capacity goal constraints were taken into consideration with equal priority. The **objective function** was formed as the sum of the negative deviations of these two constraints as shown below;

$$\min S = d_1^- + d_2^- \quad (3)$$

In accordance with these constraints and goals, the goal programming model was solved by using WINQSB program.

4.2 Modeling Bus Stop Spacing by Fuzzy Linear Programming

Fuzzy Linear Programming is assumed that objective function or the constraints can not be so crisp that instead of optimization of objective function, it is aimed to solve the problem with a satisfaction level [14].

It was seen that, especially the passenger demand and bus speed data are variable and include uncertainties so that fuzzy set theory was used for them. In accordance with these considerations, each route of the examined network was analyzed and a fuzzy linear programming model was developed for each. The total value of the access time to each stop on a route plus in-vehicle time from that stop and the global total of the access time plus in-vehicle times for all the stops on that route was obtained [15, 16].

4.2.1 The Development of Model

For each of the 26 routes, of which an optimum **Bus** Service **F**requency **Mod**el (BuFMod) was developed, a **Fu**zzy **L**inear Programming **Mod**el (FuLMod) for achieving optimal bus stop spacing was developed. On each of the 26 routes, optimum bus spacing was determined with the main object of minimizing the sum of access time and in-vehicle time.

The 26 routes were analyzed individually. The passenger access behavior assumption is given in Fig. 1 [17]. For each stop on each route, the access time to stop (Ta) and in-vehicle travel time (Tv) was computed.

Access time to bus stops (Ta)

The access time to stops (Ta) is an essential criterion for determining the level of service quality for the passengers. The spacing that does not get the passengers away from the public transport should be satisfied.

In-vehicle Travel Time (Tv)

In-vehicle travel time is the time period between getting in and out of the vehicle.

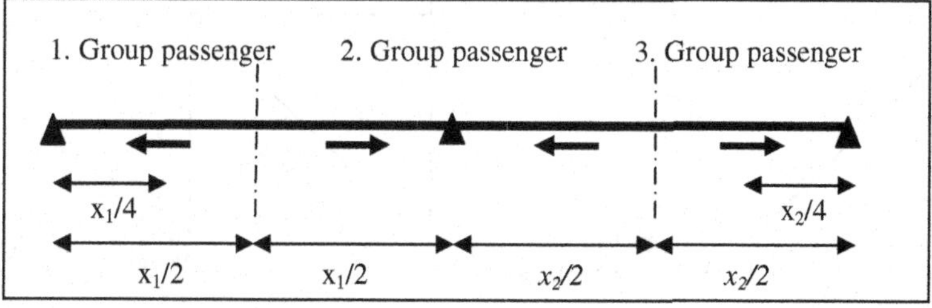

Fig. 1 Passenger arrivals to stops [17]

Pedestrian Speed (Va)

For pedestrian speed 4,4 km/h value acquired from the literature was used.

Average arriving passenger number at stops (P)

For each route, an average passenger/km value was obtained from the passenger counting data. Fuzzy set theory was used in the model for the passenger demand data's uncertainty and triangular membership functions were used. In Fig. 2, the triangular membership function for passenger parameter is given.

The access time to stops (Ta)	*In-vehicle travel time (Tv)*
$Ta_n = \frac{\left(P\cdot\frac{x_{n-1}}{2}\right)\left(\frac{x_{n-1}}{4}\right)}{V_a} + \frac{\left(P\cdot\frac{x_n}{2}\right)\left(\frac{x_n}{4}\right)}{V_a}$ (4)	$Tv_n = \left(P\cdot\frac{x_{n-1}}{2} + P\cdot\frac{x_n}{2}\right)\left(\frac{x_n}{V} + T_l\right)$ (5) $Tv = \sum_{1}^{n} Tv_n$
where; P: Average number of passengers for unit distance (pas/km) x1, x2,...,xn : Bus stop spacing (m) Va: Pedestrian Speed (km/h)	where; P: Average number of passengers for unit distance (pas/km) x1, x2,...,xn : Bus stop spacing (m) V: Average vehicle speed (km/h) Tl: Time loss due to deceleration /acceleration and passenger get in/off (sec)

Average Vehicle Travel Speed

For each route, the average vehicle speed is computed by using the data and triangular membership function is formed. In Fig. 2 triangular membership function for vehicle speed parameter is given.

Time Loss at Stops

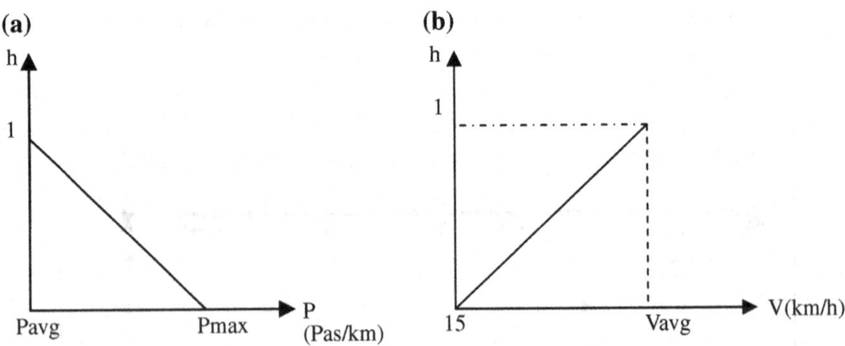

Fig. 2 Membership functions of (**a**) Average arriving passenger at stops and (**b**) average vehicle travel speed

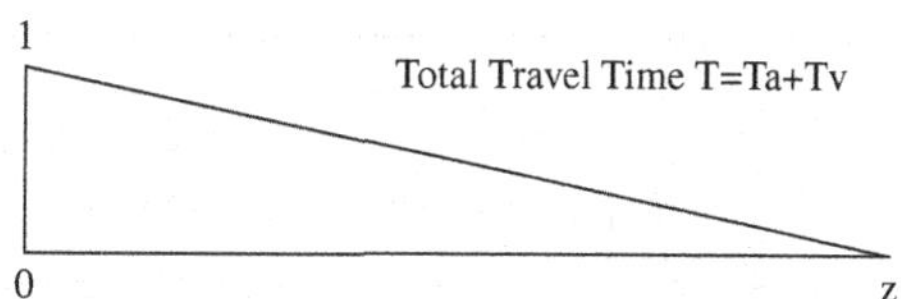

Fig. 3 Total travel time membership function

Time loss (T_l) at stops grows out of the getting in/out of the vehicles and the acceleration/deceleration maneuvers of buses while arriving and departing the stops. The getting in/out of vehicle times is computed from the data of the buses' arrival and departure time at stops. The speeds for these maneuvers are taken into consideration as 15 km/h.

$$T_l = T_l \text{ getting in/out of vehicle} + T_l \text{ deceleration/acceleration} \quad (6)$$

The deceleration and acceleration of the vehicles are calculated by using the values of the buses' specific properties used in Izmir network. The FuLMod model formed in order to minimize total travel time was solved with LINGO program and stop spacing values were obtained.

Here, a satisfaction level, h, was determined and total travel time was modeled using the triangular membership function as shown in Fig. 3. In the figure, z represents, maximum possible total travel time.

Where;

$$T = Ta + Tv$$
$$f(T) = 1 - \frac{T}{z} \geq h \quad (7)$$

Max h;

$h > 0;$

$h \leq 1;$

s.t. $x_1 + x_2 + \ldots x_n = L$

In the model, the satisfaction level is maximized and the spacing that minimize the total travel time are obtained. Differently from conventional programming, both total travel time on the route globally and total travel time on each stop are minimized. For this, a satisfaction function is formed for the route globally and for each stop. Satisfaction value (mean) of all routes is obtained as 0.99 in the study.

4.3 Comparison of Model Results with Existing Case

The results of the model BuFMod showed that the bus service frequencies are less than the existing ones. Figure 4 shows comparisons of the headways provided by BuFMod model and existing cases.

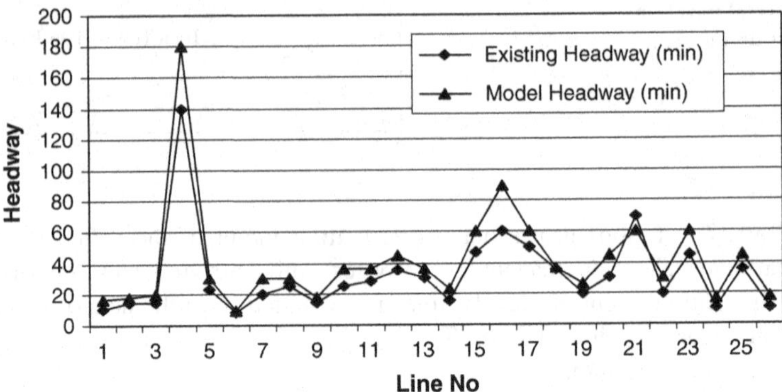

Fig. 4 BuFMod model results-existing cases for headway

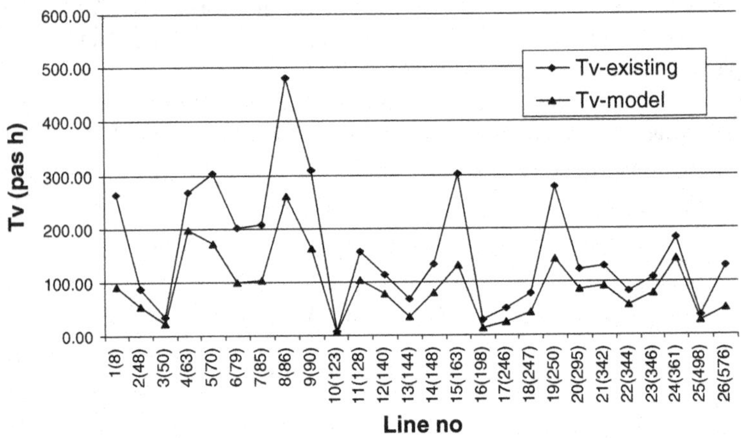

Fig. 5 FuLMod model results and existing case values of in-vehicle travel time values

The access times (Ta) and the in-vehicle travel times (Tv) of all the lines of which service frequency is examined with FuLMod model is shown in Figs. 5 and 6, with a classification of existing case and the model results.

4.4 Cost Function

The results of the two models are combined with a cost function as denoted below.

$$C_{\text{TOTAL COST}} = C_{\text{PASSENGER COST}} + C_{\text{OPERATION COST}}$$

Here, passenger benefit value was computed by using the access times of passengers to the stops and in-vehicle travel times. The time benefits of FuLMod

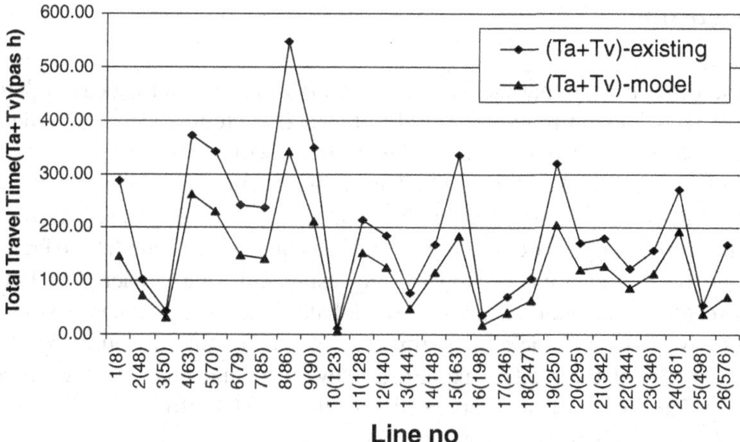

Fig. 6 FuLMod model results and existing case values for total travel time (Access Time (Ta) + In-vehicle travel time(Tv)

Table 1 Sample costs and total costs of the model results

Route no	FULMod Mode–Passenger cost			BUFMod model–Operation cost		
	Total travel time benefit (h)	Unit cost ($/h)	Total cost ($)	Trip number benefit	Unit cost ($/h)	Total cost ($)
8	141.59	5.96	549	7	33.18	4849
48	28.56	4.66	86	2	22.75	522
50	10.53	4.70	32	3	22.75	493
90	135.25	5.30	465	2	33.18	772
128	63.00	4.42	181	1	33.18	905
344	37.20	5.08	123	3	33.18	1318
361	77.48	5.36	270	7	33.18	2792
.	.	.	.	.	.	.
576	98.11	5.14	328	8	33.18	3221
FuLMod Cost			$6535	BuFMod Cost		$23391
Total cost						$29926

model results and unit passenger costs were used to compute the passenger benefit value in the equation. The unit passenger cost values ($/h) used in FuLMod Model were taken from the zoning and survey studies composed for Izmir in 2010 [17]. The unit cost values used in BuFMod Model ($/h) were calculated by using the fuel consumptions of the vehicles used on the related lines. The operation benefit value was calculated by using the BuFMod models service frequency benefit and the operation costs. The total benefit obtained is given in Table 1.

5 Conclusions

In this study, it is seen that bus frequency in the examined network is higher than the passenger demand requires. The developed BuFMod Model results decrease the frequency by a percent of 26 %. This decrease will be a benefit for operational costs and also will make a positive effect on the traffic conditions.

The results showed that, the main factors affecting the quality of service of public transit for the passengers are access time and in-vehicle travel time. These factors are decreased by optimization of stop spacing using FuLMod Model. The FuLMod Model results show that on 26 routes globally, the access time to stops (Ta) decreased by 7 %; in-vehicle travel time by 43 % and total travel time by 36 %.

The benefits of the models explained above are calculated considering a cost function. The total benefit obtained by the models for the observed 3 h (08:00–11:00) is about nearly 30000 US$.

In the study, 26 routes are examined independently from the whole network. A more comprehensive study modeling the whole network can be developed. And also for service frequency model, new variations according to the priorities of the goals can be developed.

In Fuzzy Linear programming model for determining the spacing, the behavior of passenger arrivals can have different variations according to the stop's location, being in an attraction zone or not. New fuzzy parameters can be used besides passenger demand and vehicle speed and interesting results can be obtained in future studies.

Acknowledgments This study is dedicated to Prof. Shinya Kikuchi (from Virginia Politechnic Institute and State University) who inspired many researches (including this work) on application of soft computing in transportation.

References

1. Chien, S.I., Qina, Z.: Optimization of bus stop locations for improving transit accessibility. Trans. Plann. Technol. **27**(3), 211–227 (2004)
2. Furth, P., Rahbee, A.: Optimal bus stop spacing through dynamic programming and geographic modeling. Trans. Res. Rec. **1731**, 15–22 (2000)
3. Saka, A.A.: Model for determining optimum bus-stop spacing in urban areas. J. Trans. Eng. **127**(3), 195–199 (2001)
4. Dell'Olio, L., et al.: A model of cost optimization for the location of bus-stop. In: 16th Mini–EURO Conference and 10th Meeting of EWGT (2005)
5. LeBlanc, L.J.: Transit system network design. Trans. Res. **22**, 383–390 (1988)
6. Wirasinghe, S.C.: Initial planning for urban transit systems. In: Advanced Modeling for Transit Operations and Service Planning, pp. 1–29. Elsevier Science, Oxford, (2003)
7. Alp, S.: The use of linear goal programming approach for bus transit systems, Istanbul Ticaret University. Sci. J. **1**(13), 73–91 (2008)
8. Guihaire, V., Hao, J.K.: Transit network design and scheduling: a global review. Trans. Res. A **42A**(10), 1251–1273 (2008)

9. Chien, S., Yang, Z., Hou, E.: Genetic algorithm approach for transit route planning and design. J. Trans. Eng. **127**(3), 200–207 (2001)
10. Caggiani, L., Ottomanelli, M.: Traffic equilibrium network design problem under uncertain constraints. Procedia-Soc. Behav. Sci. **20**, 372–380 (2011)
11. Murray, A.: A coverage model for improving public transit system accessibility and expanding access. Annu. Oper. Res. **123**(1), 143–156 (2003)
12. Van Nes, R., Bovy, P.H.: Importance of objectives in urban transit-network design. Trans. Res. Rec. **1735**, 25–34 (2000)
13. Furth, P.G., Wilson, W.H.M.: Setting frequencies on bus routes: theory and practice. Trans. Res. Rec. **818**, 1–7 (1981)
14. Murat Y.S., Kikuchi, S.: The fuzzy optimization approach: a comparison with the classical optimization approach using the problem of timing a traffic signal. Trans. Res. Rec. **2024**, 82–91 (Washington D.C) (2007)
15. Uludağ, N.: Modeling bus lines using fuzzy optimization and linear goal programming approaches. PhD Thesis, Pamukkale University, Denizli, p. 112 (2010)
16. Murat, Y.S., Uludag, N.: Route choice modelling in urban transportation networks using fuzzy logic and logistic regression methods. J. Sci. Ind. Res. **67**(1), 19–27 (2008)
17. Kikuchi, S.: Treatment of uncertainty in transportation analysis. Virginia Polytechnic Institute and State University, Falls Church, Lecture Notes (2006)
18. İzmir UAP İzmir Transportation Master Plan-ESHOT Report. İzmir (2010)

Index

V. Snášel et al. (eds.), *Soft Computing in Industrial Applications*,
Advances in Intelligent Systems and Computing 223, DOI: 10.1007/978-3-319-00930-8,